AF394230

Mathematical Statistics and Limit Theorems

Marc Hallin · David M. Mason
Dietmar Pfeifer · Josef G. Steinebach
Editors

Mathematical Statistics and Limit Theorems

Festschrift in Honour of Paul Deheuvels

 Springer

Editors
Marc Hallin
European Centre for Advanced Research
  in Economics and Statistics
Université Libre de Bruxelles
Bruxelles
Belgium

David M. Mason
Department of Applied Economics
  and Statistics
University of Delaware
Newark, DE
USA

Dietmar Pfeifer
School of Mathematics and Science
Carl von Ossietzky University
Oldenburg
Germany

Josef G. Steinebach
Mathematical Institute
University of Cologne
Köln
Germany

ISBN 978-3-319-12441-4   ISBN 978-3-319-12442-1   (eBook)
DOI 10.1007/978-3-319-12442-1

Library of Congress Control Number: 2015933369

Springer Cham Heidelberg New York Dordrecht London
© Springer International Publishing Switzerland 2015

Printed on acid-free paper

Springer International Publishing AG Switzerland is part of Springer Science+Business Media
(www.springer.com)

Foreword

Paul Deheuvels is well known for his work in mathematical statistics and probability, especially in the area of limit theorems. Some of the topics on which he has made significant and lasting contributions are extreme and record value theory, renewal theory, copulas, strong approximations, Erdős-Rényi laws, empirical and quantile processes, nonparametric function estimation and Karhunen-Loève expansions. Through his consulting work Paul Deheuvels has also made substantial contributions to applied statistics.

Paul Deheuvels has had a major impact on statistics in France. Most importantly, in 1980 he founded the statistics research laboratory at the Université Pierre et Marie Curie in Paris (the LSTA, Laboratoire de Statistique Théorique et Appliquée), and served as its director until 2013. He has guided a large number of doctoral students. Many of them are now placed in prominent positions in academia and industry. Among other honours, Paul Deheuvels was elected *Fellow of the Institute of Mathematical Statistics* in 1985, and *Membre de l'Académie des Sciences (France)* in 2000 (Correspondent, 1996–2000). As the only statistician in the Académie, he has been a tireless supporter and promoter of statistics in France.

An overview of Paul Deheuvels' research and role as a force in statistics is given in Adrian Raftery's contribution to this Festschrift. A list of his publications is given at the end of the volume.

Preface

During June 20–21, 2013 a committee of former students of Paul Deheuvels organized a conference on *Mathematical Statistics and Limit Theorems* in honour of his 65th birthday at the Jussieu Campus of the Université Pierre et Marie Curie—Paris VI. The committee consisted of

Salim Bouzebda (Université de Technologie de Compiègne, France)
Michel Broniatowski (Université Pierre et Marie Curie—Paris VI, France)
Sarah Ouadah (Université Pierre et Marie Curie—Paris VI, France)
Zhan Shi (Université Pierre et Marie Curie—Paris VI, France)

This volume is a collection of papers contributed by a selection of the invited speakers. Their topics are largely motivated by the research interests of Paul Deheuvels. The editorial board is grateful for the care that the contributors made in preparing their submissions, and acknowledges with thanks the efforts of the many referees who helped them in their editorial task.

Marc Hallin (ECARES, Université Libre de Bruxelles, Belgium)
David M. Mason (Department of Applied Economics and Statistics, University of Delaware, USA)
Dietmar Pfeifer (Institut für Mathematik, Carl von Ossietzky Universität Oldenburg, Germany)
Josef G. Steinebach (Mathematisches Institut, Universität zu Köln, Germany)

March 2015

Contents

Paul Deheuvels: Mentor, Advocate for Statistics, and Applied Statistician

Adrian E. Raftery

Abstract Paul Deheuvels is best known internationally as a theoretical statistician, but he has made many other contributions. Here I give a brief overview of his work as a mentor of many doctoral students, as an advocate for the discipline of statistics, particularly in the context of his work as the only statistician member of the French Académie des Sciences, and as an applied statistician.

1 Introduction

Paul Deheuvels is best known internationally as theoretical statistician and probabilist, and this volume focuses, appropriately, on these areas and his many distinguished contributions to them. However, he has been much more: a mentor to many doctoral students, an advocate for the statistical profession, particularly through his work at the Académie des Sciences, and an applied statistician. I will briefly describe some of his contributions in these areas.

2 Mentor

Paul Deheuvels was my doctoral advisor; I was one of his first doctoral students. I think his first doctoral student was Pierre Hominal, who graduated in 1979, and then Michel Broniatowski and I graduated in 1980. Since then, Deheuvels has supervised a very large number of doctoral students, close to 100, and, remarkably, 26 of those today hold academic positions, in France, Portugal, Algeria, Morocco, Senegal, and the United States.

I moved to Paris from Dublin in 1977 for graduate study. I spent my first year there (1977–1978) at the Laboratoire de Probabilités, Université Pierre et Marie Curie, where I got an education of extraordinary quality in the theory of stochastic processes.

A.E. Raftery (✉)
University of Washington, Seattle, WA, USA
e-mail: raftery@u.washington.edu

However, I quickly realized that I was more interested in statistics and wondered what to do. At that point, I met Michel Broniatowski, who urged me to talk to the young and dynamic Professor Deheuvels, who was then revitalizing Statistics at the university. I did so, and Paul welcomed me warmly and enthusiastically to study with him.

He was an excellent mentor, giving both freedom and support. He had many dissertation topics to suggest, but I ended up choosing to work on non-Gaussian time series models, which was only tangentially related to what he did. He supported my choice, and was always available to discuss progress. Often it was for five minutes in the corridor, but that was enough to set me straight if needed. I prospered under his guidance (Raftery 1980, 1981, 1982).

As one example of his intellectual support of students, he spent a great deal of time reading, critiquing, and suggesting changes to what became my first published article, "A string problem" (Raftery 1979), even though it was far from his main research interests. He rebuffed my suggestion that he become a coauthor, gruffly saying that he had done nothing to deserve it.

He has always supported his students personally as well as academically. In the 1970s in Paris, foreign graduate students sometimes had difficulties with the administration, both inside and outside the university. When he heard of some egregious administrative obstacle thrown up in the path of a foreign student, he would often go personally to the office involved and bang on the table, insisting that things be made right. He would take the time needed to solve the problem, and would generally prevail.

He continued to support his students after they graduated. In my case, he remained ready to give professional or personal advice and to write recommendation letters for decades, even though my own research interests diverged markedly from his after a few years. In another case, a student returned to her country after graduation to find herself in an untenable personal situation, and when Deheuvels learned about the situation he moved heaven and earth, using all his contacts to find her an academic position elsewhere at short notice.

3 Advocate for Statistics

Paul Deheuvels is best known for his work on probability and theoretical statistics across a stunning range of topics, starting with extreme value theory (Deheuvels 1973), and including renewal theory (Deheuvels et al. 1986), function estimation (Deheuvels and Mason 1992), and continuing work on empirical process limit theory (Deheuvels and Ouadah 2013). This has given him an international reputation, and has led to major recognition in France also. He is the only statistician who is a member of the French Académie des Sciences, and he was the first recipient of the Prix Pierre-Simon de Laplace from the Société Française de Statistique in 2007, together with Pascal Massart. This is the most prestigious prize in Statistics in France, and so far has been awarded to only four people: Deheuvels, Massart, Christian Gouriéroux, and Anestis Antoniadis.

In 1980, he created the present research laboratory of statistics at the Université Pierre et Marie Curie in Paris (the LSTA, Laboratoire de Statistique Théorique et Appliquée), and served as its Director until 2013. This has been extremely successful, now bringing together 18 faculty.

Perhaps surprisingly given his largely theoretical research, he has used his position of eminence within French science to explain, promote, and advocate for statistics as a whole in French society, including its more applied aspects.

In 1982, he published a short book in French called "Probability, Chance and Certainty," destined for the educated general public (Deheuvels 1982), now in its fourth edition (Deheuvels 2008). This is a true intellectual tour de force. In 124 small and miraculously accessible pages, he takes the reader from basic ideas of probability using games of chance, to Paul Lévy's stable distributions, and traces the history of Brownian motion from Lucretius in the first century A.D. to Einstein's rigorous development. He summarizes the controversy between objective and subjective views of probability, and leads up to the Glivenko–Cantelli theorem.

The penultimate chapter is called "The Arc Sine Law, or the Fundamental Injustice of Nature." He describes the arc sine law mathematically, and then summarizes it by saying: "in a game with two players who have equal chances, it is likely that one of them will 'win all the time'." This exemplifies the combination of mathematical rigor and engaging explanation that makes the book a model of mathematical writing for a literate public. This little work has had a huge impact, having sold over 20,000 copies. It has contributed to help France think probabilistically.

His inaugural discourse to the Académie des Sciences, "The Scientific Adventure of Statistics" (Deheuvels 2001), is a manifesto for the discipline that was widely diffused in France. It deserves also to be widely known beyond its borders for its eloquent evocation of the fruitful interplay between theory and practice that characterizes our discipline. In this era of "Big Data," his definition of statistics is worth noting:

> Statistics, you may tell me, is the science of data. But this is far from true, since such a vision confounds "data," which include all the results of observations, not always numerical, with "statistics," whose object is to construct the methodology that allows us to extract the information of interest. Statistics is to data what the lemon-squeezer is to the fruit. It works to collect all the juice without losing a single drop.

He describes the intellectual feast set before him when he first entered the discipline:

> For a while, looking at the range of statistical problems that needed solution, I was like a character in a Jorge Luis Borges novel, wandering in the infinity of an immense library.

He then talks about his work with the Compagnie Française des Pétroles on predicting extreme wave heights. I know of no better short description of the essential dialectic between application, methodology and theory in statistics:

> I remember taking chaotic helicopter flights towards platforms in the middle of powerful seas (...) These statistical problems allowed me to live through a variety of adventures in the real world of industry. I was disappointed to realize that classical mathematical tools were often totally inadequate to meet the needs of the problems. New tools were needed, and these in turn generated a whole set of theoretical problems. All this made me return with renewed enthusiasm to the conceptual beauty of statistical theory.

4 Applied Statistician

Deheuvels served as a consultant to the Compagnie Française des Pétroles (CFP, later Total) for 20 years, from 1974 to 1994. One of his early projects was the calculation of the height of the 100-year wave, for deciding how high an oil-drilling platform should be built. He did extreme value calculations which gave a value of 19 m. Initially, the engineers did not believe that such a high wave could realistically occur, but they nevertheless ended up building the platform to a height of 20 m, building in a margin over Deheuvels's calculations.

Twenty years later, on January 1, 1995 the Draupner or New Year wave occurred, with a measured height at the platform of 18.5 m (Haver 2004). See Fig. 1 for a close-up of the wave height profile. If the company had not followed Deheuvels's recommendation and built it higher than 19 m, the platform would have been destroyed, at a cost of about \$400 million.

With the CFP he also worked on planning the circulation of tankers in the Kharg Island terminal in the Persian Gulf, and on designing statistical control of catalytic cracking reactors, implemented in the OPTOR computer code. Other projects with the CFP included analyzing the risks associated with the fluctuations of financial markets (Deheuvels 1981, 1998), and analyzing slug flow processes (Deheuvels et al. 1993; Deheuvels and Bernicot 1995).

Since 1994, he has been working as a consultant with the pharmaceutical company Sanofi on problems such as estimating the differences between survival distributions under different treatments in the presence of nonproportional hazard rates. He has also developed new methods for chromatographic problems (Deheuvels et al. 2000).

A considerable amount of his work has been motivated by actuarial questions. This includes his most cited paper (Csörgő et al. 1985), where he introduced a new estimator of the tail index of a distribution, generalizing, and improving on Hill's estimator. This work has also included results in risk theory (Deheuvels and Steinebach 1990).

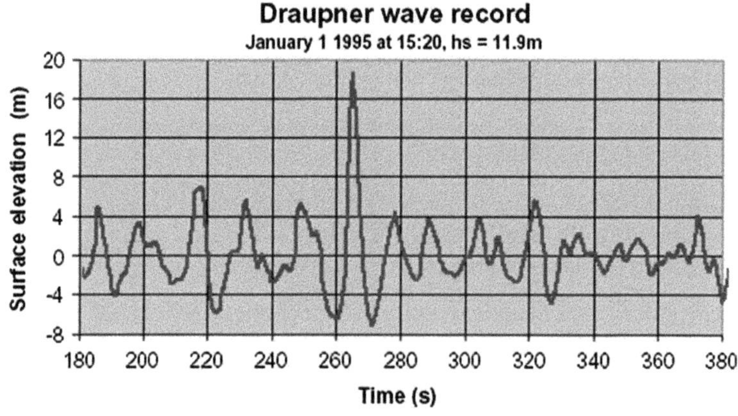

Fig. 1 Draupner wave on January 1, 1995, as measured at Draupner platform. *Source*: Haver (2004)

In his role as the only statistician member of the French Académie des Sciences, Deheuvels has spoken publicly about matters of public interest. For example, shortly after the publication of Séralini et al. (2012), which argued that genetically modified corn is harmful to rats, the Académie des Sciences convened a small committee that rapidly issued a statement criticizing the study, including its statistical underpinnings. Deheuvels legimitately complained that he had not been consulted, as the only statistician member, and he subsequently presented an alternative point of view (Deheuvels 2012, 2014). A different statistical perspective was provided by Lavielle (2013). The Séralini et al. (2012) article was subsequently retracted by the journal without the consent of its authors, a retraction that has itself been controversial. Deheuvels's position was supported by 140 French scientists in an open letter to *Le Monde* (Andalo et al. 2012).

5 Conclusion

In addition to his huge contribution to statistical theory, Paul Deheuvels has been an active mentor, advocate for the discipline of statistics, and applied statistician, developing new methods for a range of applied engineering and scientific problems.

On a more personal note, he has also done much to promote friendship among his fellow scientists around the world, many of whom have been invited to dinner by himself and his wife Joële at his home in Bourg-la-Reine outside Paris.

We wish him well in his continued activity, as his career moves into a new stage.

Acknowledgments This work was supported by a Science Foundation Ireland E.T.S. Walton visitor award, grant reference 11/W.1/I2079. I am grateful to David Mason and Paul Deheuvels for helpful comments.

References

Andalo, et al. C. (2012). Science et conscience. Le Monde, November 14, 2012. http://www.lemonde.fr/idees/article/2012/11/14/science-et-conscience_1790174_3232.html.

Csörgő, S., Deheuvels, P., & Mason, D. M. (1985). Kernel estimates of the tail index of a distribution. *Annals of Statistics, 13*, 1050–1077.

Deheuvels, P. (1973). Sur la convergence de sommes de minimums de variables aléatoires. *Comptes Rendus de l'Académie des Sciences de Paris, Série A, 276*, 309–312.

Deheuvels, P. (1981). La prévision des séries économique, une technique subjective. *Archives de l'I.S.M.E.A., 4*, 729–748.

Deheuvels, P. (1982). *La Probabilité, le Hasard et la Certitude* (1st ed.). Collection Que Sais-je? Presses Universitaires de France.

Deheuvels, P. (1998). Fluctuations du processus de risque et coefficient d'adjustement. Atti del XXI Convegno Annuale A.M.A.S.E.S. *Appendice, 21*, 41–60.

Deheuvels, P. (2001). L'aventure scientifique de la statistique. *Discours et Notice Biographiques* (pp. 47–52). Tome IV: Académie des Sciences.

Deheuvels, P. (2008). *La Probabilité, le Hasard et la Certitude* (4th ed.). Collection Que Sais-je? Presses Universitaires de France.

Deheuvels, P. (2012). Étude de Séralini sur les OGM: pourquoi sa méthodologie est statistiquement bonne. *Le Nouvel Observateur*, October 7, 2012.

Deheuvels, P. (2014). On testing stochastic dominance by exceedance, precedence and other distribution-free tests, with applications. In V. Couallier, L. Gerville-Reache, C. Huber-Carol, N. Limnios, & M. Mesbah (Eds.), *Statistical models and methods for reliability and survival analysis* (pp. 145–159). New York: Wiley.

Deheuvels, P., & Steinebach, J. (1990). On some alternative estimates of the adjustment coefficient in risk theory. *Scandinavian Actuarial Journal, 1990*, 135–159.

Deheuvels, P., & Mason, D. M. (1992). Functional laws of the iterated logarithm for the increments of empirical and quantile processes. *Annals of Probability, 20*, 1248–1287.

Deheuvels, P., & Bernicot, M. (1995). A unified model for slug flow generation. *Revue de l'Institut Français du Pétrole, 50*, 219–236.

Deheuvels, P., & Ouadah, S. (2013). Uniform-in-bandwidth functional limit laws. *Journal of Theoretical Probability, 26*, 697–721.

Deheuvels, P., Devroye, L., & Lynch, J. (1986). Exact convergence rate in the limit theorems of Erdös-Rényi and Shepp. *Annals of Probability, 14*, 209–223.

Deheuvels, P., Bernicot, M., & Dhulesia, H. (1993). On fractal dimension and modelling of bubble and slug flow processes. In A. Wilson (Ed.), *Multi phase production, BHR group conference series 4*. London: Mechanical Engineering Publications Ltd.

Deheuvels, P., Richard, J. P., & Pierre-Loti-Viaud, D. (2000). Une approche nouvelle pour l'exploitation des empreintes chromatographiques. *Annales des falsifications, de l'expertise chimique et toxicol, 92*, 455–470.

Haver, S. (2004). A possible freak wave event measured at the Draupner jacket January 1 1995. In Rogue Waves Workshop, Brest. http://www.ifremer.fr/web-com/stw2004/rw/fullpapers/walk_on_haver.pdf.

Lavielle, M. (2013). Rôle et limites de la statistique dans l'évaluation des risques sanitaires liés aux OGM. *Statistique et Société, 1*, 11–16.

Raftery, A. E. (1979). Un problème de ficelle. *Comptes Rendus de l'Académie des Sciences de Paris, Série A, 289*, 703–705.

Raftery, A. E. (1980). Estimation efficace pour un processus autorégressif exponentiel à densité discontinue. *Publications de l'Institut de Statistique des Universités de Paris, 25*, 64–91.

Raftery, A. E. (1981). Un processus autorégressif à loi marginale exponentielle: propriétés asymptotiques et estimation de maximum de vraisemblance. *Annales Scientifiques de l'Université de Clermont-Ferrand, 69*, 149–160.

Raftery, A. E. (1982). Generalised non-normal time series models. In O. D. Anderson (Ed.), *Time series analysis: Theory and practice* (pp. 621–640). Amsterdam: North-Holland.

Séralini, G. E., Clair, E., Mesnage, R., Gress, S., Defarge, N., Malatesta, M., et al. (2012). Long term toxicity of a Roundup herbicide and a Roundup-tolerant genetically modified maize. *Food and Chemical Toxicology, 50*, 4221–4231.

Lacunary Series and Stable Distributions

István Berkes and Robert Tichy

Abstract By well-known results of probability theory, any sequence of random variables with bounded second moments has a subsequence satisfying the central limit theorem and the law of the iterated logarithm in a randomized form. In this paper we give criteria for a sequence (X_n) of random variables to have a subsequence (X_{n_k}) whose weighted partial sums, suitably normalized, converge weakly to a symmetric stable distribution with parameter $0 < \alpha < 2$.

1 Introduction

It is known that sufficiently thin subsequences of general r.v. sequences behave like i.i.d. sequences. For example, Chatterji (1974a, b) and Gaposhkin (1966, 1972) proved that if a sequence (X_n) of r.v.'s satisfies $\sup_n EX_n^2 < \infty$, then one can find a subsequence (X_{n_k}) and r.v.'s X and $Y \geq 0$ such that

$$\frac{1}{\sqrt{N}} \sum_{k \leq N} (X_{n_k} - X) \xrightarrow{d} N(0, Y) \tag{1}$$

and

$$\limsup_{N \to \infty} \frac{1}{\sqrt{2N \log \log N}} \sum_{k \leq N} (X_{n_k} - X) = Y^{1/2} \quad \text{a.s.,} \tag{2}$$

Research supported by FWF grants P24302-N18, W1230 and OTKA grant K108615.

I. Berkes (✉)
Institute of Statistics, Graz University of Technology, Kopernikusgasse 24, Graz, Austria
e-mail: berkes@tugraz.at

R. Tichy
Institute of Mathematics A, Graz University of Technology, Steyrergasse 30,
8010 Graz, Austria
e-mail: tichy@tugraz.at

© Springer International Publishing Switzerland 2015
M. Hallin et al. (eds.), *Mathematical Statistics and Limit Theorems*,
DOI 10.1007/978-3-319-12442-1_2

7

where $N(0, Y)$ denotes the distribution of the r.v. $Y^{1/2}\zeta$ where ζ is an $N(0, 1)$ r.v. independent of Y. Komlós (1967) proved that under $\sup_n E|X_n| < \infty$ there exists a subsequence (X_{n_k}) and an integrable r.v. X such that

$$\lim_{N \to \infty} \frac{1}{N} \sum_{k=1}^{N} X_{n_k} = X \qquad \text{a.s.}$$

and Chatterji (1970) showed that under $\sup_n E|X_n|^p < \infty$, $0 < p < 2$ the conclusion of the previous theorem should be changed to

$$\lim_{N \to \infty} \frac{1}{N^{1/p}} \sum_{k=1}^{N} (X_{n_k} - X) = 0 \qquad \text{a.s.}$$

for some X with $E|X|^p < \infty$. Note the randomization in all these examples: the role of the mean and variance of the subsequence (X_{n_k}) is played by random variables X, Y. On the basis of these and several other examples, Chatterji (1972) formulated the following heuristic principle:

Subsequence Principle. *Let T be a probability limit theorem valid for all sequences of i.i.d. random variables belonging to an integrability class L defined by the finiteness of a norm $\| \cdot \|_L$. Then if (X_n) is an arbitrary (dependent) sequence of random variables satisfying $\sup_n \|X_n\|_L < +\infty$ then there exists a subsequence (X_{n_k}) satisfying T in a mixed form.*

In a profound paper, Aldous (1977) proved the validity of this principle for all limit theorems concerning the almost sure or distributional behavior of a sequence of functionals $f_k(X_1, X_2, \ldots)$ of a sequence (X_n) of r.v.'s. Most "usual" limit theorems belong to this class; for precise formulations, discussion and examples we refer to Aldous (1977). On the other hand, the theory does not cover functionals f_k containing parameters (as in weighted limit theorems) or allows limit theorems to involve other type of uniformities. Such uniformities play an important role in analysis. For example, if from a sequence (X_n) of r.v.'s with finite pth moments ($p \geq 1$) one can select a subsequence (X_{n_k}) such that

$$K^{-1} \left(\sum_{i=1}^{N} a_i^2 \right)^{1/2} \leq \left\| \sum_{i=1}^{N} a_i X_{n_i} \right\|_p \leq K \left(\sum_{i=1}^{N} a_i^2 \right)^{1/2}$$

for some constant $0 < K < \infty$, for every $N \geq 1$ and every $(a_1, \ldots, a_N) \in \mathbb{R}^N$, then the subspace of L^p spanned by (X_n) contains a subspace isomorphic to Hilbert space. Such embedding arguments go back to the classical paper of Kadec and Pelczynski (1962) and play an important role in Banach space theory, see e.g. Dacunha-Castelle and Krivine (1975), Aldous (1981). In the theory of orthogonal series and in Banach space theory we frequently need subsequences (f_{n_k}) of a sequence (f_n) such that

$\sum_{k=1}^{\infty} a_k f_{n_k}$ converges a.e. or in norm, after any permutation of its terms, for a class of coefficient sequences (a_k). Here we need uniformity both over a class of coefficient sequences (a_k) and over all permutations of the terms of the series. A number of uniform limit theorems for subsequences have been proved by ad hoc arguments. Révész (1965) showed that for any sequence (X_n) of r.v.'s satisfying $\sup_n E X_n^2 < \infty$ one can find a subsequence (X_{n_k}) and a r.v. X such that $\sum_{k=1}^{\infty} a_k (X_{n_k} - X)$ converges a.s. provided $\sum_{k=1}^{\infty} a_k^2 < \infty$. Under $\sup_n \|X_n\|_\infty < +\infty$, Gaposhkin (1966) showed that there exists a subsequence (X_{n_k}) and r.v.'s X and $Y \geq 0$ such that for any real sequence (a_k) satisfying the uniform asymptotic negligibility condition

$$\max_{1 \leq k \leq N} |a_k| = o(A_N), \qquad A_N = \left(\sum_{k=1}^{N} a_k^2 \right)^{1/2} \tag{3}$$

we have

$$\frac{1}{A_N} \sum_{k \leq N} a_k (X_{n_k} - X) \xrightarrow{d} N(0, Y) \tag{4}$$

and for any real sequence (a_k) satisfying the Kolmogorov condition

$$\max_{1 \leq k \leq N} |a_k| = o(A_N / (\log \log A_N)^{1/2}) \tag{5}$$

we have

$$\frac{1}{(2A_N \log \log A_N)^{1/2}} \sum_{k \leq N} a_k (X_{n_k} - X) = Y^{1/2} \qquad \text{a.s.} \tag{6}$$

For a fixed coefficient sequence (a_k) the above results follow from Aldous' general theorems, but the subsequence (X_{n_k}) provided by the proofs depends on (a_k) and to find a subsequence working for all (a_k) simultaneously requires a uniformity which is, in general, not easy to establish and it can fail in important situations. (See Guerre and Raynaud (1986) for a natural problem where uniformity is not valid.) Aldous (1977) used an equicontinuity argument to prove a permutation-invariant version of the theorem of Révész above, implying that every orthonormal system (f_n) contains a subsequence (f_{n_k}) which, using the standard terminology, is an *unconditional convergence system*. This had been a long-standing open problem in the theory of orthogonal series (see Uljanov 1964, p. 48) and was first proved by Komlós (1974). In Berkes (1989) we used the method of Aldous to prove extensions of the Kadec-Pelczynski theorem, as well as to get selection theorems for almost symmetric sequences. The purpose of the present paper is to use a similar technique to prove a uniform limit theorem of probabilistic importance, namely the analogue of Gaposhkin's uniform CLT (3)–(4) in the case when the limit distribution of the normed sums is a symmetric stable law with parameter $0 < \alpha < 2$. To formulate our result, we need some definitions. Using the terminology of Berkes and Rosenthal (1985), call the sequence (X_n) of r.v.'s *determining* if it has a limit distribution relative to any set A in the probability space with $P(A) > 0$, i.e., for any $A \subset \Omega$ with $P(A) > 0$ there exists a distribution function F_A such that

$$\lim_{n \to \infty} P(X_n < t \mid A) = F_A(t)$$

for all continuity points t of F_A. By an extension of the Helly–Bray theorem (see Berkes and Rosenthal 1985), every tight sequence of r.v.'s contains a determining subsequence. Hence in studying the asymptotic behavior of thin subsequences of general tight sequences we can assume without loss of generality that our original sequence (X_n) is determining. By Berkes and Rosenthal (1985, Proposition 2.1), for any continuity point t of the limit distribution function F_Ω, the sequence $I\{X_n \leq t\}$ converges weakly in L^∞ to some r.v. G_t; clearly $G_s \leq G_t$ a.s. for any $s \leq t$. (A sequence (ξ_n) of bounded r.v.'s is said to converge to a bounded r.v. ξ weakly in L^∞ if $E(\xi_n \eta) \longrightarrow E(\xi \eta)$ for any integrable r.v. η. To avoid confusion, we will call ordinary weak convergence of probability measures distributional convergence and denote it by $\xrightarrow{d}$. Using a standard procedure (see, e.g., Révész 1967, Lemma 6.1.4), by choosing a dense countable set D of continuity points of F_Ω, one can construct versions of $G_t, t \in D$ such that, for every fixed $\omega \in \Omega$, the function $G_t(\omega), t \in D$ extends to a distribution function. Letting μ denote the corresponding measure, μ is called the *limit random measure* of (X_n); it was introduced by Aldous (1977); for properties and applications see Aldous (1981), Berkes (1989), Berkes and Péter (1986), Berkes and Rosenthal (1985). Clearly, μ can be considered as a measurable map from the underlying probability space $(\Omega, \mathscr{F}, P)$ to the space $\mathscr{M}$ of probability measures on $\mathbb{R}$ equipped with the Prohorov metric π. It is easily seen that for any A with $P(A) > 0$ and any continuity point t of F_A we have

$$F_A(t) = E_A(\mu(-\infty, t)), \tag{7}$$

where E_A denotes conditional expectation given A. Note that μ depends on the actual r.v.'s X_n, but the distribution of μ in $(\mathscr{M}, \pi)$ depends solely on the distribution of the sequence (X_n). The situation concerning the unweighted CLT for lacunary sequences can now be summarized by the following theorem:

Theorem 1 *Let (X_n) be a determining sequence of r.v.'s with limit random measure μ. Then there exists a subsequence (X_{n_k}) satisfying, together with all of its subsequences, the CLT (1) with suitable r.v.'s X and $Y \geq 0$ if and only if*

$$\int_{-\infty}^{\infty} x^2 d\mu(x) < \infty \quad a.s. \tag{8}$$

The sufficiency part of the theorem is contained in the general subsequence theorems in Aldous (1977); the necessity was proved in Berkes and Tichy (2015). Note that the condition for the CLT for lacunary subsequences of (X_n) is given in terms of the limit random measure of (X_n) and this condition is the exact analogue of the condition in the i.i.d. case, only the common distribution of the i.i.d. variables is replaced by the limit random measure. Note also that the existence of second moments of (X_n) (or the existence of any moments) is not necessary for the conclusion of Theorem 1.

In this paper we investigate the analogous question in case of a nonnormal stable limit distribution, i.e., the question under what conditions a sequence (X_n) of r.v.'s has a subsequence (X_{n_k}) whose weighted partial sums, suitably normalized, converge weakly to an α-stable distribution, $0 < \alpha < 2$. Let, for $c > 0$ and $0 < \alpha < 2$, $G_{\alpha,c}$ denote the distribution function with characteristic function $\exp(-c|t|^\alpha)$ and let $S = S(\alpha, c)$ denote the class of symmetric distributions on $\mathbb{R}$ with characteristic function φ satisfying

$$\varphi(t) = 1 - c|t|^\alpha + o(|t|^\alpha) \qquad \text{as } t \to 0. \tag{9}$$

Our main result is

Theorem 2 *Let $0 < \alpha < 2$, $c > 0$ and let (X_n) be a determining sequence of r.v.'s with limit random measure μ. Assume that $\mu \in S(\alpha, c)$ with probability 1. Then there exists a subsequence (X_{n_k}) such that for any real sequence (a_k) satisfying*

$$\max_{1 \le k \le N} |a_k| = o(A_N), \quad A_N = \left(\sum_{k=1}^{N} |a_k|^\alpha \right)^{1/\alpha} \tag{10}$$

we have

$$A_N^{-1} \sum_{k=1}^{N} a_k X_{n_k} \xrightarrow{d} G_{\alpha,c}.$$

Condition (9) holds provided the corresponding (symmetric) distribution function F satisfies

$$1 - F(x) = c_1 x^{-\alpha} + \beta(x) x^{-\alpha}, \qquad x > 0$$

where $c_1 > 0$ is a suitable constant, $\beta(x)$ is nonincreasing for $x \ge x_0$ and $\lim_{x \to \infty} \beta(x) = 0$. (See Berkes and Dehling 1989, Lemma 3.2.) Apart from the monotonicity condition, this is equivalent to the fact that F is in the domain of normal attraction of a symmetric stable distribution. (See, e.g., Feller 1971, p. 581.) It is natural to ask if the conclusion of Theorem 2 remains valid (with a suitable centering factor) assuming only that $\mu \in S$ a.s. where S denotes the domain of normal attraction of a fixed stable distribution. From the theory in Aldous (1977) it follows that the answer is affirmative in the unweighted case $a_k = 1$, but in the uniform weighted case the question remains open. Symmetry plays no essential role in the proof of Theorem 2; it is used only in Lemma 2 and at the cost of minor changes in the proof, (9) can be replaced by a condition covering nonsymmetric distributions as well. But since we do not know the optimal condition, we restricted our investigations to the case (9) where the technical details are the simplest and the idea of the proof becomes more transparent.

Given a sequence (X_n^*) of r.v.'s and a random measure μ defined on a probability space $(\Omega, \mathscr{F}, P)$ such that X_n^* are conditionally i.i.d. given μ with conditional distribution μ, the limit random measure of (X_n^*) is easily seen to be μ. Thus in the case $\mu \in S(\alpha, c)$ a.s., (X_n^*) provides a simple example for a sequence satisfying the conditions of Theorem 2. (Since (X_n^*) is exchangeable, in this case the conclusion of Theorem 2 holds for the whole sequence (X_n^*) without passing to any subsequence.) Theorem 2 shows that any deterministic sequence (X_n) with a limit random measure μ satisfying $\mu \in S(\alpha, c)$ a.s. has a subsequence (X_{n_k}) whose weighted partial sums behave, in a uniform sense, similarly to those of (X_n^*).

2 Proof of Theorem 2

As the first step of the proof, we select a sequence $n_1 < n_2 < \ldots$ of integers such that, after a suitable discretization of (X_n), we have

$$P(X_{n_k} \in J | X_{n_1}, \ldots, X_{n_{k-1}})(\omega) \longrightarrow \mu(\omega, J) \quad \text{a.s.} \tag{11}$$

for a large class of intervals J. This step follows exactly Aldous (1977), see Proposition 11 there for details. Let (Y_n) be a sequence of r.v.'s on $(\Omega, \mathscr{F}, P)$ such that, given $\mathbf{X}$ and μ, the r.v.'s $Y_1, Y_2, \ldots$ are conditionally i.i.d. with distribution μ, i.e.,

$$P(Y_1 \in B_1, \ldots, Y_k \in B_k | \mathbf{X}, \mu) = \prod_{i=1}^{k} P(Y_i \in B_i | \mathbf{X}, \mu) \quad \text{a.s.} \tag{12}$$

$$P(Y_j \in B | \mathbf{X}, \mu) = \mu(B) \quad \text{a.s.} \tag{13}$$

for any j, k and Borel sets $B, B_1, \ldots, B_k$ on the real line. Such a sequence (Y_n) always exists after redefining (X_n) and μ on a suitable, larger probability space; for example, one can define the triple $((X_n), \mu, (Y_n))$ on the product space $\mathbb{R}^\infty \times \mathscr{M} \times \mathbb{R}^\infty$ as done in Aldous (1977, p. 72). This redefinition will not change the distribution of the sequence (X_n) and thus by Berkes and Rosenthal (1985, Proposition 2.1) it remains determining. Since the random measure μ depends on the variables X_n themselves and not only on the distribution of (X_n), this redefinition will change μ, but not the joint distribution of (X_n) and μ on which our results depend. Using (11) and a martingale argument, in Aldous (1977, Lemma 12), it is shown that

Lemma 1 *For every $\sigma(\mathbf{X})$-measurable r.v. Z and any $j \geq 1$ we have*

$$(X_{n_k}, Z) \xrightarrow{d} (Y_j, Z) \quad as \ k \to \infty.$$

We now construct a further subsequence of (X_{n_k}) satisfying the conclusion of Theorem 2. By reindexing our variables, we can assume that Lemma 1 holds with $n_k = k$. For our construction we need some auxiliary considerations. For a (nonrandom) measure $\mu \in S(\alpha, c)$, the corresponding characteristic function φ satisfies

$$\varphi(t) = 1 - c|t|^\alpha + \beta(t)|t|^\alpha, \qquad t \in \mathbb{R} \tag{14}$$

where β is a bounded continuous function on $\mathbb{R}$ with $\beta(0) = 0$. Given $\mu_1, \mu_2 \in S(\alpha, c)$ with characteristic functions φ_1, φ_2 and corresponding functions β_1, β_2 in (14), define

$$\rho(\mu_1, \mu_2) = \sup_{0 \le |t| \le 1} |\beta_1(t) - \beta_2(t)| + \sum_{k=0}^{\infty} \frac{1}{2^k} \sup_{2^k \le |t| \le 2^{k+1}} |\beta_1(t) - \beta_2(t)|. \tag{15}$$

Clearly, ρ satisfies the triangle inequality and if $\rho(\mu_1, \mu_2) = 0$, then $\beta_1(t) = \beta_2(t)$ and consequently $\varphi_1(t) = \varphi_2(t)$ for all $t \in \mathbb{R}$ and thus $\mu_1 = \mu_2$. Hence, ρ is a metric on $S(\alpha, c)$. If $\mu, \mu_1, \mu_2, \ldots \in S(\alpha, c)$ with corresponding characteristic functions $\varphi, \varphi_1, \varphi_2, \ldots$ and functions $\beta, \beta_1, \beta_2, \ldots$, then $\rho(\mu_n, \mu) \to 0$ implies that $\beta_n(t) \to \beta(t)$ and consequently $\varphi_n(t) \to \varphi(t)$ uniformly on compact intervals and thus $\mu_n \xrightarrow{d} \mu$. Conversely, if $\mu_n \xrightarrow{d} \mu$, then $\varphi_n(t) \to \varphi(t)$ uniformly on compact intervals and thus $\beta_n(t) \to \beta(t)$ uniformly on compact intervals not containing 0. Note that $\lim_{t \to 0} \beta_n(t) = 0$ for any fixed n by the definition of $S(\alpha, c)$; if this relation holds uniformly in n, then $\beta_n(t) \to \beta(t)$ will hold uniformly also on all compact intervals containing 0 and upon observing that (14) implies $|\beta(t)| \le |t|^{-\alpha}|\varphi(t) - 1| + c \le c + 2$ for $|t| \ge 1$ and thus the total contribution of the terms of the sum in (15) for $k \ge M$ is $\le 4(c+2)2^{-M}$, it follows that $\rho(\mu_n, \mu) \to 0$. Thus if for a class $H \subset S(\alpha, c)$ we have $\lim_{t \to 0} \beta(t) = 0$ uniformly for all functions β corresponding to measures in H, then in H convergence of elements in Prohorov metric and in the metric ρ are equivalent.

Let now $\varphi(t) = \varphi(t, \omega)$ denote the characteristic function of the random measure $\mu = \mu(\omega)$. By the assumption $\mu \in S(\alpha, c)$ a.s. of Theorem 2, we have

$$\varphi(t, \omega) = 1 - c|t|^\alpha + \beta(t, \omega)|t|^\alpha, \qquad t \in \mathbb{R}, \ \omega \in \Omega \tag{16}$$

where $\lim_{t \to 0} \beta(t, \omega) = 0$ a.s. Let $\xi_n(\omega) = \sup_{|t| \le 1/n} |\beta(t, \omega)|$, then we have $\lim_{n \to \infty} \xi_n(\omega) = 0$ a.s. and thus by Egorov's theorem (see Egorov 1911) for any $\varepsilon > 0$ there exists a measurable set $A \subset \Omega$ with $P(A) \ge 1 - \varepsilon$ such that $\lim_{n \to \infty} \xi_n(\omega) = 0$ and consequently $\lim_{t \to 0} \beta(t, \omega) = 0$ uniformly on A. Considering A as a new probability space, we will show that there exists a subsequence (X_{n_k}) (depending on A) satisfying the conclusion of Theorem 2 together with all its subsequences. By a diagonal argument we can get then a subsequence (X_{n_k}) satisfying the conclusion of Theorem 2 on the original Ω. Thus without loss of generality we can assume in the sequel that the function $\beta(t, \omega)$ in (16) satisfies $\lim_{t \to 0} \beta(t, \omega) = 0$ uniformly in $\omega \in \Omega$ and thus by the remarks in the previous paragraph, in the support of the random measure μ the Prohorov metric and the metric ρ generate the same convergence.

Lemma 2 *Let $\mu_1, \mu_2 \in S(\alpha, c)$ satisfy (9), let $Z_1, \ldots, Z_n$ and $Z_1^*, \ldots, Z_n^*$ be i.i.d. sequences with respective distributions μ_1, μ_2. Let $(a_1, \ldots, a_n) \in \mathbb{R}^n$, $A_n = \left(\sum_{k=1}^n |a_k|^\alpha\right)^{1/\alpha}$, $\delta_n = \max_{1 \le k \le n} |a_k|/A_n$. Then for $|t|\delta_n \le 1$ we have*

$$\left| E \exp\left(it A_n^{-1} \sum_{k=1}^n a_k Z_k\right) - E \exp\left(it A_n^{-1} \sum_{k=1}^n a_k Z_k^*\right)\right| \le |t|^\alpha \rho(\mu_1, \mu_2) \quad (17)$$

where ρ is defined by (15).

Proof Letting φ_1, φ_2 denote the characteristic function of the Z_k's resp. Z_k^*'s and using (14), (10) and the inequality

$$\left|\prod_{k=1}^n x_k - \prod_{k=1}^n y_k\right| \le \sum_{k=1}^n |x_k - y_k|,$$

valid for $|x_k| \le 1$, $|y_k| \le 1$ we get that for $|t|\delta_n \le 1$ the left-hand side of (17) equals

$$\left|\prod_{k=1}^n \varphi_1(ta_k/A_n) - \prod_{k=1}^n \varphi_2(ta_k/A_n)\right| \le \sum_{k=1}^n |\varphi_1(ta_k/A_n) - \varphi_2(ta_k/A_n)|$$

$$\le \sum_{k=1}^n |\beta_1(ta_k/A_n) - \beta_2(ta_k/A_n)||ta_k/A_n|^\alpha \le \sup_{|x| \le |t|\delta_n} |\beta_1(x) - \beta_2(x)| \sum_{k=1}^n |ta_k/A_n|^\alpha$$

$$= |t|^\alpha \sup_{|x| \le |t|\delta_n} |\beta_1(x) - \beta_2(x)| \le |t|^\alpha \rho(\mu_1, \mu_2).$$

Remark The proof of Lemma 2 shows that for any $t \in \mathbb{R}$ the left-hand side of (17) cannot exceed $|t|^\alpha \sup_{|x| \le |t|\delta_n} |\beta_1(x) - \beta_2(x)|$, a fact that will be useful in the sequel.

Given probability measures ν_n, ν on the Borel sets of a separable metric space (S, d) we say, as usual, that $\nu_n \xrightarrow{d} \nu$ if

$$\int_S f(x)d\nu_n(x) \longrightarrow \int_S f(x)d\nu(x) \quad \text{as } n \to \infty \tag{18}$$

for every bounded, real-valued continuous function f on S. (18) is clearly equivalent to

$$Ef(Z_n) \longrightarrow Ef(Z) \tag{19}$$

where Z_n, Z are r.v.'s valued in (S, d) (i.e., measurable maps from some probability space to (S, d)) with distribution ν_n, ν.

Lemma 3 (see Ranga Rao 1962). *Let (S, d) be a separable metric space and let $\nu, \nu_1, \nu_2, \ldots$ be probability measures on the Borel sets of (S, d) such that $\nu_n \xrightarrow{d} \nu$. Let $\mathscr{G}$ be a class of real-valued functions on (S, d) such that*

(a) $\mathcal{G}$ is locally equicontinuous, i.e., for every $\varepsilon > 0$ and $x \in S$ there is a $\delta = \delta(\varepsilon, x) > 0$ such that $y \in S$, $d(x, y) \le \delta$ imply $|f(x) - f(y)| \le \varepsilon$ for every $f \in \mathcal{G}$.
(b) There exists a continuous function $g \ge 0$ on S such that $|f(x)| \le g(x)$ for all $f \in \mathcal{G}$ and $x \in S$ and

$$\int_S g(x)dv_n(x) \longrightarrow \int_S g(x)dv(x) \ (< \infty) \ \text{as} \ n \to \infty. \tag{20}$$

Then

$$\int_S f(x)dv_n(x) \longrightarrow \int_S f(x)dv(x) \ \text{as} \ n \to \infty \tag{21}$$

uniformly in $f \in \mathcal{G}$.

Assume now that (X_n) satisfies the assumptions of Theorem 2, fix $t \in \mathbb{R}$ and for any $n \ge 1$, $(a_1, \ldots, a_n) \in \mathbb{R}^n$ let

$$\psi(a_1, \ldots, a_n) = E \exp\left(itA_n^{-1}\sum_{k=1}^n a_k Y_k\right), \tag{22}$$

where $A_n = (\sum_{k=1}^n |a_k|^\alpha)^{1/\alpha}$ and (Y_k) is the sequence of r.v.'s defined before Lemma 1. We show that for any $\varepsilon > 0$ there exists a sequence $n_1 < n_2 < \cdots$ of integers such that

$$(1 - \varepsilon)\psi(a_1, \ldots, a_k) \le E \exp\left(itA_k^{-1}\sum_{i=1}^k a_i X_{n_i}\right) \le (1 + \varepsilon)\psi(a_1, \ldots, a_k) \tag{23}$$

for all $k \ge 1$ and all (a_k) satisfying (10); moreover, (23) remains valid for every further subsequence of (X_{n_k}) as well. To construct n_1 we set

$$Q(\mathbf{a}, n, \ell) = \exp\left(itA_\ell^{-1}(a_1 X_n + a_2 Y_2 + \cdots + a_\ell Y_\ell)\right)$$

$$R(\mathbf{a}, \ell) = \exp\left(itA_\ell^{-1}(a_1 Y_1 + a_2 Y_2 + \cdots + a_\ell Y_\ell)\right)$$

for every $n \ge 1$, $\ell \ge 2$ and $\mathbf{a} = (a_1, \ldots, a_\ell) \in R^\ell$. We show that

$$E\left\{\frac{Q(\mathbf{a}, n, \ell)}{\psi(\mathbf{a})}\right\} \longrightarrow E\left\{\frac{R(\mathbf{a}, \ell)}{\psi(\mathbf{a})}\right\} \ \text{as} \ n \to \infty \ \text{uniformly in} \ \mathbf{a}, \ell. \tag{24}$$

(The right side of (24) equals 1.) To this end we recall that, given $\mathbf{X}$ and μ, the r.v.'s $Y_1, Y_2, \ldots$ are conditionally i.i.d. with common conditional distribution μ and thus, given $\mathbf{X}$, μ and Y_1, the r.v.'s $Y_2, Y_3, \ldots$ are conditionally i.i.d. with distribution μ. Thus

$$E\big(Q(\mathbf{a}, n, \ell)|\mathbf{X}, \mu\big) = g^{\mathbf{a}, \ell}(X_n, \mu) \tag{25}$$

16   I. Berkes and R. Tichy

and

$$E\big(R(\mathbf{a}, \ell)|\mathbf{X}, \mu, Y_1\big) = g^{\mathbf{a},\ell}(Y_1, \mu), \tag{26}$$

where

$$g^{\mathbf{a},\ell}(u, v) = E \exp\left(itA_\ell^{-1}\left(a_1 u + \sum_{i=2}^{\ell} a_i \xi_i^{(v)}\right)\right) \qquad (u \in \mathbb{R}^1, \ v \in S)$$

and $(\xi_n^{(v)})$ is an i.i.d. sequence with distribution v. Integrating (25) and (26), we get

$$E\big(Q(\mathbf{a}, n, \ell)\big) = E g^{\mathbf{a},\ell}(X_n, \mu) \tag{27}$$

$$E\big(R(\mathbf{a}, \ell)\big) = E g^{\mathbf{a},\ell}(Y_1, \mu) \tag{28}$$

and thus (24) is equivalent to

$$E\frac{g^{\mathbf{a},\ell}(X_n, \mu)}{\psi(\mathbf{a})} \longrightarrow E\frac{g^{\mathbf{a},\ell}(Y_1, \mu)}{\psi(\mathbf{a})} \quad \text{as } n \to \infty, \quad \text{uniformly in } \mathbf{a}, \ell. \tag{29}$$

We shall derive (29) from Lemmas 1–3. Recall that ρ is a metric on $S = S(\alpha, c)$; the remarks at the beginning of this section show that on the support of μ the metric ρ and the Prohorov metric π induce the same convergence and thus the same Borel σ-field; thus the limit random measure μ, which is a random variable taking values in (S, π), can be also regarded as a random variable taking values in (S, ρ). Also, μ is clearly $\sigma(\mathbf{X})$ measurable and thus $(X_n, \mu) \xrightarrow{d} (Y_1, \mu)$ by Lemma 1. (Recall that by reindexing, Lemma 1 can be assumed to hold for $n_k = k$.) Hence, (29) will follow from Lemma 3 (note the equivalence of (18) and (19)) if we show that the class of functions

$$\left\{\frac{g^{\mathbf{a},\ell}(t, v)}{\psi(\mathbf{a})}\right\} \tag{30}$$

defined on the product metric space $(\mathbb{R} \times S, \lambda \times \rho)$ (λ denotes the ordinary distance on $\mathbb{R}$) satisfies conditions (a), (b) of Lemma 3. To see the validity of (a) let us note that by (12), (13), Y_n are conditionally i.i.d. with respect to μ with conditional distribution μ, moreover, we assumed without loss of generality that the characteristic function $\varphi(t, \omega)$ of $\mu(\omega)$ satisfies (16) with $\lim_{t \to 0} \beta(t, \omega) = 0$ uniformly in ω and thus applying Lemma 2 with $\varphi_1(t) = \varphi(t, \omega)$ and $\varphi_2(t) = \exp(-c|t|^\alpha)$ and using (10) and the remark after the proof of Lemma 2 it follows that there exists an integer n_0 and a positive constant c_0 such that $\psi(\mathbf{a}) \geq c_0$ for $n \geq n_0$ and all (a_k). Thus the validity of (a) follows from Lemma 2; the validity of (b) is immediate from $|g^{\mathbf{a},\ell}(u, v)| \leq 1$. We thus proved relation (29) and thus also (24), whence it follows (note again that the right side of (24) equals 1) that

$$\psi(\mathbf{a})^{-1} E \exp\left(itA_\ell^{-1}(a_1 X_n + a_2 Y_2 + \cdots + a_\ell Y_\ell)\right) \longrightarrow 1 \tag{31}$$

as $n \to \infty$, uniformly in ℓ, $\mathbf{a}$. Hence given $\varepsilon > 0$, we can choose n_1 so large that

$$|E \exp\left(itA_\ell^{-1}(a_1 X_n + a_2 Y_2 + \cdots + a_\ell Y_\ell)\right)$$
$$- E \exp(itA_\ell^{-1}(a_1 Y_1 + a_2 Y_2 + \cdots + a_\ell Y_\ell))| \le \frac{\varepsilon}{2}\psi(a_1, \ldots, a_\ell) \qquad (32)$$

for every ℓ, $\mathbf{a}$ and $n \ge n_1$. This completes the first induction step.

Assume now that $n_1, \ldots, n_{k-1}$ have already been chosen. Exactly in the same way as we proved (31), it follows that for $\ell > k$

$$\psi(\mathbf{a})^{-1} E \exp\left(itA_\ell^{-1}(a_1 X_{n_1} + \cdots + a_{k-1} X_{n_{k-1}} + a_k X_n + a_{k+1} Y_{k+1} + \cdots + a_\ell Y_\ell)\right)$$
$$\longrightarrow \psi(\mathbf{a})^{-1} E \exp\left(itA_\ell^{-1}(a_1 X_{n_1} + \cdots + a_{k-1} X_{n_{k-1}} + a_k Y_k + \cdots + a_\ell Y_\ell)\right) \quad \text{as } n \to \infty$$

uniformly in $\mathbf{a}$ and ℓ. Hence we can choose $n_k > n_{k-1}$ so large that

$$E \exp\left(itA_\ell^{-1}(a_1 X_{n_1} + \cdots + a_{k-1} X_{n_{k-1}} + a_k X_n + a_{k+1} Y_{k+1} + \cdots + a_\ell Y_\ell)\right)$$
$$- E \exp\left(itA_\ell^{-1}(a_1 X_{n_1} + \cdots + a_{k-1} X_{n_{k-1}} + a_k Y_k + \cdots + a_\ell Y_\ell)\right)$$
$$\le \frac{\varepsilon}{2^k}\psi(a_1, \ldots, a_\ell) \qquad (33)$$

for every $(a_1, \ldots, a_\ell) \in R^\ell$, $\ell > k$ and $n \ge n_k$. This completes the kth induction step; the so constructed sequence (n_k) obviously satisfies

$$E \exp\left(itA_\ell^{-1}(a_1 X_{n_1} + \cdots + a_\ell X_{n_\ell})\right) - E \exp\left(itA_\ell^{-1}(a_1 Y_1 + \cdots + a_\ell Y_\ell)\right)$$
$$\le \varepsilon\psi(a_1, \ldots, a_\ell)$$

for every $\ell \ge 1$ and $(a_1, \ldots, a_\ell) \in R^\ell$, i.e., (23) is valid. Since in the kth induction step n_k was chosen in such a way that the corresponding inequalities (32) (for $k = 1$) and (33) (for $k > 1$) hold not only for $n = n_k$, but for all $n \ge n_k$ as well, relation (23) remains valid for any further subsequence of (X_{n_k}).

We can now easily complete the proof of Theorem 2. Letting $\psi(a_1, \ldots, a_n, t)$ denote the function defined by (22), the validity of (23) for (X_{n_k}) and its further subsequences and a diagonal argument yield a subsequence (X_{n_k}) such that for all rational t and all rational $\varepsilon > 0$ we have

$$(1 - \varepsilon)\psi(a_1, \ldots, a_k, t) \le E \exp\left(itA_k^{-1}\sum_{i=1}^{k} a_i X_{n_i}\right)$$
$$\le (1 + \varepsilon)\psi(a_1, \ldots, a_k, t) \qquad (34)$$

for $k \geq k_0(t, \varepsilon)$ and all (a_n). Recall now that without loss of generality we assumed that the characteristic function $\varphi(t, \omega)$ of $\mu(\omega)$ satisfies (16) where $\lim_{t \to 0} \beta(t, \omega) = 0$ uniformly for $\omega \in \Omega$. Applying Lemma 2 with $\varphi_1(t) = \varphi(t, \omega)$, $\varphi_2(t) = \exp(-c|t|^\alpha)$, using the Remark after the proof of the lemma and integrating with respect to ω we get

$$|\varphi(a_1, \ldots, a_k, t) - \exp(-c|t|^\alpha)| \leq |t|^\alpha \beta^*(|t|\delta_k) \tag{35}$$

for all $k \geq 1, t \in \mathbb{R}$ and all (a_k), where $\beta^*(t)$ is a function satisfying $\lim_{t \to 0} \beta^*(t) = 0$ and $\delta_k = \max_{1 \leq j \leq k} |a_j|/A_k$. Since $\delta_k \to 0$ by (10), relations (34) and (35) imply

$$E \exp\left(itA_k^{-1} \sum_{i=1}^{k} a_i X_{n_i} \right) \longrightarrow \exp(-c|t|^\alpha) \quad \text{as } k \to \infty$$

for any rational t and any (a_k) satisfying (10), and consequently

$$A_k^{-1} \sum_{i=1}^{k} a_i X_{m_i} \xrightarrow{d} G_{\alpha,c}.$$

This completes the proof of Theorem 2.

Acknowledgments We would like to thank two anonymous referees for their comments on the paper leading to a substantial improvement of the presentation.

References

Aldous, D. J. (1977). Limit theorems for subsequences of arbitrarily-dependent sequences of random variables. *Zeitschrift für Wahrscheinlichkeitstheorie und Verwandte Gebiete, 40*, 59–82.

Aldous, D. J. (1981). Subspaces of L^1 via random measures. *Transactions of the American Mathematical Society, 267*, 445–463.

Berkes, I. (1989). On almost symmetric sequences in L^p. *Acta Mathematica Hungarica, 54*, 269–278.

Berkes, I., & Dehling, H. (1989). Almost sure and weak invariance principles for random variables attracted by a stable law. *Probability Theory and Related Fields, 83*, 331–353.

Berkes, I., & Péter, E. (1986). Exchangeable r.v.'s and the subsequence principle. *Probability Theory and Related Fields, 73*, 395–413.

Berkes, I., & Rosenthal, H. P. (1985). Almost exchangeable sequences of random variables. *Zeitschrift für Wahrscheinlichkeitstheorie und Verwandte Gebiete, 70*, 473–507.

Berkes, I., & Tichy, R., On the central limit theorem for lacunary series. Preprint.

Chatterji, S. D. (1969/1970). A general strong law. *Inventiones Mathematicae, 9*, 235–245.

Chatterji, S. D. (1972) Un principe de sous-suites dans la théorie des probabilités. *Séminaire des probabilités VI, Strasbourg. Lecture Notes in Mathematics* (Vol. 258, pp. 72–89). Berlin: Springer.

Chatterji, S. D. (1974a). A principle of subsequences in probability theory: the central limit theorem. *Advances in Mathematics, 13*, 31–54.

Chatterji, S. D. (1974b). A subsequence principle in probability theory II. The law of the iterated logarithm. *Inventiones Mathematicae, 25,* 241–251.

Dacunha-Castelle, D., & Krivine, J. L. (1975). Sous-espaces de L^1. *Comptes Rendus de l'Académie des Sciences Paris Series A-B, 280,* A645–A648.

Egorov, D. F. (1911). Sur les suites de fonctions mesurables. *Comptes Rendus de l'Académie des Sciences Paris, 152,* 244–246.

Feller, W. (1971). *An introduction to probability theory and its applications* (Vol. II). New York: Wiley.

Gaposhkin, V. F. (1966). Lacunary series and independent functions. *Russian Mathematical Surveys, 21/6,* 3–82.

Gaposhkin, V. F. (1972). Convergence and limit theorems for subsequences of random variables. *(Russian) Teor. Verojatnost. i Primenen, 17,* 401–423.

Guerre, S., & Raynaud, Y. (1986) On sequences with no almost symmetric subsequence. *Texas Functional Analysis Seminar 1985–1986, Longhorn Notes, University of Texas* (pp. 83–93), Austin.

Kadec, M. I., & Pełczyński, W. (1961/1962). Bases, lacunary sequences and complemented subspaces in the spaces L_p. *Studia Mathematica, 21,* 161–176.

Komlós, J. (1967). A generalization of a problem of Steinhaus. *Acta Mathematica Academiae Scientiarum Hungaricae, 18,* 217–229.

Komlós, J. (1974). Every sequence converging to 0 weakly in L_2 contains an unconditional convergence sequence. *Arkiv för Matematik, 12,* 41–49.

Ranga Rao, R. (1962). Relations between weak and uniform convergence of measures with applications. *Annals of Mathematical Statistics, 33,* 659–680.

Révész, P. (1965). On a problem of Steinhaus. *Acta Mathematica Academiae Scientiarum Hungaricae, 16,* 311–318.

Révész, P. (1967). *The laws of large numbers.* New York: Academic Press.

Uljanov, P. (1964). Solved and unsolved problems in the theory of trigonometric and orthogonal series. *Russian Mathematical Surveys, 19*(1), 1–62.

High-Dimensional p-Norms

Gérard Biau and David M. Mason

Abstract Let $\mathbf{X} = (X_1, \ldots, X_d)$ be a $\mathbb{R}^d$-valued random vector with i.i.d. components, and let $\|\mathbf{X}\|_p = (\sum_{j=1}^d |X_j|^p)^{1/p}$ be its p-norm, for $p > 0$. The impact of letting d go to infinity on $\|\mathbf{X}\|_p$ has surprising consequences, which may dramatically affect high-dimensional data processing. This effect is usually referred to as the *distance concentration phenomenon* in the computational learning literature. Despite a growing interest in this important question, previous work has essentially characterized the problem in terms of numerical experiments and incomplete mathematical statements. In this paper, we solidify some of the arguments which previously appeared in the literature and offer new insights into the phenomenon.

1 Introduction

In what follows, for $\mathbf{x} = (x_1, \ldots, x_d)$ a vector in $\mathbb{R}^d$ and $0 < p < \infty$, we set

$$\|\mathbf{x}\|_p = \left(\sum_{j=1}^d |x_j|^p \right)^{1/p}. \tag{1}$$

Recall that for $p \geq 1$, $\|.\|_p$ is a norm on $\mathbb{R}^d$ (the L^p-norm) but for $0 < p < 1$, the triangle inequality does not hold and $\|.\|_p$ is sometimes called a prenorm. In the sequel, we take the liberty to call p-norm a norm or prenorm of the form (1), with $p > 0$.

G. Biau (✉)
Sorbonne Universités, UPMC Univ Paris 06, F-75005 Paris, France
e-mail: gerard.biau@upmc.fr

G. Biau
Institut universitaire de France, Paris, France

D.M. Mason
University of Delaware, Newark, DE 19716, USA
e-mail: davidm@udel.edu

Now, let $\mathbf{X} = (X_1, \ldots, X_d)$ be a $\mathbb{R}^d$-valued random vector with i.i.d. components. The study of the probabilistic properties of $\|\mathbf{X}\|_p$ as the dimension d tends to infinity has recently witnessed an important research effort in the computational learning community (see, e.g., François et al. 2007, for a review). This activity is easily explained by the central role played by the quantity $\|\mathbf{X}\|_p$ in the analysis of nearest neighbor search algorithms, which are currently widely used in data management and database mining. Indeed, finding the closest matching object in an L^p-sense is of significant importance for numerous applications, including pattern recognition, multimedia content retrieving (images, videos, etc.), data mining, fraud detection, and DNA sequence analysis, just to name a few. Most of these real applications involve very high-dimensional data (for example, pictures taken by a standard camera consist of several million pixels) and the curse of dimensionality (when $d \to \infty$) tends to be a major obstacle in the development of nearest neighbor-based techniques.

The effect on $\|\mathbf{X}\|_p$ of letting d go large is usually referred to as the *distance concentration phenomenon* in the computational learning literature. It is in fact a quite vague term that encompasses several interpretations. For example, it has been observed by several authors (e.g., François et al. 2007) that, under appropriate moment assumptions, the so-called *relative standard deviation* $\sqrt{\mathrm{Var}\|\mathbf{X}\|_p}/\mathbb{E}\|\mathbf{X}\|_p$ tends to zero as d tends to infinity. Consequently, by Chebyshev's inequality (this will be rigorously established in Sect. 2), for all $\varepsilon > 0$,

$$\mathbb{P}\left\{\left|\frac{\|\mathbf{X}\|_p}{\mathbb{E}\|\mathbf{X}\|_p} - 1\right| \geq \varepsilon\right\} \to 0, \text{ as } d \to \infty.$$

This simple result reveals that the relative error made as considering $\mathbb{E}\|\mathbf{X}\|_p$ instead of the random value $\|\mathbf{X}\|_p$ becomes asymptotically negligible. Therefore, high-dimensional vectors $\mathbf{X}$ appear to be distributed on a sphere of radius $\mathbb{E}\|\mathbf{X}\|_p$.

The distance concentration phenomenon is also often expressed by considering an i.i.d. $\mathbf{X}$ sample $\mathbf{X}_1, \ldots, \mathbf{X}_n$ and observing that, under certain conditions, the *relative contrast*

$$\frac{\max_{1 \leq i \leq n} \|\mathbf{X}_i\|_p - \min_{1 \leq i \leq n} \|\mathbf{X}_i\|_p}{\min_{1 \leq i \leq n} \|\mathbf{X}_i\|_p}$$

vanishes in probability as d tends to infinity, whereas the *contrast*

$$\max_{1 \leq i \leq n} \|\mathbf{X}_i\|_p - \min_{1 \leq i \leq n} \|\mathbf{X}_i\|_p$$

behaves in expectation as $d^{1/p-1/2}$ (Beyer et al. 1999; Hinneburg et al. 2000; Aggarwal et al. 2001; Kabán 2012). Thus the ratio between the largest and smallest p-distances from the sample to the origin becomes negligible as the dimension increases, and all points seem to be located at approximately the same distance. This phenomenon may dramatically affect high-dimensional data processing, analysis, retrieval, and indexing, insofar as these procedures rely on some notion of p-norm. Accordingly, serious questions are raised as to the validity of many nearest neighbor search heuristics in high dimension, a problem that can be further exacerbated by techniques that find approximate neighbors in order to improve algorithmic performance (Beyer et al. 1999).

Even if people have now a better understanding of the distance concentration phenomenon and its practical implications, it is however our belief that there is still a serious need to solidify its mathematical background. Indeed, previous work has essentially characterized the problem in terms of numerical experiments and (often) incomplete probabilistic statements, with missing assumptions and (sometimes) defective proofs. Thus, our objective in the present paper is to solidify some of the statements which previously appeared in the computational learning literature. We start in Sect. 2 by offering a thorough analysis of the behavior of the p-norm $\|\mathbf{X}\|_p$ (as a function of p and the properties of the distribution of $\mathbf{X}$) as $d \to \infty$. Section 3 is devoted to the investigation of some new asymptotic properties of the *contrast* $\max_{1 \leq i \leq n} \|\mathbf{X}_i\|_p - \min_{1 \leq i \leq n} \|\mathbf{X}_i\|_p$, both as $d \to \infty$ and $n \to \infty$. For the sake of clarity, most technical proofs are gathered in Sect. 4. The basic tools that we shall use are the law of large numbers, the central limit theorem, moment bounds for sums of i.i.d. random variables, and a coupling inequality of Yurinskiĭ (1977).

2 Asymptotic Behavior of p-Norms

2.1 Consistency

Throughout this paper, the notation $\overset{\mathbb{P}}{\to}$ and $\overset{\mathscr{D}}{\to}$ stand for convergence in probability and in distribution, respectively. The notation $u_n = \mathrm{o}(v_n)$ and $u_n = \mathrm{O}(v_n)$ mean, respectively, that $u_n/v_n \to 0$ and $u_n \leq C v_n$ for some constant C, as $n \to \infty$. The symbols $\mathrm{o}_\mathbb{P}(v_n)$ and $\mathrm{O}_\mathbb{P}(v_n)$ denote, respectively, a sequence of random variables $\{Y_n\}_{n \geq 1}$ such that $Y_n/v_n \overset{\mathbb{P}}{\to} 0$ and Y_n/v_n is bounded in probability, as $n \to \infty$.

We start this section with a general proposition that plays a key role in the analysis.

Proposition 1 *Let $\{U_d\}_{d \geq 1}$ be a sequence of random variables such that $U_d \overset{\mathbb{P}}{\to} a$, and let φ be a real-valued measurable function which is continuous at a. Assume that*

(i) φ is bounded on $[-M, M]$ for some $M > |a|$;
(ii) $\mathbb{E}|\varphi(U_d)| < \infty$ for all $d \geq 1$.

Then, as $d \to \infty$,
$$\mathbb{E}\varphi(U_d) \to \varphi(a)$$

if and only if
$$\mathbb{E}\left(\varphi\left(U_d\right) \mathbf{1}\{|U_d| > M\}\right) \to 0. \tag{2}$$

Proof The proof is easy. Condition (i) and continuity of φ at a allow us to apply the bounded convergence theorem to get

$$\mathbb{E}\left(\varphi(U_d)\mathbf{1}\{|U_d| \leq M\}\right) \to \varphi(a).$$

Since

$$\mathbb{E}\varphi(U_d) = \mathbb{E}\left(\varphi(U_d)\mathbf{1}\{|U_d| \leq M\}\right) + \mathbb{E}\left(\varphi(U_d)\mathbf{1}\{|U_d| > M\}\right),$$

the rest of the proof is obvious. $\qquad\square$

We shall now specialize the result of Proposition 1 to the case when

$$U_d = d^{-1}\sum_{j=1}^{d} Y_j := \overline{Y}_d,$$

where $\{Y_j\}_{j\geq 1}$ is a sequence of i.i.d. Y random variables with finite mean μ. In this case, by the strong law of large numbers, $U_d \to \mu$ almost surely. The following lemma gives two sufficient conditions for (2) to hold when $U_d = \overline{Y}_d$.

Lemma 1 *let φ be a real-valued measurable function. Assume that one of the following two conditions is satisfied:*

Condition 1　　*The function $|\varphi|$ is convex on $\mathbb{R}$ and $\mathbb{E}|\varphi(Y)| < \infty$.*

Condition 2　　*For some $s > 1$,*

$$\limsup_{d\to\infty} \mathbb{E}\left|\varphi(\overline{Y}_d)\right|^s < \infty.$$

Then (2) is satisfied for the sequence $\{\overline{Y}_d\}_{d\geq 1}$ with $a = \mu$ and $M > |\mu|$.

Proof Suppose that **Condition 1** is satisfied. Then note that by the convexity assumption

$$\mathbb{E}\left(\left|\varphi(\overline{Y}_d)\right|\mathbf{1}\{|\overline{Y}_d| > M\}\right) \leq d^{-1}\sum_{j=1}^{d}\mathbb{E}\left(\left|\varphi(Y_j)\right|\mathbf{1}\{|\overline{Y}_d| > M\}\right)$$
$$= \mathbb{E}\left(|\varphi(Y)|\mathbf{1}\{|\overline{Y}_d| > M\}\right).$$

Since $M > |\mu|$, we conclude that with probability one, $|\varphi(Y)|\mathbf{1}\{|\overline{Y}_d| > M\} \to 0$. Also $|\varphi(Y)|\mathbf{1}\{|\overline{Y}_d| > M\} \leq |\varphi(Y)|$. Therefore, by the dominated convergence theorem, (2) holds.

Next, notice by Hölder's inequality with $1/r = 1 - 1/s$ that

$$\mathbb{E}\left(\left|\varphi(\overline{Y}_d)\right|\mathbf{1}\{|\overline{Y}_d| > M\}\right) \leq \left(\mathbb{E}\left|\varphi(\overline{Y}_d)\right|^s\right)^{1/s}\left(\mathbb{P}\{|\overline{Y}_d| > M\}\right)^{1/r}.$$

Since $\mathbb{P}\{|\overline{Y}_d| > M\} \to 0$, (2) immediately follows from **Condition 2**. $\qquad\square$

Let us now return to the distance concentration problem, which has been discussed in the introduction. Recall that we denote by $\mathbf{X} = (X_1, \ldots, X_d)$ a $\mathbb{R}^d$-valued random vector with i.i.d. X components. Whenever for $p > 0$ $\mathbb{E}|X|^p < \infty$, we set $\mu_p = \mathbb{E}|X|^p$. Also when $\mathrm{Var}|X|^p < \infty$, we shall write $\sigma_p^2 = \mathrm{Var}|X|^p$. Proposition 1 and Lemma 1 yield the following corollary:

Corollary 1 *Fix $p > 0$ and $r > 0$.*

(i) Whenever $r/p < 1$ and $\mathbb{E}|X|^p < \infty$,

$$\frac{\mathbb{E}\|\mathbf{X}\|_p^r}{d^{r/p}} \to \mu_p^{r/p}, \ as \ d \to \infty,$$

whereas if $\mathbb{E}|X|^p = \infty$, then

$$\lim_{d\to\infty} \frac{\mathbb{E}\|\mathbf{X}\|_p^r}{d^{r/p}} = \infty.$$

(ii) Whenever $r/p \geq 1$ and $\mathbb{E}|X|^r < \infty$,

$$\frac{\mathbb{E}\|\mathbf{X}\|_p^r}{d^{r/p}} \to \mu_p^{r/p}, \ as \ d \to \infty,$$

whereas if $\mathbb{E}|X|^r = \infty$, then, for all $d \geq 1$,

$$\frac{\mathbb{E}\|\mathbf{X}\|_p^r}{d^{r/p}} = \infty.$$

Proof We shall apply Proposition 1 and Lemma 1 to $Y = |X|^p$, $Y_j = |X_j|^p$, $j \geq 1$, and $\varphi(u) = |u|^{r/p}$.

Proof of (i)
For the first part of (i), notice that with $s = p/r > 1$

$$\mathbb{E}\left|\varphi\left(\frac{\sum_{j=1}^d |X_j|^p}{d}\right)\right|^s = \frac{\sum_{j=1}^d \mathbb{E}|X_j|^p}{d} = \mathbb{E}|X|^p < \infty.$$

This shows that sufficient **Condition 2** of Lemma 1 holds, which by Proposition 1 gives the result.

For the second part of (i) observe that for any $K > 0$

$$\mathbb{E}\left(\frac{\sum_{j=1}^d |X_j|^p}{d}\right)^{r/p} \geq \mathbb{E}\left(\frac{\sum_{j=1}^d |X_j|^p \mathbf{1}\{|X_j| \leq K\}}{d}\right)^{r/p}.$$

Observing that the right-hand side of the inequality converges to $(\mathbb{E}|X|^p\mathbf{1}\{|X| \leq K\})^{r/p}$ as $d \to \infty$, we get for any $K > 0$

$$\liminf_{d\to\infty} \mathbb{E}\left(\frac{\sum_{j=1}^{d}|X_j|^p}{d}\right)^{r/p} \geq \mathbb{E}\left(|X|^p \mathbf{1}\{|X| \leq K\}\right)^{r/p}.$$

Since K can be chosen arbitrarily large and we assume that $\mathbb{E}|X|^p = \infty$, we see that the conclusion holds.

Proof of (ii)
For the first part of (ii), note that in this case $r/p \geq 1$, so φ is convex. Moreover, note that

$$\mathbb{E}\left|\varphi\left(\frac{\sum_{j=1}^{d}|X_j|^p}{d}\right)\right| = \mathbb{E}\left(\frac{\sum_{j=1}^{d}|X_j|^p}{d}\right)^{r/p}$$
$$\leq d^{-1}\mathbb{E}|X|^r$$
$$\text{(by Jensen's inequality)}$$
$$< \infty.$$

Thus sufficient **Condition 1** of Lemma 1 holds, which by Proposition 1 leads to the result.

For the second part of (ii), observe that if $\mathbb{E}|X|^r = \infty$, then, for all $d \geq 1$,

$$\mathbb{E}\left(\frac{\sum_{j=1}^{d}|X_j|^p}{d}\right)^{r/p} \geq d^{-r/p}\mathbb{E}|X|^r = \infty. \qquad \square$$

Applying Corollary 1 with $p > 0$ and $r = 2$ yields the following important result:

Proposition 2 *Fix $p > 0$ and assume that $0 < \mathbb{E}|X|^m < \infty$ for $m = \max(2, p)$. Then, as $d \to \infty$,*

$$\frac{\mathbb{E}\|\mathbf{X}\|_p}{d^{1/p}} \to \mu_p^{1/p}$$

and

$$\frac{\mathbb{E}\|\mathbf{X}\|_p^2}{d^{2/p}} \to \mu_p^{2/p},$$

which implies

$$\frac{\sqrt{\mathrm{Var}\|\mathbf{X}\|_p}}{\mathbb{E}\|\mathbf{X}\|_p} \to 0, \ \ as \ d \to \infty.$$

This result, when correctly stated, corresponds to Theorem 5 of François et al. (2007). It expresses the fact that the *relative standard deviation* converges toward zero when the dimension grows. It is known in the computational learning literature as the *p-norm concentration* in high-dimensional spaces. It is noteworthy that, by Chebyshev's inequality, for all $\varepsilon > 0$,

$$\mathbb{P}\left\{\left|\frac{\|\mathbf{X}\|_p}{\mathbb{E}\|\mathbf{X}\|_p} - 1\right| \geq \varepsilon\right\} = \mathbb{P}\left\{\left|\,\|\mathbf{X}\|_p - \mathbb{E}\|\mathbf{X}\|_p\,\right| \geq \varepsilon\mathbb{E}\|\mathbf{X}\|_p\right\}$$

$$\leq \frac{\mathrm{Var}\,\|\mathbf{X}\|_p}{\varepsilon^2\mathbb{E}^2\|\mathbf{X}\|_p} \to 0, \text{ as } d \to \infty. \tag{3}$$

That is, $\|\mathbf{X}\|_p/\mathbb{E}\|\mathbf{X}\|_p \xrightarrow{\mathbb{P}} 1$ or, in other words, the sequence $\{\|\mathbf{X}\|_p\}_{d \geq 1}$ is relatively stable (Boucheron et al. 2013). This property guarantees that the random fluctuations of $\|\mathbf{X}\|_p$ around its expectation are of negligible size when compared to the expectation, and therefore most information about the size of $\|\mathbf{X}\|_p$ is given by $\mathbb{E}\|\mathbf{X}\|_p$ as d becomes large.

2.2 Rates of Convergence

The asymptotic concentration statement of Corollary 1 can be made more precise by means of rates of convergence, at the price of stronger moment assumptions. To reach this objective, we first need a general result to control the behavior of a function of an i.i.d. empirical mean around its true value. Thus, assume that $\{Y_j\}_{j \geq 1}$ are i.i.d. Y with mean μ and variance σ^2. As before, we define

$$\overline{Y}_d = d^{-1}\sum_{j=1}^{d} Y_j.$$

Let φ be a real-valued function with derivatives φ' and φ''. Khan (2004) provides sufficient conditions for

$$\mathbb{E}\varphi(\overline{Y}_d) = \varphi(\mu) + \frac{\varphi''(\mu)\sigma^2}{2d} + \mathrm{o}(d^{-2})$$

to hold. The following lemma, whose assumptions are less restrictive, can be used in place of Khan's result (2004). For the sake of clarity, its proof is postponed to Sect. 4.

Lemma 2 *Let $\{Y_j\}_{j \geq 1}$ be a sequence of i.i.d. Y random variables with mean μ and variance σ^2, and φ be a real-valued function with continuous derivatives φ' and φ'' in a neighborhood of μ. Assume that for some $r > 1$,*

$$\mathbb{E}|Y|^{r+1} < \infty \tag{4}$$

and, with $1/s = 1 - 1/r$,

$$\limsup_{d \to \infty} \mathbb{E}\left|\varphi(\overline{Y}_d)\right|^s < \infty. \tag{5}$$

Then, as $d \to \infty$,

$$\mathbb{E}\varphi(\overline{Y}_d) = \varphi(\mu) + \frac{\varphi''(\mu)\sigma^2}{2d} + o(d^{-1}).$$

The consequences of Lemma 2 in terms of p-norm concentration are summarized in the following proposition:

Proposition 3 *Fix $p > 0$ and assume that $0 < \mathbb{E}|X|^m < \infty$ for $m = \max(4, 3p)$. Then, as $d \to \infty$,*

$$\mathbb{E}\|\mathbf{X}\|_p = d^{1/p}\mu_p^{1/p} + O(d^{1/p-1})$$

and

$$\mathrm{Var}\|\mathbf{X}\|_p = \frac{\mu_p^{2/p-2}\sigma_p^2}{d^{1-2/p}p^2} + o(d^{-1+2/p}),$$

which implies

$$\frac{\sqrt{d\,\mathrm{Var}\|\mathbf{X}\|_p}}{\mathbb{E}\|\mathbf{X}\|_p} \to \frac{\sigma_p}{p\mu_p}, \ as \ d \to \infty.$$

Proposition 3 is stated without assumptions as Theorem 6 in François et al. (2007), where it is provided with an ambiguous proof. This result shows that for a fixed large d, the *relative standard deviation* evolves with p as the ratio $\sigma_p/(p\mu_p)$. For instance, when the distribution of X is uniform,

$$\mu_p = \frac{1}{p+1} \quad \text{and} \quad \sigma_p = \frac{p}{p+1}\sqrt{\frac{1}{2p+1}}.$$

In this case, we conclude that

$$\frac{\sqrt{d\,\mathrm{Var}\|\mathbf{X}\|_p}}{\mathbb{E}\|\mathbf{X}\|_p} \to \sqrt{\frac{1}{2p+1}}.$$

Thus, in the uniform setting, the limiting *relative standard deviation* is a strictly decreasing function of p. This observation is often interpreted by saying that p-norms are more concentrated for larger values of p. There are, however, distributions for which this is not the case. A counterexample is given by a balanced mixture of two-standard Gaussian random variables with mean 1 and -1, respectively (see François et al. 2007, p. 881). In that case, it can be seen that the asymptotic *relative standard deviation* with $p \leq 1$ is smaller than for values of $p \in [8, 30]$, making fractional norms more concentrated.

Proof (*Proposition* 3) Fix $p > 0$ and introduce the functions on $\mathbb{R}$

$$\varphi_1(u) = |u|^{1/p} \quad \text{and} \quad \varphi_2(u) = |u|^{2/p}.$$

Assume that $\mathbb{E}|X|^{\max(4,p)} < \infty$. Applying Corollary 1 we get that, as $d \to \infty$,

$$\mathbb{E}\left(\frac{\sum_{j=1}^d |X_j|^p}{d}\right)^{2/p} \to \mu_p^{2/p}$$

and

$$\mathbb{E}\left(\frac{\sum_{j=1}^d |X_j|^p}{d}\right)^{4/p} \to \mu_p^{4/p}.$$

This says that with $s = 2$, for $i = 1, 2$,

$$\limsup_{d\to\infty} \mathbb{E}\left|\varphi_i\left(\frac{\sum_{j=1}^d |X_j|^p}{d}\right)\right|^s < \infty.$$

Now, let $Y = |X|^p$ and set $r = 2$. If we also assume that $\mathbb{E}|Y|^{r+1} = \mathbb{E}|Y|^3 = \mathbb{E}|X|^{3p} < \infty$, we get by applying Lemma 2 to φ_1 and φ_2 that for $i = 1, 2$

$$\mathbb{E}\varphi_i(\overline{Y}_d) = \varphi_i(\mu_p) + \frac{\varphi_i''(\mu_p)\sigma_p^2}{2d} + \mathrm{o}(d^{-1}).$$

Thus, whenever $\mathbb{E}|X|^m < \infty$, where $m = \max(4, 3p)$,

$$\mathbb{E}|\overline{Y}_d|^{1/p} = \mu_p^{1/p} + \frac{1}{p}\left(\frac{1-p}{p}\right)\frac{\mu_p^{1/p-2}\sigma_p^2}{2d} + \mathrm{o}(d^{-1})$$

and

$$\mathbb{E}|\overline{Y}_d|^{2/p} = \mu_p^{2/p} + \frac{1}{p}\left(\frac{2-p}{p}\right)\frac{\mu_p^{2/p-2}\sigma_p^2}{d} + \mathrm{o}\left(d^{-1}\right).$$

Therefore, we see that

$$\mathrm{Var}|\overline{Y}_d|^{1/p} = \mathbb{E}|\overline{Y}_d|^{2/p} - \mathbb{E}^2|\overline{Y}_d|^{1/p}$$

$$= \frac{\mu_p^{2/p-2}\sigma_p^2}{dp^2} + \mathrm{o}\left(d^{-1}\right).$$

The identity $\overline{Y}_d = d^{-1}\sum_{j=1}^d |X_j|^p$ yields the desired results. $\qquad\square$

We conclude the section with a corollary, which specifies inequality (3).

Corollary 2 *Fix $p > 0$.*

(i) If $0 < \mathbb{E}|X|^m < \infty$ for $m = \max(4, 3p)$, then, for all $\varepsilon > 0$,

$$\mathbb{P}\left\{ \left| \frac{\|\mathbf{X}\|_p}{\mathbb{E}\|\mathbf{X}\|_p} - 1 \right| \geq \varepsilon \right\} \leq \frac{\sigma_p^2}{\varepsilon^2 d p^2 \mu_p^2} + \mathrm{o}(d^{-1}).$$

(ii) If for some positive constant C, $0 < |X| \leq C$ almost surely, then, for $p \geq 1$ and all $\varepsilon > 0$,

$$\mathbb{P}\left\{ \left| \frac{\|\mathbf{X}\|_p}{\mathbb{E}\|\mathbf{X}\|_p} - 1 \right| \geq \varepsilon \right\} \leq 2 \exp\left(-\varepsilon^2 \left(\frac{d^{2/p-1} \mu_p^{2/p}}{2C^2} + \mathrm{o}(d^{2/p-1}) \right) \right).$$

Proof Statement (i) is an immediate consequence of Proposition 3 and Chebyshev's inequality. Now, assume that $p \geq 1$, and let $A = [-C, C]$. For $\mathbf{x} = (x_1, \ldots, x_d) \in \mathbb{R}^d$, let $g : A^d \to \mathbb{R}$ be defined by

$$g(\mathbf{x}) = \|\mathbf{x}\|_p = \left(\sum_{j=1}^d |x_j|^p \right)^{1/p}.$$

Clearly, for each $1 \leq j \leq d$,

$$\sup_{\substack{(x_1,\ldots,x_d)\in A^d \\ x_j' \in A}} \left| g(x_1, \ldots, x_d) - g(x_1, \ldots, x_{j-1}, x_j', x_{j+1}, \ldots, x_d) \right|$$

$$= \sup_{\mathbf{x}\in A^d, x_j' \in A} \left| \|\mathbf{x}\|_p - \|\mathbf{x}'\|_p \right|,$$

where $\mathbf{x}'$ is identical to $\mathbf{x}$, except on the jth coordinate where it takes the value x_j'. It follows, by Minkowski inequality (which is valid here since $p \geq 1$), that

$$\sup_{\substack{(x_1,\ldots,x_d)\in A^d \\ x_j' \in A}} \left| g(x_1, \ldots, x_d) - g(x_1, \ldots, x_{j-1}, x_j', x_{j+1}, \ldots, x_d) \right|$$

$$\leq \sup_{\substack{\mathbf{x}\in A^d \\ x_j' \in A}} \|\mathbf{x} - \mathbf{x}'\|_p$$

$$= \sup_{(x,x')\in A^2} |x - x'| \leq 2C.$$

Consequently, using the bounded difference inequality (McDiarmid 1989), we obtain

$$\mathbb{P}\left\{\left|\frac{\|\mathbf{X}\|_p}{\mathbb{E}\|\mathbf{X}\|_p} - 1\right| \geq \varepsilon\right\} = \mathbb{P}\left\{\left|\|\mathbf{X}\|_p - \mathbb{E}\|\mathbf{X}\|_p\right| \geq \varepsilon\mathbb{E}\|\mathbf{X}\|_p\right\}$$

$$\leq 2\exp\left(-\frac{2(\varepsilon\mathbb{E}\|\mathbf{X}\|_p)^2}{4dC^2}\right)$$

$$= 2\exp\left(-\varepsilon^2\left(\frac{d^{2/p-1}\mu_p^{2/p}}{2C^2} + \mathrm{o}(d^{2/p-1})\right)\right),$$

where, in the last inequality, we used Proposition 3. This concludes the proof. $\qquad\square$

3 Minima and Maxima

Another important question arising in high-dimensional nearest neighbor search analysis concerns the relative asymptotic behavior of the minimum and maximum p-distances to the origin within a random sample. To be precise, let $\mathbf{X}_1, \ldots, \mathbf{X}_n$ be an i.i.d. $\mathbf{X}$ sample, where $\mathbf{X} = (X_1, \ldots, X_d)$ is as usual a $\mathbb{R}^d$-valued random vector with i.i.d. X components. We will be primarily interested in this section in the asymptotic properties of the difference (the *contrast*) $\max_{1\leq i\leq n}\|\mathbf{X}_i\|_p - \min_{1\leq i\leq d}\|\mathbf{X}_i\|_p$.

Assume, to start with, that n is fixed and only d is allowed to grow. Then an immediate application of the law of large numbers shows that, whenever $\mu_p = \mathbb{E}|X|^p < \infty$, almost surely as $d \to \infty$,

$$d^{-1/p}\left(\max_{1\leq i\leq n}\|\mathbf{X}_i\|_p - \min_{1\leq i\leq n}\|\mathbf{X}_i\|_p\right) \xrightarrow{\mathbb{P}} 0.$$

Moreover, if $0 < \mu_p < \infty$, then

$$\frac{\max_{1\leq i\leq n}\|\mathbf{X}_i\|_p}{\min_{1\leq i\leq n}\|\mathbf{X}_i\|_p} \xrightarrow{\mathbb{P}} 1.$$

The above ratio is sometimes called the *relative contrast* in the computational learning literature. Thus, as d becomes large, all observations seem to be distributed at approximately the same p-distance from the origin. The concept of nearest neighbor (measured by p-norms) in high dimension is therefore less clear than in small dimension, with resulting computational difficulties and algorithmic inefficiencies.

These consistency results can be specified by means of asymptotic distributions. Recall that if $Z_1, \ldots, Z_n$ are i.i.d standard normal random variables, the sample range is defined to be

$$M_n = \max_{1\leq i\leq n} Z_i - \min_{1\leq i\leq n} Z_i.$$

32   G. Biau and D.M. Mason

The asymptotic distribution of M_n is well known (see, e.g., David 1981). Namely, for any x one has

$$\lim_{n\to\infty} \mathbb{P}\left\{\sqrt{2\log n}\left(M_n - 2\sqrt{2\log n} + \frac{\log\log n + \log 4\pi}{2\sqrt{2\log n}}\right) \le x\right\}$$
$$= \int_{-\infty}^{\infty} \exp\left(-t - e^{-t} - e^{-(x-t)}\right) dt.$$

For future reference, we shall sketch the proof of this fact here. It is well known that with

$$a_n = \sqrt{2\log n} \quad \text{and} \quad b_n = \sqrt{2\log n} - \frac{1}{2}\frac{(\log\log n + \log 4\pi)}{\sqrt{2\log n}} \tag{6}$$

we have

$$\left(a_n(\max_{1\le i\le n} Z_i - b_n), a_n(\min_{1\le i\le n} Z_i + b_n)\right) \to (E, -E'), \tag{7}$$

where E and E' are independent, $E = E'$ and $\mathbb{P}\{E \le x\} = \exp(-\exp(-x))$, $-\infty < x < \infty$. (The asymptotic independence of the maximum and minimum part can be inferred from Theorem 4.2.8 of Reiss 1989, and the asymptotic distribution part from Example 2 on p. 71 of Resnick 1987.) From (7) we get

$$a_n(\max_{1\le i\le n} Z_i - \min_{1\le i\le n} Z_i) - 2a_n b_n \overset{\mathscr{D}}{\to} E + E'.$$

Clearly,

$$\mathbb{P}\{E + E' \le x\} = \int_{-\infty}^{\infty} \exp\left(-e^{-(x-t)}\right) \exp(-e^{-t})e^{-t} dt$$
$$= \int_{-\infty}^{\infty} \exp\left(-t - e^{-t} - e^{-(x-t)}\right) dt.$$

Our first result treats the case when n is fixed and $d \to \infty$.

Proposition 4 *Fix $p > 0$, and assume that $0 < \mathbb{E}|X|^p < \infty$ and $0 < \sigma_p < \infty$. Then, for fixed n, as $d \to \infty$,*

$$d^{1/2-1/p}\left(\max_{1\le i\le n} \|\mathbf{X}_i\|_p - \min_{1\le i\le n} \|\mathbf{X}_i\|_p\right) \overset{\mathscr{D}}{\to} \frac{\sigma_p \mu_p^{1/p-1}}{p} M_n.$$

To our knowledge, this is the first statement of this type in the analysis of high-dimensional nearest neighbor problems. In fact, most of the existing results merely bound the asymptotic expectation of the (normalized) difference and ratio between the max and the min, but with bounds which are unfortunately not of the same order in n as soon as $n \ge 3$ (see, e.g.,Theorem 3 in Hinneburg et al. 2000).

One of the consequences of Proposition 4 is that, for fixed n, the difference between the farthest and nearest neighbors does not necessarily go to zero in probability as d tends to infinity. Indeed, we see that the size of

$$\max_{1 \le i \le n} \|\mathbf{X}_i\|_p - \min_{1 \le i \le n} \|\mathbf{X}_i\|_p$$

grows as $d^{1/p-1/2}$. For example, this difference increases with dimensionality as $\sqrt{d}$ for the L^1 (Manhattan) metric and remains stable in distribution for the L^2 (Euclidean) metric. It tends to infinity in probability for $p < 2$ and to zero for $p > 2$. This observation is in line with the conclusions of Hinneburg et al. (2000), who argue that nearest neighbor search in a high-dimensional space tends to be meaningless for norms with larger exponents, since the maximum observed distance tends toward the minimum one. It should be noted, however, that the variance of the limiting distribution depends on the value of p.

Remark 1 Let $Z_1, \ldots, Z_n$ be i.i.d standard normal random variables, and let

$$R_n = \frac{\max_{1 \le i \le n} Z_i}{\min_{1 \le i \le n} Z_i}.$$

Assuming $\mu_p > 0$ and $0 < \sigma_p < \infty$, one can prove, using the same technique, that

$$\frac{\max_{1 \le i \le n} \|\mathbf{X}_i\|_p - d^{1/p}\mu_p}{\min_{1 \le i \le n} \|\mathbf{X}_i\|_p - d^{1/p}\mu_p} \xrightarrow{\mathcal{D}} R_n.$$

Proof (Proposition 4) Denote by $\mathbf{Z}_n$ a centered Gaussian random vector in $\mathbb{R}^n$, with identity covariance matrix. By the central limit theorem, as $d \to \infty$,

$$\sqrt{d}\left[\left(\frac{\|\mathbf{X}_1\|_p^p}{d}, \ldots, \frac{\|\mathbf{X}_n\|_p^p}{d}\right) - (\mu_p, \ldots, \mu_p)\right] \xrightarrow{\mathcal{D}} \sigma_p \mathbf{Z}_n.$$

Applying the delta method with the mapping $f(x_1, \ldots, x_n) = (x_1^{1/p}, \ldots, x_n^{1/p})$ (which is differentiable at $(\mu_p, \ldots, \mu_p)$ since $\mu_p > 0$), we obtain

$$\sqrt{d}\left[\left(\frac{\|\mathbf{X}_1\|_p}{d^{1/p}}, \ldots, \frac{\|\mathbf{X}_n\|_p}{d^{1/p}}\right) - (\mu_p^{1/p}, \ldots, \mu_p^{1/p})\right] \xrightarrow{\mathcal{D}} \frac{\sigma_p \mu_p^{1/p-1}}{p} \mathbf{Z}_n.$$

Thus, by continuity of the maximum and minimum functions,

$$d^{1/2-1/p}\left(\max_{1 \le i \le n} \|\mathbf{X}_i\|_p - \min_{1 \le i \le n} \|\mathbf{X}_i\|_p\right) \xrightarrow{\mathcal{D}} \frac{\sigma_p \mu_p^{1/p-1}}{p} M_n. \qquad \square$$

In the previous analysis, n (the sample size) was fixed whereas d (the dimension) was allowed to grow to infinity. A natural question that arises concerns the impact of letting n be a function of d such that n tends to infinity as $d \to \infty$ (Mallows 1972). Proposition 5 below offers a first answer.

Proposition 5 *Fix $p \geq 1$, and assume that $0 < \mathbb{E}|X|^{3p} < \infty$ and $\sigma_p > 0$. For any sequence of positive integers $\{n(d)\}_{d \geq 1}$ converging to infinity and satisfying*

$$n(d) = o\left(\frac{d^{1/5}}{\log^{6/5} d}\right), \quad \text{as } d \to \infty, \tag{8}$$

we have

$$\frac{p a_{n(d)} d^{1/2-1/p}}{\mu_p^{1/p-1} \sigma_p} \left(\max_{1 \leq i \leq n(d)} \|\mathbf{X}_i\|_p - \min_{1 \leq i \leq n(d)} \|\mathbf{X}_i\|_p \right) - 2 a_{n(d)} b_{n(d)} \xrightarrow{\mathscr{D}} E + E',$$

where a_n and b_n are as in (6), and E and E' are as in (7).

Proof In the following, we let $\delta(d) = 1/\log d$. For future use note that

$$\delta^2(d) \log n(d) \to 0 \quad \text{and} \quad \frac{n^5(d)}{d \delta^6(d)} \to 0, \text{ as } d \to \infty. \tag{9}$$

In the proof, we shall often suppress the dependence of n and δ on d. For $1 \leq i \leq n$, we set

$$\mathbf{X}_i = (X_{1,i}, \ldots, X_{d,i}) \quad \text{and} \quad \|\mathbf{X}_i\|_p^p = \sum_{j=1}^d |X_{j,i}|^p.$$

We see that for $n \geq 1$,

$$\left(\frac{\|\mathbf{X}_1\|_p^p - d\mu_p}{\sqrt{d}\sigma_p}, \ldots, \frac{\|\mathbf{X}_n\|_p^p - d\mu_p}{\sqrt{d}\sigma_p} \right)$$
$$= \left(\frac{\sum_{j=1}^d |X_{j,1}|^p - d\mu_p}{\sqrt{d}\sigma_p}, \ldots, \frac{\sum_{j=1}^d |X_{j,n}|^p - d\mu_p}{\sqrt{d}\sigma_p} \right)$$
$$:= (Y_1, \ldots, Y_n) = \mathbf{Y}_n \in \mathbb{R}^n.$$

As above, let $\mathbf{Z}_n = (Z_1, \ldots, Z_n)$ be a centered Gaussian random vector in $\mathbb{R}^n$, with identity covariance matrix. Write, for $1 \leq j \leq d$,

$$\boldsymbol{\xi}_j = \left(\frac{|X_{j,1}|^p - \mu_p}{\sqrt{d}\sigma_p}, \ldots, \frac{|X_{j,n}|^p - \mu_p}{\sqrt{d}\sigma_p} \right)$$

and note that $\sum_{j=1}^{d} \boldsymbol{\xi}_j = \mathbf{Y}_n$. Set $\beta = \sum_{j=1}^{d} \mathbb{E}\|\boldsymbol{\xi}_j\|_2^3$. Then, by Jensen's inequality,

$$\mathbb{E}\|\boldsymbol{\xi}_j\|_2^3 = \mathbb{E}\left(\frac{\sum_{i=1}^{n}\left(|X_{j,i}|^p - \mu_p\right)^2}{d\sigma_p^2}\right)^{3/2} \leq \left(\frac{n}{d\sigma_p^2}\right)^{3/2} \mathbb{E}\left|\,|X|^p - \mu_p\,\right|^3.$$

This gives that for any $\delta > 0$, possibly depending upon n,

$$B := \beta n \delta^{-3} \leq \frac{n^{5/2}}{\sqrt{d}\sigma_p^3}\mathbb{E}\left|\,|X|^p - \mu_p\,\right|^3 \delta^{-3}.$$

Applying a result of Yurinskiĭ (1977) as formulated in Sect. 4 of Chap. 10 of Pollard (2001) we get, on a suitable probability space depending on $\delta > 0$ and $n \geq 1$, there exist random vectors $\mathbf{Y}'_n$ and $\mathbf{Z}'_n$ satisfying $\mathbf{Y}'_n \overset{\mathscr{D}}{=} \mathbf{Y}_n$ and $\mathbf{Z}'_n \overset{\mathscr{D}}{=} \mathbf{Z}_n$ such that

$$\mathbb{P}\left\{\|\mathbf{Y}'_n - \mathbf{Z}'_n\|_2 > 3\delta\right\} \leq CB\left(1 + \frac{|\log(B)|}{n}\right), \tag{10}$$

where C is a universal constant. To avoid the use of primes, we shall from now on drop them from the notation and write $\mathbf{Y}_n \overset{\mathscr{D}}{=} \mathbf{Y}'_n$ and $\mathbf{Z}_n \overset{\mathscr{D}}{=} \mathbf{Z}'_n$, where it is understood that the pair $(\mathbf{Y}_n, \mathbf{Z}_n)$ satisfies inequality (10) for the given $\delta > 0$.

Using the fact that

$$\left|\max_{1 \leq i \leq n} x_i - \max_{1 \leq i \leq n} y_i\right| \leq \sqrt{\sum_{i=1}^{n}(x_i - y_i)^2},$$

we get, for all $\varepsilon > 0$,

$$\mathbb{P}\left\{a_n|\max_{1 \leq i \leq n} Y_i - \max_{1 \leq i \leq n} Z_i| > \varepsilon\right\} \leq \mathbb{P}\left\{\sqrt{2\log n}\,\|\mathbf{Y}_n - \mathbf{Z}_n\|_2 > \varepsilon\right\}.$$

Thus, for all d large enough,

$$\mathbb{P}\left\{a_n|\max_{1 \leq i \leq n} Y_i - \max_{1 \leq i \leq n} Z_i| > \varepsilon\right\} \leq \mathbb{P}\left\{\sqrt{2\log n}\,\|\mathbf{Y}_n - \mathbf{Z}_n\|_2 > 3\delta\sqrt{2\log n}\right\}$$

$$\text{(since } \delta\sqrt{\log n} \to 0 \text{ as } d \to \infty\text{)}$$

$$= \mathbb{P}\left\{\|\mathbf{Y}_n - \mathbf{Z}_n\|_2 > 3\delta\right\}.$$

From (10), we deduce that for all $\varepsilon > 0$ and all d large enough,

$$\mathbb{P}\left\{a_n|\max_{1 \leq i \leq n} Y_i - \max_{1 \leq i \leq n} Z_i| > \varepsilon\right\} \leq CB\left(1 + \frac{|\log(B)|}{n}\right).$$

But, by our choice of $\delta(d)$ and (9),

$$B\left(1 + \frac{|\log(B)|}{n}\right) \to 0,$$

so that

$$a_n\left|\max_{1\le i\le n} Y_i - \max_{1\le i\le n} Z_i\right| = o_{\mathbb{P}}(1).$$

Similarly, one proves that

$$a_n\left|\min_{1\le i\le n} Y_i - \min_{1\le i\le n} Z_i\right| = o_{\mathbb{P}}(1).$$

Thus, by (7), we conclude that

$$\left(a_n(\max_{1\le i\le n} Y_i - b_n), a_n(\min_{1\le i\le n} Y_i + b_n)\right) \xrightarrow{\mathscr{D}} (E, -E'). \tag{11}$$

Next, we have

$$\left(a_n(\max_{1\le i\le n} Y_i - b_n), a_n(\min_{1\le i\le n} Y_i + b_n)\right)$$
$$= \left(a_n\left(\frac{\max_{1\le i\le n}\|\mathbf{X}_i\|_p^p}{\sqrt{d}\sigma_p} - \frac{\sqrt{d}\mu_p}{\sigma_p} - b_n\right),\right.$$
$$\left. a_n\left(\frac{\min_{1\le i\le n}\|\mathbf{X}_i\|_p^p}{\sqrt{d}\sigma_p} - \frac{\sqrt{d}\mu_p}{\sigma_p} + b_n\right)\right)$$
$$= \left(a_n\left(\frac{\max_{1\le i\le n}\|\mathbf{X}_i\|_p^p}{\sqrt{d}\sigma_p} - \beta_n\right), a_n\left(\frac{\min_{1\le i\le n}\|\mathbf{X}_i\|_p^p}{\sqrt{d}\sigma_p} - \beta_n'\right)\right),$$

where $\beta_n = \frac{\sqrt{d}\mu_p}{\sigma_p} + b_n$ and $\beta_n' = \frac{\sqrt{d}\mu_p}{\sigma_p} - b_n$. Note that $a_n \to \infty$ and (11) imply that both

$$\frac{\max_{1\le i\le n}\|\mathbf{X}_i\|_p^p}{\sqrt{d}\sigma_p} - \beta_n \xrightarrow{\mathbb{P}} 0 \quad \text{and} \quad \frac{\min_{1\le i\le n}\|\mathbf{X}_i\|_p^p}{\sqrt{d}\sigma_p} - \beta_n' \xrightarrow{\mathbb{P}} 0. \tag{12}$$

Observe also that by a two term Taylor expansion, for a suitable $\tilde{\beta}_n$ between β_n and $(\max_{1\le i\le n}\|\mathbf{X}_i\|_p^p)/(\sqrt{d}\sigma_p)$,

$$\frac{pa_n}{\beta_n^{1/p-1}}\left(\left(\frac{\max_{1\le i\le n}\|\mathbf{X}_i\|_p^p}{\sqrt{d}\sigma_p}\right)^{1/p}-\beta_n^{1/p}\right)$$

$$= a_n\left(\frac{\max_{1\le i\le n}\|\mathbf{X}_i\|_p^p}{\sqrt{d}\sigma_p}-\beta_n\right)$$

$$+\frac{a_n}{\beta_n^{1/p-1}}\frac{1-p}{2p}\widetilde{\beta}_n^{1/p-2}\left(\frac{\max_{1\le i\le n}\|\mathbf{X}_i\|_p^p}{\sqrt{d}\sigma_p}-\beta_n\right)^2.$$

We obtain by (11) and (12) that

$$a_n^2\left(\frac{\max_{1\le i\le n}\|\mathbf{X}_i\|_p^p}{\sqrt{d}\sigma_p}-\beta_n\right)^2\frac{\widetilde{\beta}_n^{1/p-2}}{a_n\beta_n^{1/p-1}}=O_{\mathbb{P}}\left(\frac{1}{a_n\beta_n}\right)=o_{\mathbb{P}}(1).$$

Similarly,

$$\frac{pa_n}{(\beta_n')^{1/p-1}}\left(\left(\frac{\min_{1\le i\le n}\|\mathbf{X}_i\|_p^p}{\sqrt{d}\sigma_p}\right)^{1/p}-(\beta_n')^{1/p}\right)$$

$$= a_n\left(\frac{\min_{1\le i\le n}\|\mathbf{X}_i\|_p^p}{\sqrt{d}\sigma_p}-\beta_n'\right)+o_{\mathbb{P}}(1).$$

Keeping in mind that $\beta_n/\beta_n'\to 1$, we get

$$\frac{pa_n}{\beta_n^{1/p-1}}\left(\left(\frac{\max_{1\le i\le n}\|\mathbf{X}_i\|_p^p}{\sqrt{d}\sigma_p}\right)^{1/p}-\beta_n^{1/p},\left(\frac{\min_{1\le i\le n}\|\mathbf{X}_i\|_p^p}{\sqrt{d}\sigma_p}\right)^{1/p}-(\beta_n')^{1/p}\right)$$

$$\overset{\mathscr{D}}{\to}(E,-E')$$

and hence

$$\frac{pa_n}{\beta_n^{1/p-1}}\left(\frac{\max_{1\le i\le n}\|\mathbf{X}_i\|_p}{(\sqrt{d}\sigma_p)^{1/p}}-\frac{\min_{1\le i\le n}\|\mathbf{X}_i\|_p}{(\sqrt{d}\sigma_p)^{1/p}}-\beta_n^{1/p}+(\beta_n')^{1/p}\right)\overset{\mathscr{D}}{\to}E+E'.$$

Next notice that (8) implies that $b_n/\sqrt{d}\to 0$, as $d\to\infty$. Thus, recalling

$$\frac{\beta_n}{\sqrt{d}u_p/\sigma_p}=1+\frac{b_n}{\sqrt{d}\mu_p/\sigma_p}\quad\text{and}\quad\frac{\beta_n'}{\sqrt{d}u_p/\sigma_p}=1-\frac{b_n}{\sqrt{d}\mu_p/\sigma_p},$$

we are led to

$$\frac{pa_n}{\beta_n^{1/p-1}}\left(\beta_n^{1/p} - (\beta_n')^{1/p}\right) = 2a_nb_n + O(a_nb_n^2\beta_n^{-1}) = 2a_nb_n + o(1).$$

Therefore we get

$$\frac{pa_{n(d)}d^{1/2-1/p}}{\mu_p^{1/p-1}\sigma_p}\left(\max_{1\le i\le n(d)}\|\mathbf{X}_i\|_p - \min_{1\le i\le n(d)}\|\mathbf{X}_i\|_p\right)$$
$$-2a_{n(d)}b_{n(d)} \xrightarrow{\mathscr{D}} E + E'. \qquad \square$$

4 Proof of Lemma 2

In the sequel, to lighten notation a bit, we set $\overline{Y} = \overline{Y}_d$. Choose any $\varepsilon > 0$ and $\delta > 0$ such that φ has continuous derivatives φ' and φ'' on $I_\delta = [\mu - \delta, \mu + \delta]$ and $|\varphi''(\mu) - \varphi''(x)| \le \varepsilon$ for all $x \in I_\delta$. We see that by Taylor's theorem that for $\overline{Y} \in I_\delta$

$$\varphi(\overline{Y}) = \varphi(\mu) + \varphi'(\mu)(\overline{Y} - \mu) + 2^{-1}\varphi''(\widetilde{\mu})(\overline{Y} - \mu)^2, \qquad (13)$$

where $\widetilde{\mu}$ lies between $\overline{Y}$ and μ. Clearly,

$$\left|\mathbb{E}\varphi(\overline{Y}) - \varphi(\mu) - \frac{\sigma^2\varphi''(\mu)}{2d}\right|$$
$$= \left|\mathbb{E}\left(\varphi(\overline{Y}) - \left(\varphi(\mu) + \varphi'(\mu)(\overline{Y} - \mu) + 2^{-1}\varphi''(\mu)(\overline{Y} - \mu)^2\right)\right)\right|$$
$$\le \left|\mathbb{E}\left(\left\{\varphi(\overline{Y}) - \left(\varphi(\mu) + \varphi'(\mu)(\overline{Y} - \mu) + 2^{-1}\varphi''(\mu)(\overline{Y} - \mu)^2\right)\right\}\mathbf{1}\{\overline{Y} \in I_\delta\}\right)\right|$$
$$+ \mathbb{E}\left(\left|\varphi(\overline{Y})\right|\mathbf{1}\{\overline{Y} \notin I_\delta\}\right) + \mathbb{E}\left(\left|P(\overline{Y})\right|\mathbf{1}\{\overline{Y} \notin I_\delta\}\right),$$

where
$$P(y) = \varphi(\mu) + \varphi'(\mu)(y - \mu) + 2^{-1}\varphi''(\mu)(y - \mu)^2.$$

Now using (13) and $|\varphi''(\mu) - \varphi''(x)| \le \varepsilon$ for all $x \in I_\delta$, we may write

$$\left|\mathbb{E}\left(\left\{\varphi(\overline{Y}) - \left(\varphi(\mu) + \varphi'(\mu)(\overline{Y} - \mu) + 2^{-1}\varphi''(\mu)(\overline{Y} - \mu)^2\right)\right\}\mathbf{1}\{\overline{Y} \in I_\delta\}\right)\right|$$
$$\le \frac{\varepsilon}{2}\mathbb{E}(\overline{Y} - \mu)^2 = \frac{\varepsilon\sigma^2}{2d}.$$

Next, we shall bound

$$\mathbb{E}\left(\left|\varphi(\overline{Y})\right|\mathbf{1}\{\overline{Y} \notin I_\delta\}\right) + \mathbb{E}\left(\left|P(\overline{Y})\right|\mathbf{1}\{\overline{Y} \notin I_\delta\}\right) := \Delta_d^{(1)} + \Delta_d^{(2)}.$$

Recall that we assume that for some $r > 1$, condition (4) holds. In this case, by Theorem 28 on p. 286 of Petrov (1975) applied with "r" replaced by "$r + 1$", for all $\delta > 0$,

$$\mathbb{P}\left\{|\overline{Y} - \mu| \geq \delta\right\} = \mathrm{o}(d^{-r}). \tag{14}$$

Then, by using Hölder's inequality, (5) and (14), we get

$$\Delta_d^{(1)} \leq \left(\mathbb{E}\left|\varphi(\overline{Y})\right|^s\right)^{1/s}\left(\mathbb{P}\{\overline{Y} \notin I_\delta\}\right)^{1/r} = \mathrm{o}(d^{-1}).$$

We shall next bound $\Delta_d^{(2)}$. Obviously from (14)

$$|\varphi(\mu)|\,\mathbb{P}\{\overline{Y} \notin I_\delta\} = \mathrm{o}(d^{-1}).$$

Furthermore, by the Cauchy-Schwarz inequality and (14),

$$\mathbb{E}\left|\varphi'(\mu)(\overline{Y} - \mu)\mathbf{1}\{\overline{Y} \notin I_\delta\}\right| \leq \left|\varphi'(\mu)\right|\sigma d^{-1/2}\mathrm{o}(d^{-r/2}) = \mathrm{o}(d^{-1}),$$

and by Hölder's inequality with $p = (r + 1)/2$ and

$$q^{-1} = 1 - p^{-1} = 1 - 2/(r + 1) = (r - 1)/(r + 1),$$

we have

$$2^{-1}\left|\varphi''(\mu)\right|\mathbb{E}\left((\overline{Y} - \mu)^2\mathbf{1}\{\overline{Y} \notin I_\delta\}\right)$$
$$\leq 2^{-1}\left|\varphi''(\mu)\right|\left(\mathbb{E}|\overline{Y} - \mu|^{r+1}\right)^{2/(r+1)}\left(\mathbb{P}\{\overline{Y} \notin I_\delta\}\right)^{1/q}.$$

Applying Rosenthal's inequality (see Eq. 3 in Giné and Mason 2003) we obtain

$$\mathbb{E}|\overline{Y} - \mu|^{r+1} = \mathbb{E}\left|d^{-1}\sum_{i=1}^{d}(Y_i - \mu)\right|^{r+1}$$
$$\leq \left(\frac{15(r + 1)}{\log(r + 1)}\right)^{r+1}\max\left(d^{-(r+1)/2}\left(\mathbb{E}Y^2\right)^{(r+1)/2}, d^{-r}\mathbb{E}|Y|^{r+1}\right).$$

Thus

$$\left(\mathbb{E}|\overline{Y} - \mu|^{r+1}\right)^{2/(r+1)} = \mathrm{O}(d^{-1}),$$

which when combined with (14) gives

$$2^{-1}\left|\varphi''(\mu)\right|\left(\mathbb{E}|\overline{Y} - \mu|^{r+1}\right)^{2/(r+1)}\left(\mathbb{P}\{\overline{Y} \notin I_\delta\}\right)^{(r-1)/(r+1)} = \mathrm{o}(d^{-1}).$$

Thus

$$\Delta_d^{(2)} = o(d^{-1}).$$

Putting everything together, we conclude that for any $\varepsilon > 0$

$$\limsup_{d \to \infty} d \left| \mathbb{E}\varphi(\overline{Y}_d) - \varphi(\mu) - \frac{\sigma^2 \varphi''(\mu)}{2d} \right| \le \frac{\varepsilon \sigma^2}{2}.$$

Since $\varepsilon > 0$ can be chosen arbitrarily small, this completes the proof.

Acknowledgments The authors thank the referee for pointing out a misstatement in the original version of the paper.

References

Aggarwal, C. C., Hinneburg, A., & Keim, D. A. (2001). On the surprising behavior of distance metrics in high dimensional space. In *Proceedings of the 8th International Conference on Database Theory* (pp. 420–434). Berlin: Springer.

Beyer, K. S., Goldstein, J., Ramakrishnan, R., Shaft, U. (1999). When is nearest neighbor meaningful? In *Proceedings of the 7th International Conference on Database Theory* (pp. 217–235). Berlin: Springer.

Boucheron, S., Lugosi, G., & Massart, P. (2013). *Concentration inequalities: A nonasymptotic theory of independence*. Oxford: Oxford University Press.

David, H. (1981). *Order statistics* (2nd ed.). New York: Wiley.

François, D., Wertz, V., & Verleysen, M. (2007). The concentration of fractional distances. *IEEE Transactions on Knowledge and Data Engineering, 19*, 873–886.

Giné, E., Mason, D. M., & Zaitsev, A. Y. (2003). The L_1-norm density estimator process. *The Annals of Probability, 31*, 719–768.

Hinneburg, A., Aggarwal, C. C., & Keim, D. A. (2000). What is the nearest neighbor in high dimensional spaces. In *Proceedings of the 26th International Conference on Very Large Data Bases* (pp. 506–515). San Francisco: Morgan Kaufmann.

Kabán, A. (2012). Non-parametric detection of meaningless distances in high dimensional data. *Statistics and Computing, 22*, 375–385.

Khan, R. A. (2004). Approximation for the expectation of a function of the sample mean. *Statistics, 38*, 117–122.

Mallows, C. L. (1972). A note on asymptotic joint normality. *The Annals of Mathematical Statistics, 43*, 508–515.

McDiarmid, C. (1989). On the method of bounded differences. In J. Siemons (Ed.), *Surveys in combinatorics, 1989. London Mathematical Society Lecture Note Series* (Vol. 141, pp. 148–188). Cambridge: Cambridge University Press.

Petrov, V. V. (1975). *Sums of independent random variables. Ergebnisse der Mathematik und ihrer Grenzgebiete.* (Vol. 82), New York: Springer.

Pollard, D. (2001). *A user's guide to measure theoretic probability*. Cambridge: Cambridge University Press.

Reiss, R.-D. (1989). *Approximate distributions of order statistics. With Applications to Nonparametric Statistics*. New York: Springer.

Resnick, S. I. (1987). *Extreme values, regular variation, and point processes*. New York: Springer.

Yurinskiĭ, V. V. (1977). On the error of the Gaussian approximation for convolutions. *Teoriya Veroyatnostei i ee Primeneniya, 22*, 242–253.

Estimating and Detecting Jumps.
Applications to $D[0, 1]$-Valued
Linear Processes

Denis Bosq

Abstract This paper is devoted to the estimation of the intensity and the density of jumps for $D[0, 1]$-valued random variables and the construction of detectors for constant or random jumps. Limit theorems are obtained in the context of continuous observations or high-frequency data. Applications to jumps for $D[0, 1]$-valued moving average and autoregressive processes are considered. We also study the special case where there is an infinity of jumps. Thus, our approach is somewhat different from that which consists of studying jumps in semimartingales.

Keywords Jumps · Functional linear processes · Cadlag · Limit theorems · Estimation · Detection

1 Introduction

Prediction of functional autoregressive processes in Banach spaces and Hilbert spaces has been extensively studied, see Bosq (2000, 2014), Bosq and Blanke (2007), Ferraty and Romain (2011) and Horváth and Kokoszka (2012) for results and references. In general, the context is continuous random functions with application to electricity consumption, electrocardiograms, variation of temperature, etc. The associated time interval being a day, a week, one year, etc.

This paper is devoted to the case where some jumps appear in the context of finance, economy, climatic variations, electricity price modelling, etc.; also note that big continuous variations may be considered as jumps since, in that situation, prediction accuracy is poor.

The natural space for studying jumps is $D = D[0, 1]$, i.e. the space of real cadlag (continuous on the right and with a limit on the left) functions defined on $[0, 1]$. It is well known that, if D is equipped with the uniform norm, it becomes a non-separable space and that this infers measurability problems. Thus it is better to replace it with the Skorohod metric, cf. Billingsley (1999).

D. Bosq (✉)
LSTA, Université Pierre et Marie Curie - Paris 6, Paris, France
e-mail: denis.bosq@upmc.fr

© Springer International Publishing Switzerland 2015

M. Hallin et al. (eds.), *Mathematical Statistics and Limit Theorems*,
DOI 10.1007/978-3-319-12442-1_4

Now, one observes independent or dependent copies of the D-valued random variable X and wants to estimate the jump's intensity and jump's density, and, if data are collected in discrete time, to detect the position of jumps. Concerning the dependent copies we will consider the moving average model (MAD) and the autoregressive model (ARD). Our approach is somewhat different from that which consists of studying jumps in diffusions or in semimartingales. It explains why we do not indicate references concerning that topic (note that there are at least 547 titles containing the words "jump diffusion", cf. Math. Sci. Net).

In Sect. 2, one supposes that there exists a jump at $t_0 \in]0, 1[$, where t_0 is constant, and that observations are collected in continuous time. Then, under suitable conditions, it is easy to obtain limit theorems for the empirical estimators of $E(X(t_0) - X(t_0-))$ and $E|X(t_0) - X(t_0-)|$, namely the strong law of large numbers (SLLN) and the central limit theorem (CLT).

Section 3 is devoted to high-frequency data (HFD). The situation is more intricate since one has to detect t_0 before estimating the jump's intensity; we construct a statistical detector of t_0. If the sample paths of X satisfy a Hölder condition on $[0, t_0[$ and on $[t_0, 1]$ we obtain a consistent detector of t_0 with a convergence rate. Insertion of the detector allows us to obtain the SLLN and CLT.

In Sect. 4 we study the case where the jump's position T is random. Again we obtain limit theorems in continuous and discrete time. The asymptotic behaviour of the kernel density estimator of T from HFD is considered in Sect. 5: a result from Deheuvels (2000) allows us to obtain sharp rates of convergence.

Section 6 deals with the case where X possesses two random jumps S and T. The associated scheme is observations of the copies $X_1, \ldots, X_n$ with positions of jumps $S_1, T_1, \ldots, S_n, T_n$. The problem is to distinguish between the "S-jump" and "T-jump". We present a simple discrimination method that leads to consistent estimators of the jumps intensities as well in continuous time as in discrete time.

In Sect. 7 we apply the above results to the case where $(X_n, n \in \mathbb{Z})$ is a D-valued linear process. First we consider the moving average of order one (MAD(1)):

$$X_n = Z_n - a(Z_{n-1}), \ n \in \mathbb{Z}$$

where (Z_n) is a D-strong white noise and $a : D \longrightarrow D$ a continuous linear operator. If (Z_n) has only one jump, (X_n) has in general two jumps and one may apply results in Sect. 6. Another important model is the autoregressive process of order 1 (ARD(1)) defined by:

$$X_n = \rho(X_{n-1}) + Z_n, \ n \in \mathbb{Z}.$$

If ρ is linear continuous with $\rho(D) \subseteq C = C[0, 1]$, the space of real continuous functions defined on $[0, 1]$, and if (Z_n) has only one jump, then X_n has the same jump and one may apply results of Sect. 4. In the general situation X_n has an infinity of jumps and a different method is considered in a special case: it allows us to obtain consistent estimators of the jumps intensities.

2 Jump at a Fixed Point with Continuous Data

In order to study jumps we consider the space $D = D\,[0, 1]$ of cadlag real functions defined on $[0, 1]$. We denote $D\,(\|.\|)$ the space D equipped with its uniform norm but, since it is not separable (cf. Billingsley (1999, p. 157)), it is better to replace it with the space $D(d)$ equipped with the Skorohod metric (cf. Billingsley (1999, p. 123)). Note that uniform convergence implies d-convergence, but the converse is not true except if the limit is a continuous function (cf. Billingsley (1999, p. 124)). Finally, note that $x \longmapsto x(t_0) - x(t_0-)$, $x \in D$ is a continuous linear form on $D\,(\|.\|)$ and that it is $\mathscr{D}\text{-}\mathscr{B}_{\mathbb{R}}$ measurable, where $\mathscr{B}_{\mathbb{R}}$ is the $\sigma-$algebra of Borel sets over $\mathbb{R}$ (cf. Billingsley (1995, Theorems 1 and 2)). These properties will be used below.

Now let X be a random variable, defined on a probability space $(\Omega, \mathscr{A}, P)$ and with values in $(D, \mathscr{D})$ where $\mathscr{D}$ denotes the $\sigma-$algebra associated with the Skorohod metric d (cf. Billingsley (1999, p. 157)). We suppose that $E\,\|X\| < \infty$ and set

$$\Delta(t_0) = X(t_0) - X(t_0-), \ (t_0 \in\,]0, 1[),$$

where $X(t_0-) = \lim_{s \to t_0-} X(s)$. If X has a jump at t_0, then $\Delta(t_0)$ is an integrable, non-degenerate random variable.

One observes $X_1, \ldots, X_n$ that are copies (possibly dependent) of X. Since $X_1, \ldots, X_n$ are equidistributed and $x \longmapsto x(t_0) - x(t_0-)$, $x \in D$ is a measurable linear form on $D\,(\|.\|)$, then the random variables

$$\Delta_i(t_0) = X_i(t_0) - X_i(t_0-), \ 1 \le i \le n$$

have the same distribution as $\Delta(t_0)$. Now the empirical unbiased estimator of $E\,\Delta(t_0)$ is

$$\overline{\Delta}_n(t_0) = \frac{1}{n} \sum_{i=1}^{n} \Delta_i(t_0), \ n \ge 1,$$

and the following simple statement gives consistency:

Proposition 1 *If the process $(X_n, n \in \mathbb{Z})$ satisfies the SLLN on $D(d)$, i.e.:*

$$d\left(0, \frac{1}{n} \sum_{i=1}^{n} X_i - EX\right) \xrightarrow[n \to \infty]{} 0, \ a.s. \tag{1}$$

then

$$\overline{\Delta}_n(t_0) \to E\,\Delta(t_0), \ a.s., \tag{2}$$

furthermore

$$\frac{1}{n} \sum_{i=1}^{n} X_i(t_0) \to EX(t_0), \ a.s. \tag{3}$$

and

$$\frac{1}{n} \sum_{i=1}^{n} X_i(t_0-) \to EX(t_0-), \ a.s. \tag{4}$$

Proof First, since 0 is a continuous function, it follows that d-convergence and uniform convergence are equivalent, thus (1) implies

$$\left\| \frac{1}{n} \sum_{i=1}^{n} X_i - EX \right\| \to 0, \ \text{a.s.} \tag{5}$$

Now consider the continuous linear form on D defined by

$$\varphi_{t_0}(x) = x(t_0) - x(t_0-), \ x \in D.$$

From (5) and $D\,(\|.\|)$ continuity we get

$$\varphi_{t_0}\left(\frac{1}{n} \sum_{i=1}^{n} X_i - EX \right) \to 0, \ \text{a.s.}$$

and, by linearity

$$\frac{1}{n} \sum_{i=1}^{n} \Delta_i(t_0) \to \varphi_{t_0}(EX), \ \text{a.s.}$$

since $\varphi_{t_0}(EX) = E\varphi_{t_0}(X)$, (2) follows.

Finally (5) entails (3), and (4) comes from

$$\frac{1}{n} \sum_{i=1}^{n} X_i(t_0-) = \overline{\Delta}_n(t_0) + \frac{1}{n} \sum_{i=1}^{n} X_i(t_0). \qquad \square$$

Note that (1) cannot be replaced with $d(EX, \frac{1}{n} \sum_{i=1}^{n} X_i) \to 0$, a.s. since d is not translation invariant.

Concerning the strong law of large numbers for independent D-valued random variables it appears in Daffer and Taylor (1979): they study the case where X is convex tight and the case where X belongs to the cone of non-decreasing elements of D. For linear processes in D observed in continuous time, see El Hajj (2011). See also Schiopu-Kratina and Daffer (1999), Bezandry (2006) and Ranga Rao (1963).

Note that (3) holds even if t_0 is not a jump point. Now it is easy to realise that one may have $E\Delta(t_0) = 0$ although there is a non-degenerated jump at t_0. For example take

$$X(t) = U\, I_{[0,t_0]}(t) + V\, I_{[t_0,1]}(t), \ 0 \le t \le 1 \ (0 < t_0 < 1),$$

where U and V are independent with Bernoulli distribution of parameter $1/2$. Then $E\Delta(t_0) = 0$ while $E|\Delta(t_0)| = \frac{1}{2}$. In such a situation one must estimate $E|\Delta(t_0)|$; the following statement gives consistency.

Proposition 2 *If one of the following assumptions holds:*

$B_1 - (|\Delta_n(t_0)|)$ *is* m—*dependent and integrable.*

$B_2 - (|\Delta_n(t_0)|)$ *is geometrically strongly mixing and satisfies the Cramer's condition:*

$$E\left||\Delta_n(t_0)| - E|\Delta_n(t_0)|\right|^k \leq c^{k-2} k! \operatorname{Var}|\Delta_n(t_0)|, \ k \geq 3, \ n \geq 1, \ (c > 0),$$

then $\frac{1}{n} \sum_{i=1}^{n} |\Delta_i(t_0)| \to E|\Delta(t_0)|$, *a.s.*

Proof If B_1 holds, the proof is straightforward. Under B_2 one may adapt the proof in Bosq (1998, p. 35). Details are omitted. $\qquad\qquad\square$

We now turn to limit in distribution:

Proposition 3 *If $E\|X\|^2 < \infty$ and if $(X_n - EX_n)$ is a process which satisfies the central limit theorem in $D(\|.\|)$, then*

$$\frac{1}{\sqrt{n}} \sum_{i=1}^{n} (\Delta_i(t_0) - E\Delta_i(t_0)) \Rightarrow N_{t_0}, \tag{6}$$

where N_{t_0} is centred Gaussian. If, in addition, (X_n) is iid, it is clear that $N_{t_0} \sim \mathcal{N}(0, \operatorname{Var}\Delta(t_0))$.

Proof Since

$$\frac{1}{\sqrt{n}} \sum_{i=1}^{n} (X_i - EX_i) \Rightarrow N$$

where N is a D-valued Gaussian random variable, the mapping theorem (Billingsley 1999, p. 21) and continuity of φ_{t_0} for $D(\|.\|)$ yields

$$\frac{1}{\sqrt{n}} \sum_{i=1}^{n} \varphi_{t_0}(X_i - EX_i) \Rightarrow \varphi_{t_0}(N),$$

hence (6). $\qquad\qquad\square$

Concerning the central limit theorem in D we refer to Hahn (1978) and Bloznelis and Paulauskas (1993). Bézandry and Fernique (1992) have obtained the CLT in $D(\|.\|)$. The case of linear processes appears in El Hajj (2011); see also Norzhigitov and Sharipov (2010). Now, if $E\Delta(t_0) = 0$, one may prefer to derive a central limit theorem for $(|\Delta_n(t_0)|)$, see for example Merlevède et al. (1999, Theorem 1.1).

3 High-Frequency Data

We now study the case where one observes data in discrete time but with high frequency, namely $X_i(j\,k_n^{-1})$, $k_n \geq 1$, $0 \leq j \leq k_n$, $1 \leq i \leq n$, with $k_n \to \infty$ as $n \to \infty$. This scheme is classical but somewhat artificial, however, it allows us to obtain informations concerning asymptotics. We suppose that X has a single jump at $t_0 \in \,]0, 1[$, with t_0 constant and, since data are discrete, we need a regularity assumption for the sample paths:

A_α—*There exists a real random variable M with $EM^2 < \infty$; and $\alpha \in]0, 1]$ such that*

$$|X(t) - X(s)| \leq M\,|t - s|^\alpha, \quad (s, t) \in I_{t_0}$$

where $I_{t_0} = [0, t_0[\,^2 \cup [t_0, 1]^2$.

Example 1 (Fractional Brownian motions)

Consider two fractional Brownian motions $B_1 = (B_1(t), 0 \leq t \leq t_0)$ and $B_2 = (B_2(t), t_0 \leq t \leq 1)$, possibly dependent, with covariance $\frac{c_\beta}{2}(|s|^{2\beta} + |t|^{2\beta} - |t - s|^{2\beta})$ where $0 < \beta < 1$. Then, one may verify that A_α holds as soon as $0 < \alpha < \beta$. Note that $\beta = \frac{1}{2}$ corresponds with Wiener processes.

Example 2 (Ornstein–Uhlenbeck process with trend)

Define the model

$$X(t) = \mu(t) + \int_0^t e^{-\theta(t-s)} dW(s), \ 0 \leq t \leq 1, \tag{7}$$

where $\mu(.)$ is a non-random real function, with a jump at t_0 $(\mu(t_0) \neq \mu(t_0-))$, such that $\mu(0) = \mu(1)$, and continuously differentiable, except at t_0. The integral is taken in Ito's sense, W is a Wiener process and θ is positive. Then, if $\alpha < \frac{1}{2}$, A_α holds. The interpretation of (7) is periodic trend plus noise.

Now, in order to study the behaviour of the detector of t_0 and the jump's estimator we need some preliminaries. First, for every integer $n \geq 1$, there exists an integer $j_0(n)$ such that

$$t'_{0,n} := \frac{j_0(n) - 1}{k_n} < t_0 \leq \frac{j_0(n)}{k_n} := t_{0,n}. \tag{8}$$

Second, if $X_1, \ldots, X_n$ are copies of X, possibly dependent, by enlarging the probability space one can construct associated copies of M such that

$$|X_i(t) - X_i(s)| \leq M_i\,|t - s|^\alpha, \quad (s, t) \in I_{t_0}, 1 \leq i \leq n. \tag{9}$$

Now we need two lemmas:

Lemma 1 *Let $Y_{1,n}, \ldots, Y_{n,n}$ be real independent centered random variables such that, for every n, $E(Y_{i,n}^4) \leq c$, $1 \leq i \leq n$, where c is a constant not depending on n, then*

$$P\left(\left|\frac{1}{n}\sum_{i=1}^{n}Y_{i,n}\right|\geq\eta\right)\leq\frac{3c}{n^2\eta^4},\ \ \eta>0,\ n\geq 1, \tag{10}$$

hence

$$\frac{1}{n}\sum_{i=1}^{n}Y_{i,n}\to 0,\ \ a.s.\ and\ in\ L^2. \tag{11}$$

Proof Similar to the proof of theorem 1 pp. 388–389 in Shiryaev (1996).   □

In the next lemma $\Delta_{i,n}=X_i(t_{0,n})-X_i(t'_{0,n})$, $1\leq i\leq n$ and $\Delta_i=X_i(t_0)-X_i(t_0-)$, $1\leq i\leq n$. Moreover, $a_n\asymp b_n$ means $a_n=\mathcal{O}(b_n)$ and $b_n=\mathcal{O}(a_n)$.

Lemma 2 *Suppose that A_α holds, then:*
1. If $E\,\|X\|^2<\infty$ and if $(X_n,\ n\geq 1)$ satisfies the L^2-law of large numbers, then

$$\frac{1}{n}\sum_{i=1}^{n}\Delta_{i,n}\to E\Delta(t_0),\ (L^2). \tag{12}$$

In addition, if (Δ_n) is iid and $k_n\asymp n^{1/2\alpha}$, we have

$$E\left(\frac{1}{n}\sum_{i=1}^{n}\Delta_{i,n}-E\Delta(t_0)\right)^2=\mathcal{O}\left(\frac{1}{n}\right). \tag{13}$$

2. Suppose that $E\,\|X\|<\infty$, that $(\Delta_n(t_0),\ n\geq 1)$ satisfies the SLLN and that the following assumption holds:

   A_q *— There exists $q\geq 1$, such that $EM^q<\infty$ and $\displaystyle\sum_n k_n^{-\alpha q}<\infty$*

   then

$$\frac{1}{n}\sum_{i=1}^{n}\Delta_{i,n}\to E\Delta(t_0),\ a.s. \tag{14}$$

   If (Δ_n) is iid, one may remove A_q.

Proof 1. Since

$$\frac{1}{n}\sum_{i=1}^{n}(\Delta_{i,n}-\Delta_i(t_0))=\frac{1}{n}\sum_{i=1}^{n}(X_i(t_{0,n})-X_i(t_0))+\frac{1}{n}\sum_{i=1}^{n}\left(X_i(t_0-)-X_i\left(t'_{0,n}\right)\right),$$

A_α and (9) imply

$$\delta_n:=\left|\frac{1}{n}\sum_{i=1}^{n}(\Delta_{i,n}-\Delta_i(t_0))\right|\leq 2\left(\frac{1}{n}\sum_{i=1}^{n}M_i\right)k_n^{-\alpha}. \tag{15}$$

Now we have $(\frac{1}{n}\sum_{i=1}^{n} M_i)^2 \le \frac{1}{n}(\sum_{i=1}^{n} M_i^2)$, thus $E(\frac{1}{n}\sum_{i=1}^{n} M_i)^2 \le EM^2$, hence

$$E\delta_n^2 \le 4\,EM^2\,k_n^{-2\alpha} \to 0 \tag{16}$$

and since $(\Delta_n(t_0))$ satisfies the $L^2 - LLN$, (12) follows. Now Eq. (13) is an easy consequence of (16).

2. Note that Jensen's inequality gives

$$\left(\frac{1}{n}\sum_{i=1}^{n} M_i\right)^q \le \frac{1}{n}\sum_{i=1}^{n} M_i^q,$$

hence $E(\frac{1}{n}\sum_{i=1}^{n} M_i)^q \le EM^q$ and, for $\eta > 0$, (15) yields

$$P(\delta_n \ge \eta) \le P\left(\frac{2}{n}\sum_{i=1}^{n} M_i\,k_n^{-\alpha} \ge \eta\right) \le \frac{2^q}{\eta^q k_n^{\alpha q}}\,E\left(\frac{1}{n}\sum_{i=1}^{n} M_i\right)^q$$

$$\le \left(\frac{2}{\eta}\right)^q EM^q k_n^{-\alpha q},$$

thus $\sum_n P(\delta_n \ge \eta) < \infty$, $\eta > 0$, it follows that $\delta_n \to 0$ a.s. and since $(\Delta_n(t_0))$ satisfies the SLLN one obtains (14).

Now, if (Δ_n) is iid (M_n) is also iid, it follows that $\frac{1}{n}\sum_{i=1}^{n} M_i \to EM$ a.s., thus $\frac{1}{n}\sum_{i=1}^{n} M_i\,k_n^{-\alpha} \to 0$ a.s. hence the result from (15). $\qquad\square$

Since $t_{0,n}$ and $t'_{0,n}$ are not observed it is necessary to construct a *detector* of t_0. To this aim we set

$$\hat{t}_{0,n} = \frac{1}{k_n}\,\arg\max_{1 \le j \le k_n}\,S_{j,n},$$

where

$$S_{j,n} = \frac{1}{n}\sum_{i=1}^{n}\left[X_i\left(\frac{j}{k_n}\right) - X_i\left(\frac{j-1}{k_n}\right)\right],\ 1 \le j \le k_n,$$

with an arbitrary choice of j if there is a tie.

Proposition 4 (SLLN) *Suppose that $E\,\|X\| < \infty$ and, $E(X(t_0) - X(t_0-)) > 0$. Then, under A_α and A_q, and if $(\Delta_n(t_0))$ satisfies the SLLN, we have*

$$\hat{t}_{0,n} \to t_0+,\ a.s., \tag{17}$$

$$\hat{t}_{0,n} - \frac{1}{k_n} \to t_0-,\ a.s., \tag{18}$$

and

$$\frac{1}{n}\sum_{i=1}^{n}\left(X_i(\hat{t}_{0,n}) - X_i\left(\hat{t}_{0,n} - \frac{1}{k_n}\right)\right) \to E(X(t_0) - X(t_0-)), \ a.s.. \qquad (19)$$

Again, if (X_n) is iid one may remove A_q.

Proof Put $U_n = \max_{j \neq j_0(n)} S_{j,n}$ and $V_n = S_{j_0(n),n}$ where $j_0(n)$ is defined in (8). Now A_α entails

$$|U_n| \le \max_{j \neq j_0(n)} \frac{1}{n}\sum_{i=1}^{n}\left|X_i\left(\frac{j}{k_n}\right) - X_i\left(\frac{j-1}{k_n}\right)\right| \le \left(\frac{1}{n}\sum_{i=1}^{n} M_i\right) k_n^{-\alpha} \to 0, \ a.s.,$$

$$(20)$$

from A_q and similarly as in the proof of Lemma 2.

On the other hand

$$\left|V_n - \frac{1}{n}\sum_{i=1}^{n}\Delta_i(t_0)\right| \le \left|\frac{1}{n}\sum_{i=1}^{n}(X_i(t_{0,n}) - X_i(t_0))\right|$$

$$+ \left|\frac{1}{n}\sum_{i=1}^{n}\left(X_i(t_0-) - X_i\left(t_{0,n} - \frac{1}{k_n}\right)\right)\right| \le \left(\frac{2}{n}\sum_{i=1}^{n} M_i\right) k_n^{-\alpha}$$

and the bound tends to 0 almost surely. Now, since $(\Delta_n(t_0))$ satisfies the SLLN, it follows that

$$V_n \to E(X(t_0) - X(t_0-)) > 0 \ a.s.. \qquad (21)$$

Then, (20) and (21) imply that there exists $\Omega_0 \in \mathscr{A}$ with $P(\Omega_0) = 1$ such that, for every $\omega \in \Omega_0$, there exists an integer $N(\omega)$ such that $n \ge N(\omega)$ gives

$$U_n < \frac{1}{2} E(X(t_0) - X(t_0-)) < V_n,$$

therefore

$$\hat{t}_{0,n}(\omega) = t_{0,n}, \ n \ge N(\omega), \ \omega \in \Omega_0,$$

and, since $t_{0,n} \to t_0+$, and $t'_{0,n} = t_{0,n} - \frac{1}{k_n} \to t_0-$, one obtains (17) and (18).

Finally,

$$\frac{1}{n}\sum_{i=1}^{n}\left(X_i\left(\hat{t}_{0,n}(\omega)\right) - X_i\left(\hat{t}_{0,n}(\omega) - \frac{1}{k_n}\right)\right)$$

$$= \frac{1}{n}\sum_{i=1}^{n}\left(X_i(t_{0,n}) - X_i\left(t'_{0,n}\right)\right), \ n \ge N(\omega),$$

$\omega \in \Omega_0$, hence (19) from Lemma 2. $\qquad \square$

In the case where $E(X(t_0) - X(t_0-)) < 0$ one may put

$$\tilde{t}_{0,n} = \frac{1}{k_n} \arg \min_{1 \le j \le k_n} S_{j,n},$$ (22)

for obtaining similar results.

If $E(X(t_0) - X(t_0-)) = 0$ or if one does not know the sign of $E(X(t_0) - X(t_0-))$ the study will be performed by using

$$\check{t}_{0,n} = \frac{1}{k_n} \arg \max_{1 \le j \le k_n} \sum_{i=1}^{n} \left| X_i\left(\frac{j}{k_n}\right) - X_i\left(\frac{j-1}{k_n}\right) \right|$$

and one can obtain analogous results, the details are omitted. This method holds if one wants to estimate the jump in Example 1.

We now turn to limit in distribution for the High-Frequency Data case. For convenience we set $Y(t) = X(t) - EX(t)$ and $Y(t-) = X(t-) - EX(t-)$; a similar notation will be used for $X_1, \ldots, X_n$. We need some lemmas:

Lemma 3 *Under A_α, $E\|X\|^2 < \infty$ and if $X_1, \ldots, X_n$ are iid, then*

$$\frac{1}{\sqrt{n}} \sum_{i=1}^{n} \left[Y_i(t_{0,n}), \ Y_i\left(t'_{0,n}\right) \right] \Rightarrow N \sim \mathcal{N}(0, \Gamma_{t_0}),$$ (23)

where Γ_{t_0} is the covariance matrix of the vector $(Y(t_0), Y(t_0-))$.

Proof We have

$$d_n := E\left[\frac{1}{\sqrt{n}} \sum_{i=1}^{n} \left[Y_i(t_{0,n}) - Y_i(t_0) \right] \right]^2 = E\left[Y(t_{0,n}) - Y(t_0) \right]^2,$$

then, by using A_α one obtains

$$d_n \le 4\,EM^2 \left| t_0 - t_{0,n} \right|^{2\alpha} \le 4\,EM^2 \, k_n^{-2\alpha} \to 0.$$

The same property holds for $t'_{0,n}$, hence

$$\frac{1}{n} E \left\| \sum_{i=1}^{n} \left[Y_i(t_{0,n}) - Y_i(t_0), \ Y_i\left(t'_{0,n}\right) - Y_i(t_0-) \right] \right\|_{\mathbb{R}^2}^2 \to 0,$$

now it is straightforward to show that

$$\frac{1}{\sqrt{n}} \sum_{i=1}^{n} [Y_i(t_0),\; Y_i(t_0-)] \Rightarrow N \sim \mathcal{N}(0,\; \Gamma_{t_0}),$$

hence (23). $\qquad\square$

Lemma 4 *If A_α holds with $EM^4 < \infty$, $E\,\|X\|^4 < \infty$, $X_1,\ldots,X_n$ are iid and $E(X(t_0) - X(t_0-)) > 0$, then*

$$P\left(\hat{t}_{0,n} \neq t_{0,n}\right) = \mathcal{O}\left(\frac{1}{n^2}\right). \tag{24}$$

Proof Recall that $U_n = \max_{j \neq j_0(n)} S_{j,n}$ and $V_n = S_{j_0(n),n}$ and note that $\hat{t}_{0,n} \neq t_{0,n}$ entails $V_n \leq U_n$. Then, from (20), $\hat{t}_{0,n} \neq t_{0,n}$ implies $V_n \leq \bar{M}_n k_n^{-\alpha}$ where $\bar{M}_n = \frac{1}{n}\sum_{i=1}^{n} M_i$. Now, for all $\eta > 0$,

$$P(V_n \leq \bar{M}_n\, k_n^{-\alpha})$$
$$= P\left(V_n \leq \bar{M}_n\, k_n^{-\alpha},\, \bar{M}_n - EM > \eta\right) + P\left(V_n \leq \bar{M}_n\, k_n^{-\alpha},\, \bar{M}_n - EM \leq \eta\right)$$
$$\leq P\left(\left|\bar{M}_n - EM\right| > \eta\right) + P\left(V_n \leq (\eta + EM)\, k_n^{-\alpha},\, \bar{M}_n - EM \leq \eta\right)$$

and since

$$EV_n = E\left(X(t_{0,n}) - X\left(t'_{0,n}\right)\right) \to E\left(X(t_0) - X(t_0-)\right) > 0,$$

then, for n large enough (not random), one has $(\eta + EM)\, k_n^{-\alpha} - EV_n < -\gamma,\, (\gamma > 0)$ hence

$$P\left(V_n \leq (\eta + EM)\, k_n^{-\alpha}\right) = P\left(V_n - EV_n \leq (\eta + EM)\, k_n^{-\alpha} - EV_n\right)$$
$$\leq P\left(|V_n - EV_n| > \gamma\right),$$

finally

$$P\left(\hat{t}_{0,n} \neq t_{0,n}\right) \leq P\left(\left|\bar{M}_n - EM\right| > \eta\right) + P\left(|V_n - EV_n| > \gamma\right),$$

using twice the bound (10) one obtains the desired result. $\qquad\square$

Remark 1 Note that one may slightly improve the detector by setting $\widetilde{t}_{0n} = \hat{t}_{0n} - \frac{1}{2k_n}$ for obtaining $P\left[\left|\widetilde{t}_{0n} - t_0\right| > \frac{1}{2k_n}\right] = \mathcal{O}\left(\frac{1}{n^2}\right)$.

We are now in a position to get limit in distribution:

Proposition 5 (CLT) *Under assumptions made in Lemma 4, we have*

$$\frac{1}{\sqrt{n}} \sum_{i=1}^{n} \left[Y_i\left(\hat{t}_{0,n}\right), \; Y_i\left(\hat{t}_{0,n} - \frac{1}{k_n}\right) \right] \Rightarrow N \sim \mathcal{N}\left(0, \Gamma_{t_0}\right), \tag{25}$$

Proof From Lemma 3 it suffices to show that

$$d_n' = E \left\| \frac{1}{\sqrt{n}} \sum_{i=1}^{n} \left[Y_i\left(\hat{t}_{0,n}\right), Y_i\left(\hat{t}_{0,n} - \frac{1}{k_n}\right) \right] - \frac{1}{\sqrt{n}} \sum_{i=1}^{n} \left[Y_i(t_{0,n}), Y_i\left(t_{0,n}'\right) \right] \right\|_{\mathbb{R}^2}^2 \to 0.$$

Now, write

$$E \left(\frac{1}{\sqrt{n}} \sum_{i=1}^{n} \left[Y_i\left(\hat{t}_{0,n}\right) - Y_i(t_{0,n}) \right] \right)^2$$

$$= E \left(\frac{1}{\sqrt{n}} \sum_{i=1}^{n} \left[Y_i\left(\hat{t}_{0,n}\right) - Y(t_{0,n}) \right] I_{\hat{t}_{0,n} \neq t_{0,n}} \right)^2$$

$$\leq \left(E \left(\frac{1}{\sqrt{n}} \sum_{i=1}^{n} \left[Y_i\left(\hat{t}_{0,n}\right) - Y(t_{0,n}) \right] \right)^4 \right)^{1/2} \left(E \left(I_{\hat{t}_{0,n} \neq t_{0,n}} \right)^4 \right)^{1/2}.$$

From A_α and the condition $E \|X\|^4 < \infty$ we infer that

$$E \left(\frac{1}{\sqrt{n}} \sum_{i=1}^{n} \left[Y_i\left(\hat{t}_{0,n}\right) - Y_i(t_{0,n}) \right] \right)^2 = \mathcal{O}\left(P\left(\hat{t}_{0,n} \neq t_{0,n}\right)^{1/2} \right)$$

which tends to 0 from Lemma 4.

A similar property holds for $E \left(\frac{1}{\sqrt{n}} \sum_{i=1}^{n} \left[Y_i\left(\hat{t}_{0,n} - \frac{1}{k_n}\right) - Y_i\left(t_{0,n}'\right) \right] \right)^2$ and (25) follows. $\qquad\square$

Corollary 1 *We have*

$$\frac{1}{\sqrt{n}} \sum_{i=1}^{n} \left[Y_i\left(\hat{t}_{0,n}\right) - Y\left(\hat{t}_{0,n} - \frac{1}{k_n}\right) \right] \Rightarrow N \sim \mathcal{N}(0, \operatorname{Var}(X(t_0) - X(t_0-))).$$

Proof Clear. $\qquad\square$

Of course it is again possible to adapt the above results to cases where $E(X(t_0) - X(t_0-)) < 0$, and $E(X(t_0) - X(t_0-)) = 0$ with $E|X(t_0) - X(t_0-)| > 0$.

4 Random Jumps

We now suppose that X has one and only one jump at T, where T is a real random variable such that $P(T \in \,]0, 1[) = 1$. In the continuous case, if $(X_n, n \geq 1)$ are copies of X, one may associate copies $(T_n, n \geq 1)$ of T. Then, it is not difficult to state propositions concerning T since, if $(X_n, n \geq 1)$ are iid, m-dependent or strongly mixing, the same properties hold for $(T_n, n \geq 1)$. Thus, under regularity conditions we have:

$$\frac{1}{n} \sum_{i=1}^{n} T_i \rightarrow ET, \text{ a.s.,} \tag{26}$$

$$\frac{1}{\sqrt{n}} \sum_{i=1}^{n} (T_i - ET_i) \Rightarrow N \sim \mathcal{N}\left(0, \sum_{h=-\infty}^{+\infty} \mathrm{Cov}(T_0, T_h)\right), \tag{27}$$

and similar results for $(X(T) - X(T-))$ and $|(X(T) - X(T-)|$.

For the discrete case observations, the situation is more intricate. Again we take data of the form $X_i(jk_n^{-1})$, $k_n \geq 1$, $0 \leq j \leq k_n$, $1 \leq i \leq n$, and define a random integer $J(i, n)$ such that

$$(J(i, n) - 1) k_n^{-1} < T_i \leq J(i, n) k_n^{-1} := T_{in}, \; 1 \leq i \leq n. \tag{28}$$

Now, for every (i, j), we put

$$d_{ijn} = \left| X_i\left(jk_n^{-1}\right) - X_i\left((j-1)k_n^{-1}\right) \right|,$$

and

$$D_{in} = \max_{1 \leq j \leq k_n} d_{ijn}.$$

An empirical estimator of $E\Delta(T) = E\,|(X(T) - X(T-)|$ is

$$\bar{D}_n = \frac{1}{n} \sum_{i=1}^{n} D_{in},$$

and the assumption corresponding to A_α is

$A'_\alpha - \exists M : EM^2 < \infty, \exists \alpha \in \,]0, 1] : |X(t) - X(s)| \leq M\,|t - s|^\alpha,$
$\quad (s, t) \in I_T := [0, T[^2 \cup [T, 1]^2.$

Example 3 Consider two Wiener processes $W_k = (W_k(t), 0 \leq t \leq 1)$, $k = 1, 2$ and suppose that T and (W_1, W_2) are independent. In order to construct X one may put

$$X(t) = W_1(t)\, I_{[0,T]}(t) + W_2(t)\, I_{[T,1]}(t), \; 0 \leq t \leq 1.$$

Then, if $0 < \alpha < \frac{1}{2}$ and

$$M = \sup_{0 \le s \ne t \le 1} \frac{|W_1(t) - W_1(s)|}{|t - s|^\alpha} + \sup_{0 \le s \ne t \le 1} \frac{|W_2(t) - W_2(s)|}{|t - s|^\alpha},$$

A'_α is satisfied.

Proposition 6 *If* $E \, \|X\|^4 < \infty$, *A'_α holds and (X_n) is an iid sequence, then*

$$\bar{D}_n \to E \Delta(T), \quad a.s. \text{ and in } L^2.$$

Proof For $1 \le i \le n$, set $U_{in} = \max_{j \ne J(i,n)} d_{ijn}$ and $V_{in} = |X_i(T_{in}) - X_i (T_{in} - k_n^{-1})|$, we have

$$\bar{D}_n = \frac{1}{n} \sum_{i=1}^{n} \max(U_{in}, V_{in}),$$

hence

$$0 \le \frac{1}{n} \sum_{i=1}^{n} V_{in} \le \bar{D}_n \le \frac{1}{n} \sum_{i=1}^{n} U_{in} + \frac{1}{n} \sum_{i=1}^{n} V_{in}, \tag{29}$$

and A'_α yields

$$0 \le \frac{1}{n} \sum_{i=1}^{n} U_{in} \le \bar{M}_n \, k_n^{-\alpha} \to 0, \quad a.s. \tag{30}$$

since $\bar{M}_n$ converges almost surely. It follows that $\bar{D}_n$ and $\frac{1}{n} \sum_{i=1}^{n} V_{in}$ have the same a.s. behaviour. Now, since $E \, \|X\|^4 < \infty$, $(V_{in} - E V_{in})$ satisfy the conditions in Lemma 1, therefore

$$\frac{1}{n} \sum_{i=1}^{n} V_{in} - E V_{1n} \to 0 \quad a.s.$$

and the dominated convergence theorem entails a.s. convergence. The proof is similar for L^2-convergence. $\qquad\square$

Proposition 7 *Under A'_α, (X_n) iid, $E \, \|X\|^2 < \infty$ and $n^{1/2} k_n^{-\alpha} \to 0$, it is true that*

$$\sqrt{n} \left(\bar{D}_n - E \Delta(T) \right) \Rightarrow N \sim \mathcal{N} \left(0, \, \mathrm{Var}(\Delta(T)) \right), \tag{31}$$

Proof From (29) and (30) and using $n^{1/2} k_n^{-\alpha} \to 0$, it is easy to see that $\sqrt{n} \bar{D}_n$ and $\frac{1}{\sqrt{n}} \sum_{i=1}^{n} V_{in}$ have the same asymptotic behaviour in distribution. Now let us set $V_i = |X_i(T_i) - X_i(T_i -)|$. We clearly have

$$a_n := \frac{1}{\sqrt{n}} \sum_{i=1}^{n} (V_i - E V_i) \Rightarrow N \sim \mathcal{N}(0, \sigma^2).$$

We want to compare a_n with

$$b_n := \frac{1}{\sqrt{n}} \sum_{i=1}^{n} (V_{in} - EV_{in}).$$

Noting that

$$V_i \leq |X_i(T_i) - X_i(T_{in})| + V_{in} + \left| X_i(T_i-) - X_i\left(T_{in} - \frac{1}{k_n}\right) \right|,$$

A'_α gives $V_i \leq 2M_i k_n^{-\alpha} + V_{in}$, similarly $V_{in} \leq 2M_i k_n^{-\alpha} + V_i$ and also $|EV_{in} - EV_i| \leq 2EM_i k_n^{-\alpha}$ hence $n^{1/2}|EV_{in} - EV_i| \to 0$ and finally

$$|a_n - b_n| \leq 2\, n^{1/2} k_n^{-\alpha} \left[\bar{M}_n + EM \right] \to 0 \ \ \text{a.s.}$$

and (31) follows. $\qquad\qquad\qquad\qquad\qquad\qquad\qquad\qquad\qquad\qquad\qquad\qquad\quad\square$

5 Estimating the Density of T

Suppose that T has a continuous and strictly positive density f on $[0, 1]$ and that the observations $X_1, \ldots, X_n$ are iid. and in continuous time, then one may define the density estimator $f_n(t)$ by setting

$$f_n(t) = \frac{1}{nh_n} \sum_{i=1}^{n} K\left(\frac{t - T_i}{h_n} \right), \quad 0 \leq t \leq 1, \tag{32}$$

where K is a Lipschitzian symmetric density with bounded support. Then, if f is of class C_2 and if $h_n \asymp (\frac{\ln n}{n})^{1/5}$, it follows that

$$\|f_n - f\| = \mathcal{O}\left(\left(\frac{\ln n}{n} \right)^{2/5} \right) \ \ \text{a.s.} \tag{33}$$

see, for example, Deheuvels (2000).

Now, if $X_i(jk_n^{-1})$, $k_n \geq 1$, $0 \leq j \leq k_n$, $1 \leq i \leq n$ are observed, the density estimator takes the form

$$\widehat{f_n}(t) = \frac{1}{nh_n} \sum_{i=1}^{n} K\left(\frac{t - \widehat{T}_{in}}{h_n} \right), \quad t \in [0, 1],$$

where

$$\widehat{T}_{in} = k_n^{-1} \arg \max_{1 \le j \le k_n} \left| X_i \left(j k_n^{-1} \right) - X_i \left((j-1) k_n^{-1} \right) \right|, \ 1 \le i \le n,$$

and if there is a tie one may choose the greatest j (for example).

Moreover, to $\widehat{f}_n$ one can associate f_n in (32) and the pseudo-estimator

$$\varphi_n(t) = \frac{1}{n h_n} \sum_{i=1}^{n} K \left(\frac{t - T_{in}}{h_n} \right), \ t \in [0, 1],$$

where $T_{in} = J(i, n) k_n^{-1}$, cf. (28). We now make the following assumption

$$A'_q - \exists q \ge 2: \ EM^q < \infty, \ E \left(\frac{M}{|X(T) - X(T-)|} \right)^q < \infty.$$

Proposition 8 *Under* A'_α, A'_q, *and*

$$\| f_n - f \| \to 0 \ a.s. \tag{34}$$

the conditions $\sum n k_n^{-\alpha q} < \infty$ *and* $h_n^2 k_n \to \infty$ *imply*

$$\left\| \widehat{f}_n - f \right\| \to 0 \ a.s..$$

Proof We define U_{in} and V_{in} similarly as in Proposition 6 and note that

$$\widehat{T}_{in} \ne T_{in} \implies V_{in} \le U_{in}. \tag{35}$$

As in Proposition 7 we have $V_i - V_{in} \le 2 M_i k_n^{-\alpha}$ and $0 \le U_{in} \le M_i k_n^{-\alpha}$, hence

$$U_{in} - V_{in} \le 3 M_i k_n^{-\alpha} - V_i,$$

from (35) it follows that

$$P \left(\widehat{T}_{in} \ne T_{in} \right) \le P(U_{in} - V_{in} \ge 0) \le P \left(3 M_i k_n^{-\alpha} - V_i \ge 0 \right) \le P \left(\frac{M_i}{V_i} \ge \frac{k_n^{\alpha}}{3} \right).$$

Now A'_q implies

$$P \left(\widehat{T}_{in} \ne T_{in} \right) \le E \left(\frac{M}{|X(T) - X(T-)|} \right)^q \frac{3^q}{k_n^{\alpha q}},$$

thus

$$P\left(\exists i \in \{1, \ldots, n\} : \widehat{T}_{in} \neq T_{in}\right) \leq \sum_{i=1}^{n} P\left(\widehat{T}_{in} \neq T_{in}\right) = \mathcal{O}\left(nk_n^{-\alpha q}\right), \tag{36}$$

since $\sum nk_n^{-\alpha q} < \infty$, we have $\widehat{T}_{in} = T_{in}$, $1 \leq i \leq n$ almost surely for n large enough and consequently $\widehat{f}_n = \varphi_n$. Now, since K satisfies a Lipschitz condition, there exists a constant $c(K)$ such that

$$|f_n(t) - \varphi_n(t)| \leq \frac{1}{nh_n^2} c(K) \sum_{i=1}^{n} |T_{in} - T_i| \leq \frac{c(K)}{h_n^2 k_n}. \tag{37}$$

Next, since
$$\left\|\widehat{f}_n - f\right\| \leq \left\|\widehat{f}_n - \varphi_n\right\| + \left\|\varphi_n - f_n\right\| + \left\|f_n - f\right\|, \tag{38}$$

the result follows from (34), (37) and (38). $\qquad\square$

Example 4 If $h_n \asymp n^{-\gamma}$, $(0 < \gamma < 1)$, $k_n \asymp n^{\beta}$, $\beta > 0$, then, the condition $\beta > \max(\frac{2}{\alpha q}, 2\gamma)$ entails uniform consistency of $\widehat{f}_n$.

Corollary 2 *If* $h_n \asymp \left(\frac{\ln n}{n}\right)^{1/5}$ *and* $\beta > \max\left(\frac{2}{\alpha q}, \frac{4}{5}\right)$, *then*

$$\left\|\widehat{f}_n - f\right\| = \mathcal{O}\left(\left(\frac{\ln n}{n}\right)^{2/5}\right) \quad a.s.$$

Proof It is a consequence of (33), (36) and (37). $\qquad\square$

We now turn to L^2-convergence.

Proposition 9 *If* A'_{α}, A'_{q} *hold,* (X_n) *is iid, and if* $h_n \to 0$, $nh_n \to \infty$, $h_n k_n^{\alpha q} \to \infty$, $h_n^2 k_n \to \infty$, *then*
$$E\left(\widehat{f}_n(t) - f(t)\right)^2 \to 0, \ 0 < t < 1. \tag{39}$$

If, in addition, $f \in C_2[0, 1]$, $h_n \asymp n^{-1/5}$, $k_n \asymp n^{4/5}$, $\alpha q \geq \frac{3}{4}$, *we have*

$$E\left(\widehat{f}_n(t) - f(t)\right)^2 = \mathcal{O}\left(n^{-4/5}\right), \ 0 < t < 1. \tag{40}$$

Proof Set $c_{in} = K(\frac{t-\widehat{T}_{in}}{h_n}) - K(\frac{t-T_{in}}{h_n})$, we have

$$b_n = \left| E\left(\widehat{f}_n(t) - \varphi_n(t)\right) \right| = \frac{1}{nh_n} \left| E\left[\sum_{i=1}^{n} c_{in} I_{\widehat{T}_{in} \neq T_{in}} \right] \right|$$

$$\leq \frac{2\|K\|}{nh_n} \sum_{i=1}^{n} P\left(\widehat{T}_{in} \neq T_{in}\right).$$

From (36) it follows that $b_n = \mathcal{O}(\frac{1}{h_n k_n^{\alpha q}}) \to 0$. On the other hand

$$v_n := \mathrm{Var}\left(\widehat{f}_n(t) - \varphi_n(t)\right) = \frac{1}{n^2 h_n^2} \sum_{i=1}^{n} \mathrm{Var}\left(c_{in} I_{\widehat{T}_{in} \neq T_{in}}\right)$$

$$\leq \frac{1}{n^2 h_n^2} \sum_{i=1}^{n} E\left(c_{in} I_{\widehat{T}_{in} \neq T_{in}}\right)^2.$$

Thus $v_n \leq \frac{1}{n^2 h_n^2} \sum_{i=1}^{n} 4\|K\|_\infty^2 P(\widehat{T}_{in} \neq T_{in}) = \mathcal{O}(\frac{1}{nh_n^2 k_n^{\alpha q}})$, and therefore

$$E\left(\widehat{f}_n(t) - \varphi_n(t)\right)^2 = b_n^2 + v_n = \mathcal{O}\left(\frac{1}{h_n^2 k_n^{2\alpha q}}\right) + \mathcal{O}\left(\frac{1}{nh_n^2 k_n^{\alpha q}}\right).$$

Now from (37), one obtains

$$E(\varphi_n(t) - f_n(t))^2 = \mathcal{O}\left(\frac{1}{h_n^4 k_n^2}\right).$$

Finally

$$h_n \to 0,\ nh_n \to \infty \implies E\left(f_n(t) - f(t)\right)^2 \to 0,$$
$$h_n^2 k_n \to \infty \implies E(\varphi_n(t) - f_n(t))^2 \to 0,$$
$$nh_n \to \infty,\ h_n k_n^{\alpha q} \to \infty \implies E\left(\widehat{f}_n(t) - \varphi_n(t)\right)^2 \to 0,$$

and (39) follows.

Now if $f \in C_2[0,1]$, $h_n \asymp n^{-1/5}$, it is well known that $E(f_n(t) - f(t))^2 = O(n^{-4/5})$ (see for example Rosenblatt (1985)) then, it is easy to verify that the choice $k_n \asymp n^{4/5}$ and the condition $\alpha q \geq \frac{3}{4}$ give (40). $\qquad\square$

Proposition 10 (limit in distribution)

Under the conditions in Proposition 8 and if $h_n \to 0$, $nh_n^5 \to \infty$, $\frac{\sqrt{n}}{h_n^{3/2} k_n} \to 0$, $\sum n\, k_n^{-\alpha q} < \infty$ then

$$\sqrt{nh_n}\left(\widehat{f_n}(t) - f(t)\right) \Rightarrow N \sim \mathcal{N}\left(0,\; f(t)\int K^2\right). \tag{41}$$

Proof First

$$\sqrt{nh_n}(\varphi_n(t) - f_n(t)) \le \frac{\|K\|\sqrt{n}}{h_n^{3/2}k_n} \to 0.$$

Now, we have seen that, if $\sum n\, k_n^{-\alpha q} < \infty$ we have $\widehat{f_n}(t) = \varphi_n(t)$ for n large enough. Then

$$\sqrt{nh_n}\left(\widehat{f_n}(t) - f(t)\right) = \sqrt{nh_n}\left(\widehat{f_n}(t) - \varphi_n(t)\right)$$
$$+ \sqrt{nh_n}(\varphi_n(t) - f_n(t)) + \sqrt{nh_n}(f_n(t) - f(t))$$

and since $\sqrt{nh_n}(\widehat{f_n}(t) - f_n(t)) \to 0$ in probability and $\sqrt{nh_n}(f_n(t) - f(t))$ is asymptotically Gaussian (see, for example Rosenblatt (1985)), one obtains the desired result. $\qquad\square$

Example 5 Take $h_n = \frac{n^{-1/5}}{\log n}$, $k_n \asymp n^{\beta}$ with $\beta > \max(\frac{4}{5},\, \frac{2}{\alpha q})$ then (41) holds.

Finally, if $T_1, \ldots, T_n$ are correlated, one may use results in Bosq and Blanke (2007) for obtaining consistent density estimators with sharp rates.

6 The Case of More than One Jump

If X admits several independent random jumps, one does not know their respective positions. Thus, it is necessary to modify the method used in Sect. 4. The principle is to use relations between the coefficients and the roots of a polynomial equation. For convenience we only study the case of two jumps.

We now suppose that X admits two independent random jumps S and T with values in $]0, 1[$ such that $P(S = T) = 0$ and

$$E\Delta := E\,|X(T) - X(T-)| > E\delta := E\,|X(S) - X(S-)| > 0. \tag{42}$$

Constructing X: For example, suppose that S and T are equidistributed with density f over $[0, 1]$ and that (S, T, W_1, W_2, W_3) are globally independent, where the W_i's are Wiener processes on $[0, 1]$ with parameters σ_i^2 such that $\sigma_3^2 > \sigma_1^2$. Construction of X can be performed as follows; if $S \le T$ set

$$X(t) = W_1(t)\, I_{0 \le t < S} + W_2(t)\, I_{S \le t < T} + W_3(t) I_{T \le t \le 1},$$

and if $T < S$

$$X(t) = W_3(t)\, I_{0 \le t < T} + W_2(t)\, I_{T \le t < S} + W_1(t) I_{S \le t \le 1}.$$

It is easy to verify that

$$EΔ = \int_0^1 E\,|W_2(s) - W_3(s)|\,f(s)ds = \sqrt{\frac{2}{\pi}}\,\sqrt{\sigma_2^2 + \sigma_3^2}\int_0^1 \sqrt{s}\,f(s)ds$$

and that $E\delta$ has a similar expression (replace σ_3^2 by σ_1^2), thus (42) is satisfied.

The main point is that, given X, we do not know the respective positions of S and T. Now, since $\max(\Delta, \delta)$ and $\min(\Delta, \delta)$ are observed, we may set

$$\xi = \max(\Delta, \delta) + \min(\Delta, \delta) = \Delta + \delta$$

$$\eta = \max(\Delta, \delta)\min(\Delta, \delta) = \Delta\delta$$

Now consider the following assumption:
B-Δ and δ are independent and square integrable.

For example if $W_2 = 0$ in the construction of X condition B holds. From B it follows that

$$E\Delta\,E\delta = \frac{1}{4}\left(E\xi^2 - E\eta^2\right).$$

The associated discriminant is $[E(\Delta - \delta)]^2 = E\eta^2 - \mathrm{Var}(\xi) > 0$. Thus

$$E\delta = \frac{E\xi - \sqrt{E\eta^2 - \mathrm{Var}(\xi)}}{2}, \tag{43}$$

and

$$E\Delta = \frac{E\xi + \sqrt{E\eta^2 - \mathrm{Var}(\xi)}}{2}. \tag{44}$$

Now, if $X_1, \ldots, X_n$ are observed, the corresponding variables $\xi_1, \eta_1, \ldots, \xi_n, \eta_n$ are also observed and one may use (43) and (44) for constructing empirical estimators δ_n and Δ_n of $E\delta$ and $E\Delta$. Note that, if v_n denotes the empirical estimator of $E\eta^2 - \mathrm{Var}(\xi)$, the estimator of the discriminant must be $\max(0, v_n)$.

Proposition 11 (continuous case)
If $X_1, \ldots, X_n$ are iid and if B holds, then

$$\delta_n \to E\delta \ a.s., \ \Delta_n \to E\Delta, \ a.s.$$

Similar results hold for the other kinds of jumps and if (X_n) is strongly mixing. The details are omitted.

We now consider the same family of discretized data as above and make the following assumption:

A_α''—*There exists M such that $EM^q < \infty$, $(q \geq 2)$ and $\alpha \in]0, 1]$ such that*

$$|X(t) - X(s)| \leq M\,|t - s|^\alpha,$$

where $(s, t) \in I_{S,T} := [0, \min(S, T)[^2 \cup [\min(S, T), \max(S, T)[^2 \cup [\max(S, T), 1]^2$.

Now there exist integers $J_{in}(S)$ and $J_{in}(T)$ such that

$$\frac{J_{in}(S) - 1}{k_n} < S_i \leq \frac{J_{in}(S)}{k_n} := S_{in}, \quad 1 \leq i \leq n,$$

and

$$\frac{J_{in}(T) - 1}{k_n} < T_i \leq \frac{J_{in}(T)}{k_n} := T_{in}, \quad 1 \leq i \leq n,$$

where S_i and T_i are the positions of the jumps associated with X_i.

In order to construct an estimator of the jumps intensity we set

$$D_{in} = \max_{1 \leq j \leq k_n} \left| X_i\left(\frac{j}{k_n}\right) - X_i\left(\frac{j-1}{k_n}\right) \right|$$

and we denote J_{in}' an integer that maximises D_{in}. Then we put

$$d_{in} = \max_{1 \leq j \leq k_n,\, j \neq J_{in}'} \left| X_i\left(\frac{j}{k_n}\right) - X_i\left(\frac{j-1}{k_n}\right) \right|.$$

Now, similarly as in the continuous case we set $\xi_{in} = D_{in} + d_{in}$ and $\eta_{in} = |D_{in} - d_{in}|$. Finally we consider the random variable

$$U_{in}' = \max_{1 \leq j \leq k_n,\, j \neq \{J_{in}(S), J_{in}(T)\}} \left| X_i\left(\frac{j}{k_n}\right) - X_i\left(\frac{j-1}{k_n}\right) \right|, \quad k_n \geq 3.$$

Lemma 5 *If A_α'' holds, $\min(\Delta, \delta) \geq a > 0$, (X_n) is iid, $E\,\|X\|^4 < \infty$ and $\sum nk_n^{-\alpha q} < \infty$, then*

$$\frac{1}{n}\sum_{i=1}^{n} \xi_{in} \to E\Delta + E\delta, \quad a.s. \tag{45}$$

and

$$\frac{1}{n}\sum_{i=1}^{n} \eta_{in} \to E\Delta E\delta, \quad a.s. \tag{46}$$

Proof A''_α yields $0 \leq U'_{in} \leq M_i k_n^{-\alpha}$ and similarly as above we deduce that

$$P\left(\exists i \in \{1, \ldots, n\} : U'_{in} > \frac{a}{2}\right) = \mathscr{O}\left(n k_n^{-\alpha q}\right),$$

then, for n large enough, $U'_{in} \leq \frac{a}{2}$ $(a.s.)$, $1 \leq i \leq n$. Now, since $\min(\Delta, \delta) \geq a$, it follows that, a.s. for n large enough we have

$$D_{in} = \max(\Delta_{in}, \delta_{in}), \quad d_{in} = \min(\Delta_{in}, \delta_{in}),$$

where $\Delta_{in} = \left|X_i(T_{in}) - X_i(T_{in} - \frac{1}{k_n})\right|$ and $\delta_{in} = \left|X_i(S_{in}) - X_i(S_{in} - \frac{1}{k_n})\right|$. Now it suffices to apply the law of large numbers to (ξ_{in}) and (η_{in}) via Lemma 1 and the dominated convergence theorem for obtaining (45) and (46). $\qquad\square$

Finally we denote by $\hat{\delta}_n$ and $\hat{\Delta}_n$ the associated empirical estimators of $E\delta$ and $E\Delta$ (cf. (43) and (44)) and we obtain

Proposition 12 (HFD case)
> *Under the conditions in Lemma 5 and condition B, we get*

$$\hat{\delta}_n \to E\delta, \quad \hat{\Delta}_n \to E\Delta \quad a.s..$$

Proof Clear. $\qquad\square$

Note that it is possible to adapt the above method to the case of more than two jumps, however, the computations are rather intricate.

7 Applications to D-Valued Linear Processes

7.1 Definitions and Examples

A D- strong white noise is a sequence $(Z_n, \ n \in \mathbb{Z})$ of iid centred D-valued random variables such that $0 < E\|Z_0\|^2 < \infty$. Now, in order to define linear processes in D we consider the space $\mathscr{L} = \mathscr{L}(D, D)$ of continuous linear operators with respect to $D(\|.\|)$. The linear norm is denoted by $\|.\|_{\mathscr{L}}$.

The $D-$moving average process of order 1 (MAD(1)) is given by

$$X_n = \mu + Z_n - a(Z_{n-1}), \quad n \in \mathbb{Z}, \tag{47}$$

where $\mu \in D$ and $a \in \mathscr{L}$.

The D-autoregressive process (ARD) is solution of

$$X_n = \nu + \rho(X_{n-1}) + Z_n, \quad n \in \mathbb{Z} \tag{48}$$

with $v \in D$ and $\rho \in \mathscr{L}$. If there exists an integer j_0 such that $\left\| \rho^{j_0} \right\|_{\mathscr{L}} < 1$, then it may be shown that (48) has a unique solution given by

$$X_n = v + \sum_{j=0}^{\infty} \rho^j (Z_{n-j}), \quad n \in \mathbb{Z}, \tag{49}$$

where convergence takes place with probability one in the uniform norm (cf. El Hajj (2013)).

Examples of continuous linear operators in $D(\|.\|)$ appear below:

Example 6 Set $\rho(x)(t) = \int_0^1 r(s, t)\, x(s)ds$, $x \in D$ where r is (uniformly) continuous on $[0, 1]^2$, then $\rho \in \mathscr{L}$, and $\rho(D) \subset C$.

Example 7 Set $a(x)(t) = a_0(t)\, x(t)$, $0 \le t \le 1$, $x \in D$, where a_0 is continuous and $0 < |a_0(t)| \le c < 1, 0 \le t \le 1$, and c is constant. Then $a \in \mathscr{L}$ and x and $a(x)$ have the same jumps.

7.2 Some Results

For lack of place we only give indications concerning applications. More complete results will appear elsewhere.

First, since $(Z_n, n \in \mathbb{Z})$ is iid, the results appearing in the above sections are valid, providing suitable assumptions. Concerning the MAD (1) process, if $\mu \in C$ and $a(D) \subset C$, it follows that Z_n and X_n have the same jumps with the same intensity:

$$Z_n(T_n) - Z_n(T_n-) = X_n(T_n) - X_n(T_n-).$$

Also, if μ has a jump point at t_0, it is possible to consider couples of jumps (t_0, T_n) by using methods similar as in Sect. 6.

Another special case is Example 7: Z_n has a single jump at T_n and T_n admits a density, it follows that X_n admits two independent jumps with respective intensities $E\,|Z_n(T_n) - Z_n(T_n-)|$ and $E\,|a(T_{n-1})|\,|Z_{n-1}(T_{n-1}) - Z_{n-1}(T_{n-1}-)|$ and results in Sect. 6 can be easily adapted since (X_n) is stationary and 1-dependent.

We turn to the ARD (1) process. Again, if $\rho(D) \subset C$, Z_n and X_n have the same jumps and one may apply results concerning t_0, T and (S, T). Now, if one uses the operator in Example 7 and if $Z_n(t_0) - Z_n(t_0-)$ admits a density and satisfies the Cramer condition, then $X_n(t_0) - X_n(t_0-)$ is geometrically strongly mixing, Bradley (1986), and Proposition 2 applies.

7.3 The Case of Infinitely Many Jumps

The general case is much more difficult since it depends how ρ^j transforms the jumps. We only consider the special case where ρ is a number belonging to $]-1, 1[$. Then

$$X_n = \sum_{i=1}^{n} \rho^j Z_{n-j}, \quad n \in \mathbb{Z}. \tag{50}$$

Again we suppose that Z_n has a single jump at T_n where T_n has a density. Then X_n has the jumps $(T_{n-j} \; j \geq 0)$ with intensities $(\lambda \, |\rho|^j \, , j \geq 0)$, where $\lambda = E \, |Z_n(T_n) - Z_n(T_n-)|$. First we construct an estimator of ρ. For this purpose we consider the relation

$$\int_0^1 X_n(t)dt = \rho \int_0^1 X_{n-1}(t)dt + \int_0^1 Z_n(t)dt$$

For convenience we write it in the form $Y_n = \rho Y_{n-1} + E_n$. Hence an estimator of ρ, defined as the empirical autocorrelation coefficient based on $Y_1, \ldots, Y_n$, is given by:

$$\hat{\rho}_n = \frac{\sum_{i=1}^{n-1} Y_i Y_{i+1}}{\sum_{i=1}^{n} Y_i^2}.$$

Now consider the series

$$R_i = \sum_{j=0}^{\infty} \left| X_i(T_{i-j}) - X_i(T_{i-j}-) \right|.$$

We have $E R_i = \frac{\lambda}{1-|\rho|}$. It follows that R_i is almost surely finite. If $R_1, \ldots, R_n$ are observable (even if, in practice, the jumps are negligible if j is large enough), then putting

$$\bar{R}_n = \frac{R_1 + \cdots + R_n}{n}$$

one obtains an estimator of $\frac{\lambda}{1-|\rho|}$, thus, an empirical estimator of λ is given by

$$\hat{\lambda}_n = \left(1 - \left|\hat{\rho}_n\right|\right) \bar{R}_n.$$

Proposition 13 *Suppose that $(Z_n, \, n \in \mathbb{Z})$ is a D-white noise such that we have $c := E \, \|Z_n - E Z_n\|^4 < \infty$, then $\hat{\lambda}_n \, |\hat{\rho}_n|^j \to \lambda \, |\rho|^j \; j \geq 0 \; a.s..$*

Proof For convenience we suppose that ρ is strictly positive. Since $\hat{\rho}_n \to \rho$ a.s. (cf. Brockwell and Davis (1991)), it suffices to show that $\bar{R}_n \to E R_1$ a.s.. Set

$$\bar{S}_{jn} = \rho^j \frac{1}{n} \sum_{i=1}^{n} (W_{ij} - E W_{ij}),$$

where $W_{ij} = \left| Z_i(T_{i-j}) - Z_i(T_{i-j}-) \right|$, then we have

$$\bar{R}_n - ER_1 = \sum_{j=0}^{\infty} \bar{S}_{jn}.$$

Now choose $\rho' \in]\rho, 1[$, we obtain

$$P\left(\left|\bar{R}_n - ER_1\right| > \eta\right) \leq P\left(\exists j \geq 0 : \left|\bar{S}_{jn}\right| > \eta \frac{\rho'^j}{1 - \rho'}\right) \quad \eta > 0.$$

Therefore

$$P\left(\left|\bar{R}_n - ER_1\right| > \eta\right) \leq \sum_{j=0}^{\infty} P\left(\left|\bar{S}_{jn}\right| > \eta \frac{\rho'^j}{1 - \rho'}\right)$$

and Lemma 1 entails

$$P\left(\left|\bar{R}_n - ER_1\right| > \eta\right) \leq \frac{1}{n^2} \left(1 - \rho'\right)^4 \frac{3c}{\eta^4} \sum_{j=0}^{\infty} \left(\frac{\rho}{\rho'}\right)^{4j}$$

and the Borel-Cantelli lemma gives the result. $\qquad\square$

Some simulations illustrating the theoretical results appear in the thesis by El Hajj (2013).

Acknowledgments I want to thank the associate editor and the two referees for their very useful comments which allowed me to notably improve the original manuscript.

References

Bezandry, P. H. (2006). Laws of large numbers for $D[0, 1]$. *Statistics and Probability Letters, 76*(10), 981–985.

Bézandry, P. H., & Fernique, X. (1992). Sur la propriété de la limite centrale dans $D[0, 1]$. *Annales de l'Institut Henri Poincaré Probability and Statistics, 28*(1), 31–46.

Billingsley, P. (1999). *Convergence of probability measures* (2nd ed.). *Wiley Series in Probability and Statistics: Probability and Statistics*. New York: Wiley.

Bloznelis, M., & Paulauskas, V. (1993). On the central limit theorem in $D[0, 1]$. *Statistics and Probability Letters, 17*(2), 105–111.

Bosq, D. (1998). Nonparametric statistics for stochastic processes. *Estimation and prediction* (2nd ed.,). *Lecture Notes in Statistics (Vol. 110)*. New York: Springer.

Bosq, D. (2000). *Linear processes in function spaces. Lecture Notes in Statistics (Vol. 149)*. New York: Springer.

Bosq, D. (2014). Computing the best linear predictor in a Hilbert space. Applications to general ARMAH processes. *Journal of Multivariate Analysis, 124*, 436–450.

Bosq, D., & Blanke, D. (2007). *Prediction and inference in large dimensions. Wiley Series in Probability and Statistics*. New York: Wiley-Dunod.

Bradley, R. C. (1986). Basic properties of strong mixing conditions. In: *Dependence in probability and statistics (Oberwolfach, 1985). Programme Probability and Statistics* (vol. 11, pp. 165–192). Boston: Birkhäuser.

Brockwell, P. J., & Davis, R. A. (1991). *Time series: Theory and methods* (2nd ed.). New York: Springer.

Daffer, P. Z., & Taylor, R. L. (1979). Laws of large numbers for $D[0, 1]$. *Annals of Probability, 7*(1), 85–95.

Deheuvels, P. (2000). Uniform limit laws for kernel density estimators on possibly unbounded intervals. Recent advances in reliability theory, Statistics in Indiana Technology (Bordeaux, 2000) (pp. 477–492). Boston: Birkhäuser.

El Hajj, L. (2011). Théorèmes limites pour les processus autorégressifs à valeurs dans $D[0, 1]$. *Comptes Rendus Mathematics Academy Sciences Paris, 349*(13–14), 821–825.

El Hajj, L. (2013). Inférence statistique pour des variables fonctionnelles à sauts. Thèse de doctorat, Université Pierre et Marie Curie-Paris 6, Décembre 2013.

Ferraty, F., & Romain, Y. (Eds.). (2011). *The Oxford handbook of functional data analysis*. Oxford: Oxford University Press.

Hahn, M. G. (1978). Central limit theorems in $D[0, 1]$. *Zeitschrift für Wahrscheinlichkeitstheorie und verwandte Gebiete, 44*(2), 89–101.

Horváth, L., & Kokoszka, P. (2012). *Inference for functional data with applications. Springer Series in Statistics*. New York: Springer.

Merlevède, F., Peligrad, M., & Bosq, D. (1999). Asymptotic normality for density kernel estimators in discrete and continuous time. *Journal of Multivariate Analysis, 68*(1), 78–95.

Norzhigitov, A. F., & Sharipov, O. S. (2010). The central limit theorem for weakly dependent random variables with values in the space $D[0, 1]$. *Uzbekskii Matematicheskii Zhurnal, 2*, 103–112.

Pestman, W. R. (1995). Measurability of linear operators in the Skorokhod topology. *Bulletin of the Belgian Mathematical Society—Simon Stevin, 2*(4), 381–388.

Ranga Rao, R. (1963). The law of large numbers for $D[0, 1]$-valued random variables. *Theory of Probability and its Applications, 8*, 70–74.

Rosenblatt, M. (1985). *Stationary sequences and random fields*. Boston: Birkhäuser Boston Inc.

Şchiopu-Kratina, I., & Daffer, P. (1999). Tightness and the law of large numbers in $D[0, 1]$. *Revue Roumaine de Mathématique Pures et Appliquées, 44*(3), 473–480. (2000).

Shiryaev, A. N. (1996). In R. P. Boas (Ed.), *Probability, graduate texts in mathematics* (2nd ed., Vol. 95). New York: Springer. translated from the first (1980).

A Sharp Abelian Theorem for the Laplace Transform

Maëva Biret, Michel Broniatowski and Zansheng Cao

Abstract This paper states asymptotic equivalents for the moments of the Esscher transform of a distribution on $\mathbb{R}$ with smooth density in the upper tail. As a by-product it provides a tail approximation for its moment generating function, and shows that the Esscher transforms have a Gaussian behavior for large values of the parameter.

1 Introduction

Let X denote a real-valued random variable with support $\mathbb{R}$ and distribution P_X with density p.

The moment generating function of X

$$\Phi(t) := \mathbb{E}[\exp(tX)] \tag{1}$$

is supposed to be finite in a nonvoid neighborhood $\mathcal{N}$ of 0. This hypothesis is usually referred to as a Cramér type condition.

The tilted density of X (or Esscher transform of its distribution) with parameter t in $\mathcal{N}$ is defined on $\mathbb{R}$ by

$$\pi_t(x) := \frac{\exp(tx)}{\Phi(t)} p(x).$$

M. Biret
Snecma Villaroche, Rond-Point René Ravaud-Réau, Moissy-cramayel, France
e-mail: maeva.biret@snecma.fr

M. Broniatowski (✉)
LSTA, Université Pierre et Marie Curie, 5 Place Jussieu, Paris, France
e-mail: michel.broniatowski@upmc.fr

Z. Cao
IRMA, Université de Strasbourg, 7 Rue René Descartes, Strasbourg, France
e-mail: cao@math.unistra.fr

© Springer International Publishing Switzerland 2015
M. Hallin et al. (eds.), *Mathematical Statistics and Limit Theorems*,
DOI 10.1007/978-3-319-12442-1_5

For $t \in \mathcal{N}$, the functions

$$t \to m(t) := \frac{d}{dt} \log \Phi(t), \tag{2}$$

$$t \to s^2(t) := \frac{d^2}{dt^2} \log \Phi(t), \tag{3}$$

$$t \to \mu_j(t) := \frac{d^j}{dt^j} \log \Phi(t), \, j \in (2, \infty). \tag{4}$$

are, respectively, the expectation and the centered moments of a random variable with density π_t.

When Φ is steep, meaning that

$$\lim_{t \to t^+} m(t) = \infty \tag{5}$$

and

$$\lim_{t \to t^-} m(t) = -\infty$$

where $t^+ := ess \sup \mathcal{N}$ and $t^- := ess \inf \mathcal{N}$ then m parametrizes $\mathbb{R}$ (this is steepness, see Barndorff-Nielsen (1978)). We will only require (5) to hold.

This paper presents sharp approximations for the moments of the tilted density π_t under conditions pertaining to the shape of p in its upper tail, when t tends to the upper bound of $\mathcal{N}$.

Such expansions are relevant in the context of extreme value theory as well as in approximations of very large deviation probabilities for the empirical mean of independent and identically distributed summands. We refer to Feigin and Yashchin (1983) in the first case, where convergence in type to the Gumbel extreme distribution follows from the self-neglecting property of the function s^2, and to Broniatowski and Mason (1994) in relation with extreme deviation probabilities. The fact that up to a normalization, and under the natural regularity conditions assumed in this paper, the tilted distribution with density $\pi_t(x)$ converges to a standard Gaussian law as t tends to the essential supremum of the set $\mathcal{N}$ is also of some interest.

2 Notation and Hypotheses

Thereafter we will use indifferently the notation $f(t) \underset{t \to \infty}{\sim} g(t)$ and $f(t) \underset{t \to \infty}{=} g(t)(1 + o(1))$ to specify that f and g are asymptotically equivalent functions.

The density p is assumed to be of the form

$$p(x) = \exp(-(g(x) - q(x))), \qquad x \in \mathbb{R}_+. \tag{6}$$

For the sake of this paper, only the form of p for positive x matters.

The function g is positive, convex, four times differentiable and satisfies

$$\frac{g(x)}{x} \xrightarrow[x\to\infty]{} \infty. \tag{7}$$

Define

$$h(x) := g'(x). \tag{8}$$

In the present context, due to (7) and the assumed conditions on q to be stated hereunder, $t^+ = +\infty$.

Not all positive convex g's satisfying (7) are adapted to our purpose. We follow the line of Juszczak and Nagaev (2004) to describe the assumed regularity conditions of h. See also Balkema et al. (1993) for somehow similar conditions.

We firstly assume that the function h, which is a positive function defined on $\mathbb{R}_+$, is either regularly or rapidly varying in a neighborhood of infinity; the function h is monotone and, by (7), $h(x) \to \infty$ when $x \to \infty$.

The following notation is adopted:

$RV(\alpha)$ designates the class of regularly varying functions of index α defined on $\mathbb{R}_+$,
$\psi(t) := h^{\leftarrow}(t)$ designates the inverse of h. Hence ψ is monotone for large t and $\psi(t) \to \infty$ when $t \to \infty$,
$\sigma^2(x) := 1/h'(x)$,
$\hat{x} := \hat{x}(t) = \psi(t)$,
$\hat{\sigma} := \sigma(\hat{x}) = \sigma(\psi(t))$.

The two cases considered for h, the regularly varying case and the rapidly varying case, are described below. The first one is adapted to regularly varying functions g, whose smoothness is described through the following condition pertaining to h.

Case 1 (The Regularly varying case) It will be assumed that h belongs to the subclass of $RV(\beta)$, $\beta > 0$, with

$$h(x) = x^\beta l(x),$$

where

$$l(x) = c \exp \int_1^x \frac{\varepsilon(u)}{u} du \tag{9}$$

for some positive c. We assume that $x \mapsto \varepsilon(x)$ is twice differentiable and satisfies

$$\begin{cases} \varepsilon(x) \underset{x\to\infty}{=} o(1), \\ x|\varepsilon'(x)| \underset{x\to\infty}{=} O(1), \\ x^2|\varepsilon^{(2)}(x)| \underset{x\to\infty}{=} O(1). \end{cases} \tag{10}$$

It will also be assumed that

$$|h^{(2)}(x)| \in RV(\theta) \tag{11}$$

where θ is a real number such that $\theta \leq \beta - 2$.

Remark 1 Under (9), when $\beta \neq 1$ then, under (11), $\theta = \beta - 2$. Whereas, when $\beta = 1$ then $\theta \leq \beta - 2$. A sufficient condition for the last assumption (11) is that $\varepsilon'(t) \in RV(\gamma)$, for some $\gamma < -1$. Also in this case when $\beta = 1$, then $\theta = \beta + \gamma - 1$.

Example 1 (*Weibull density*) Let p be a Weibull density with shape parameter $k > 1$ and scale parameter 1, namely

$$p(x) = kx^{k-1} \exp(-x^k), \qquad x \geq 0$$
$$= k \exp(-(x^k - (k-1)\log x)).$$

Take $g(x) = x^k - (k-1)\log x$ and $q(x) = 0$. Then it holds

$$h(x) = kx^{k-1} - \frac{k-1}{x} = x^{k-1}\left(k - \frac{k-1}{x^k}\right).$$

Set $l(x) = k - (k-1)/x^k$, $x \geq 1$, which verifies

$$l'(x) = \frac{k(k-1)}{x^{k+1}} = \frac{l(x)\varepsilon(x)}{x}$$

with

$$\varepsilon(x) = \frac{k(k-1)}{kx^k - (k-1)}.$$

Since the function $\varepsilon(x)$ satisfies the three conditions in (10), then $h(x) \in RV(k-1)$.

Case 2 (*The Rapidly varying case*) Here we have $h^{\leftarrow}(t) = \psi(t) \in RV(0)$ and

$$\psi(t) = c \exp \int_1^t \frac{\varepsilon(u)}{u} du \tag{12}$$

for some positive c, and $t \mapsto \varepsilon(t)$ is twice differentiable with

$$\begin{cases} \varepsilon(t) \underset{t\to\infty}{=} o(1), \\ \dfrac{t\varepsilon'(t)}{\varepsilon(t)} \underset{t\to\infty}{\longrightarrow} 0, \\ \dfrac{t^2\varepsilon^{(2)}(t)}{\varepsilon(t)} \underset{t\to\infty}{\longrightarrow} 0. \end{cases} \tag{13}$$

Note that these assumptions imply that $\varepsilon(t) \in RV(0)$.

Example 2 (*A rapidly varying density*) Define p through

$$p(x) = c \exp(-e^{x-1}), x \geq 0.$$

Then $g(x) = h(x) = e^{x-1}$ and $q(x) = 0$ for all nonnegative x. We show that $h(x)$ is a rapidly varying function. It holds $\psi(t) = \log t + 1$. Since $\psi'(t) = 1/t$, let $\varepsilon(t) = 1/(\log t + 1)$ such that $\psi'(t) = \psi(t)\varepsilon(t)/t$. Moreover, the three conditions of (13) are satisfied. Thus $\psi(t) \in RV(0)$ and $h(x)$ is a rapidly varying function.

Denote by $\mathscr{R}$ the class of functions with either regular variation defined as in Case 1 or with rapid variation defined as in Case 2.

We now state hypotheses pertaining to the bounded function q in (6). We assume that

$$|q(x)| \in RV(\eta), \text{ for some } \eta < \theta - \tfrac{3\beta}{2} - \tfrac{3}{2} \text{ if } h \in RV(\beta) \tag{14}$$

and

$$|q(\psi(t))| \in RV(\eta), \text{ for some } \eta < -\tfrac{1}{2} \text{ if } h \text{ is rapidly varying.} \tag{15}$$

3 An Abelian-Type Theorem

We have

Theorem 1 *Let $p(x)$ be defined as in (6) and $h(x)$ belong to $\mathscr{R}$. Denote by $m(t)$, $s^2(t)$ and $\mu_j(t)$ for $j = 3, 4, \ldots$ the functions defined in (2), (3) and (4). Then it holds*

$$m(t) \underset{t\to\infty}{=} \psi(t)(1 + o(1)),$$

$$s^2(t) \underset{t\to\infty}{=} \psi'(t)(1 + o(1)),$$

$$\mu_3(t) \underset{t\to\infty}{=} \psi^{(2)}(t)(1 + o(1)),$$

$$\mu_j(t) \underset{t\to\infty}{=} \begin{cases} M_j s^j(t)(1 + o(1)), & \text{for even } j > 3 \\ \frac{(M_{j+3} - 3jM_{j-1})\mu_3(t)s^{j-3}(t)}{6}(1 + o(1)), & \text{for odd } j > 3 \end{cases},$$

where M_i, $i > 0$, denotes the ith order moment of standard normal distribution.

Using (6), the moment generating function $\Phi(t)$ defined in (1) takes on the form

$$\Phi(t) = \int_0^\infty e^{tx} p(x)dx = c \int_0^\infty \exp(K(x,t) + q(x))dx, \qquad t \in (0, \infty)$$

where

$$K(x,t) = tx - g(x). \tag{16}$$

If $h \in \mathscr{R}$, then for fixed t, $x \mapsto K(x,t)$ is a concave function and takes its maximum value at $\hat{x} = h^{\leftarrow}(t)$.

As a direct by-product of Theorem 1 we obtain the following Abel-type result.

Theorem 2 *Under the same hypotheses as in Theorem 1, we have*

$$\Phi(t) = \sqrt{2\pi}\,\hat{\sigma}\,e^{K(\hat{x},t)}(1 + o(1)).$$

Remark 2 It is easily verified that this result is in accordance with Theorem 4.12.11 of Bingham et al. (1987), Theorem 3 of Borovkov (2008), and Theorem 4.2 of Juszczak and Nagaev (2004). Some classical consequence of Kasahara's Tauberian theorem can be paralleled with Theorem 2. Following Theorem 4.2.10 in Bingham et al. (1987), with f defined as g above, it follows that $-\log \int_x^\infty p(v)dv \sim g(x)$ as $x \to \infty$ under Case 1, a stronger assumption than required in Theorem 4.2.10 of Bingham et al. (1987). Theorem 4.12.7 in Bingham et al. (1987) hence applies and provides an asymptotic equivalent for $\log \Phi(t)$ as $t \to \infty$; Theorem 2 improves on this result, at the cost of the additional regularity assumptions of Case 1. Furthermore, these results complement those in Broniatowski and Fuchs (1995) Sect. 3.2, in Case 2.

We also derive the following consequence of Theorem 1.

Theorem 3 *Under the present hypotheses, denote $\mathscr{X}_t$ a random variable with density $\pi_t(x)$. Then as $t \to \infty$, the family of random variables*

$$\frac{\mathscr{X}_t - m(t)}{s(t)}$$

converges in distribution to a standard normal distribution.

Remark 3 This result holds under various hypotheses, as developed, for example, in Balkema et al. (1993) or Feigin and Yashchin (1983). Under log-concavity of p it also holds locally; namely the family of densities π_t converges pointwise to the standard gaussian density; this yields asymptotic results for the extreme deviations of the empirical mean of i.i.d. summands with light tails (see Broniatowski and Mason (1994)), and also provides sufficient conditions for P_X to belong to the domain of attraction of the Gumbel distribution for the maximum through criterions pertaining to the Mill's ratio (see Feigin and Yashchin (1983)).

Remark 4 That g is four times derivable can be relaxed; in Case 1 with $\beta > 2$ or in Case 2, g a three times derivable function, together with the two first lines in (10) and (13), provides Theorems 1, 2, and 3. Also it may be seen that the order of differentiability of g in Case 1 with $0 < \beta \leq 2$ is related to the order of the moment of the tilted distribution for which an asymptotic equivalent is obtained. This will be developed in the forthcoming paper.

The proofs of the above results rely on Lemma 5–9. Lemma 5 is instrumental for Lemma 9.

Acknowledgments The authors thank an anonymous referee for comments and suggestions which helped greatly to the improvement on a previous version of this paper.

Appendix: Proofs

The following lemma provides a simple argument for the local uniform convergence of regularly varying functions.

Lemma 1 *Consider $l(t) \in RV(\alpha)$, $\alpha \in \mathbb{R}$. For any function f such that $f(t) \underset{t \to \infty}{=} o(t)$, it holds*

$$\sup_{|x| \le f(t)} |l(t+x)| \underset{t \to \infty}{\sim} |l(t)|. \tag{17}$$

If $f(t) = at$ with $0 < a < 1$, then it holds

$$\sup_{|x| \le at} |l(t+x)| \underset{t \to \infty}{\sim} (1+a)^\alpha |l(t)|. \tag{18}$$

Proof By Theorem 1.5.2 of Bingham et al. (1987), if $l(t) \in RV(\alpha)$, then for all I

$$\sup_{\lambda \in I} \left| \frac{l(\lambda t)}{l(t)} - \lambda^\alpha \right| \underset{t \to \infty}{\longrightarrow} 0,$$

with $I = [A, B]$ $(0 < A \le B < \infty)$ if $\alpha = 0$, $I = (0, B]$ $(0 < B < \infty)$ if $\alpha > 0$ and $I = [A, \infty)$ $(0 < A < \infty)$ if $\alpha < 0$.

Putting $\lambda = 1 + x/t$ with $f(t) \underset{t \to \infty}{=} o(t)$, we obtain

$$\sup_{|x| \le f(t)} \left| \frac{l(t+x)}{l(t)} - \left(1 + \frac{f(t)}{t} \right)^\alpha \right| \underset{t \to \infty}{\longrightarrow} 0,$$

which implies (17).

When $f(t) = at$ with $0 < a < 1$, we get

$$\sup_{|x| \le at} \left| \frac{l(t+x)}{l(t)} - (1+a)^\alpha \right| \underset{t \to \infty}{\longrightarrow} 0,$$

which implies (18). $\qquad\square$

Now we quote some simple expansions pertaining to the function h under the two cases considered in the above Sect. 2.

Lemma 2 *We have under Case 1,*

$$h'(x) = \frac{h(x)}{x}[\beta + \varepsilon(x)],$$

$$h^{(2)}(x) = \frac{h(x)}{x^2}[\beta(\beta - 1) + a\varepsilon(x) + \varepsilon^2(x) + x\varepsilon'(x)],$$

$$h^{(3)}(x) = \frac{h(x)}{x^3}[\beta(\beta - 1)(\beta - 2) + b\varepsilon(x) + c\varepsilon^2(x) + \varepsilon^3(x)$$

$$+ x\varepsilon'(x)(d + e\varepsilon(x)) + x^2\varepsilon^{(2)}(x)].$$

where a, b, c, d, e are some real constants.

Corollary 1 *We have under Case 1, $h'(x) \underset{x\to\infty}{\sim} \beta h(x)/x$ and $|h^{(i)}(x)| \le C_i h(x)/x^i$, $i = 1, 2, 3$, for some constants C_i and for large x.*

Corollary 2 *We have under Case 1, $\hat{x}(t) = \psi(t) \in RV(1/\beta)$ (see Theorem (1.5.15) of Bingham et al. (1987)) and $\hat{\sigma}^2(t) = \psi'(t) \sim \beta^{-1}\psi(t)/t \in RV(1/\beta - 1)$.*

It also holds

Lemma 3 *We have under Case 2,*

$$\psi^{(2)}(t) \underset{t\to\infty}{\sim} -\frac{\psi(t)\varepsilon(t)}{t^2} \text{ and } \psi^{(3)}(t) \underset{t\to\infty}{\sim} 2\frac{\psi(t)\varepsilon(t)}{t^3}.$$

Lemma 4 *We have under Case 2,*

$$h'(\psi(t)) = \frac{1}{\psi'(t)} = \frac{t}{\psi(t)\varepsilon(t)},$$

$$h^{(2)}(\psi(t)) = -\frac{\psi^{(2)}(t)}{(\psi'(t))^3} \underset{t\to\infty}{\sim} \frac{t}{\psi^2(t)\varepsilon^2(t)},$$

$$h^{(3)}(\psi(t)) = \frac{3(\psi^{(2)}(t))^2 - \psi^{(3)}(t)\psi'(t)}{(\psi'(t))^5} \underset{t\to\infty}{\sim} \frac{t}{\psi^3(t)\varepsilon^3(t)}.$$

Corollary 3 *We have under Case 2, $\hat{x}(t) = \psi(t) \in RV(0)$ and $\hat{\sigma}^2(t) = \psi'(t) = \psi(t)\varepsilon(t)/t \in RV(-1)$. Moreover, we have $h^{(i)}(\psi(t)) \in RV(1), i = 1, 2, 3$.*

Before beginning the proofs of our results we quote that the regularity conditions (9) and (12) pertaining to the function h allow for the above simple expansions. Substituting the constant c in (9) and (12) by functions $x \to c(x)$ which converge smoothly to some positive constant c adds noticeable complexity.

We now come to the proofs of five lemmas which provide the asymptotics leading to Theorems 1 and 2.

Lemma 5 *It holds*

$$\frac{\log \hat{\sigma}}{\int_1^t \psi(u)du} \underset{t\to\infty}{\longrightarrow} 0.$$

Proof By Corollaries 2 and 3, we have that $\psi(t) \in RV(1/\beta)$ in Case 1 and $\psi(t) \in RV(0)$ in Case 2. Using Theorem 1 of Feller (1971), Sect. 8.9 or Proposition 1.5.8 of Bingham et al. (1987), we obtain

$$\int_1^t \psi(u)du \underset{t\to\infty}{\sim} \begin{cases} t\psi(t)/(1+1/\beta) \in RV(1+1/\beta) & \text{if } h \in RV(\beta) \\ t\psi(t) \in RV(1) & \text{if } h \text{ is rapidly varying} \end{cases} . \tag{19}$$

Also by Corollaries 2 and 3, we have that $\hat\sigma^2 \in RV(1/\beta - 1)$ in Case 1 and $\hat\sigma^2 \in RV(-1)$ in Case 2. Thus $t \mapsto \log\hat\sigma \in RV(0)$ by composition and

$$\frac{\log\hat\sigma}{\int_1^t \psi(u)du} \underset{t\to\infty}{\sim} \begin{cases} \frac{\beta+1}{\beta} \times \frac{\log\hat\sigma}{t\psi(t)} \in RV\left(-1-\frac{1}{\beta}\right) & \text{if } h \in RV(\beta) \\ \frac{\log\hat\sigma}{t\psi(t)} \in RV(-1) & \text{if } h \text{ is rapidly varying} \end{cases} ,$$

which proves the claim. $\qquad\square$

The next steps of the proof make use of the function

$$L(t) := (\log t)^3 .$$

Lemma 6 *We have*

$$\sup_{|x|\le\hat\sigma L(t)} \left| \frac{h^{(3)}(\hat x + x)}{h^{(2)}(\hat x)} \right| \hat\sigma L^4(t) \xrightarrow[t\to\infty]{} 0.$$

Proof Case 1. By Corollary 1 and by (11) we have

$$|h^{(3)}(x)| \le C\frac{|h^{(2)}(x)|}{x},$$

for some constant C and x large. Since, by Corollary 2, $\hat x \in RV(1/\beta)$ and $\hat\sigma^2 \in RV(1/\beta - 1)$, we have

$$\frac{|x|}{\hat x} \le \frac{\hat\sigma L(t)}{\hat x} \in RV\left(-\frac{1}{2} - \frac{1}{2\beta}\right)$$

and $|x|/\hat x \xrightarrow[t\to\infty]{} 0$ uniformly in $\{x : |x| \le \hat\sigma L(t)\}$. For large t and all x such that $|x| \le \hat\sigma L(t)$, we have

$$|h^{(3)}(\hat x + x)| \le C\frac{|h^{(2)}(\hat x + x)|}{\hat x + x} \le C \sup_{|x|\le\hat\sigma L(t)} \frac{|h^{(2)}(\hat x + x)|}{\hat x + x}$$

whence

$$\sup_{|x|\le\hat\sigma L(t)} |h^{(3)}(\hat x + x)| \le C \sup_{|x|\le\hat\sigma L(t)} \frac{|h^{(2)}(\hat x + x)|}{\hat x + x}$$

where

$$\sup_{|x|\le\hat\sigma L(t)} \frac{|h^{(2)}(\hat x + x)|}{\hat x + x} \underset{t\to\infty}{\sim} \frac{|h^{(2)}(\hat x)|}{\hat x},$$

using (17) for the regularly varying function $|h^{(2)}(\hat x)| \in RV(\theta/\beta)$, with $f(t) = \hat\sigma L(t) \underset{t\to\infty}{=} o(\hat x)$. Thus for t large enough and for all $\delta > 0$

$$\sup_{|x|\le\hat\sigma L^4(t)} \left|\frac{h^{(3)}(\hat x + x)}{h^{(2)}(\hat x)}\right| \hat\sigma L^4(t) \le C\frac{\hat\sigma L^4(t)}{\hat x}(1 + \delta) \in RV\left(\frac{1}{2\beta} - \frac{1}{2} - \frac{1}{\beta}\right),$$

which proves Lemma 6 in Case 1.

Case 2. By Lemma 4, we have that $h^{(3)}(\psi(t)) \in RV(1)$. Using (18), we have for $0 < a < 1$ and t large enough

$$\sup_{|v|\le at} |h^{(3)}(\psi(t + v))| \underset{t\to\infty}{\sim} (1 + a)h^{(3)}(\psi(t)).$$

In the present case $\hat x \in RV(0)$ and $\hat\sigma^2 \in RV(-1)$. Setting $\psi(t + v) = \hat x + x = \psi(t) + x$, we have $x = \psi(t + v) - \psi(t)$ and $A := \psi(t - at) - \psi(t) \le x \le \psi(t + at) - \psi(t) =: B$, since $t \mapsto \psi(t)$ is an increasing function. It follows that

$$\sup_{|v|\le at} h^{(3)}(\psi(t + v)) = \sup_{A\le x\le B} h^{(3)}(\hat x + x).$$

Now note that (cf. p. 127 in Bingham et al. (1987))

$$B = \psi(t + at) - \psi(t) = \int_t^{t+at} \psi'(z)dz$$
$$= \int_t^{t+at} \frac{\psi(z)\varepsilon(z)}{z}dz \underset{t\to\infty}{\sim} \psi(t)\varepsilon(t)\log(1 + a),$$

since $\psi(t)\varepsilon(t) \in RV(0)$. Moreover, we have

$$\frac{\hat\sigma L(t)}{\psi(t)\varepsilon(t)} \in RV(-1) \text{ and } \frac{\hat\sigma L(t)}{\psi(t)\varepsilon(t)} \underset{t\to\infty}{\longrightarrow} 0.$$

It follows that $\hat\sigma L(t) \underset{t\to\infty}{=} o(B)$ and in a similar way, we have $\hat\sigma L(t) \underset{t\to\infty}{=} o(A)$. Using Lemma 4 and since $\hat\sigma L^4(t) \in RV(-1/2)$, it follows that for t large enough and for all $\delta > 0$

$$\sup_{|x|\le\hat\sigma L(t)} \frac{|h^{(3)}(\psi(t+v))|}{|h^{(2)}(\psi(t))|}\hat\sigma L^4(t) \le \sup_{A\le x\le B} \frac{|h^{(3)}(\psi(t+v))|}{|h^{(2)}(\psi(t))|}\hat\sigma L^4(t)$$

$$\le (1+a)\frac{\hat\sigma L^4(t)}{\psi(t)\varepsilon(t)}(1+\delta) \in RV\left(-\frac{1}{2}\right),$$

which concludes the proof of Lemma 6 in Case 2. $\qquad\square$

Lemma 7 *We have*

$$|h^{(2)}(\hat x)|\hat\sigma^4 \xrightarrow[t\to\infty]{} 0,$$

$$|h^{(2)}(\hat x)|\hat\sigma^3 L(t) \xrightarrow[t\to\infty]{} 0.$$

Proof Case 1. By Corollaries 1 and 2, we have

$$|h^{(2)}(\hat x)|\hat\sigma^4 \le \frac{C_2}{\beta^2 t} \in RV(-1)$$

and

$$|h^{(2)}(\hat x)|\hat\sigma^3 L(t) \le \frac{C_2}{\beta^{3/2}}\frac{L(t)}{\sqrt{t\psi(t)}} \in RV\left(-\frac{1}{2\beta}-\frac{1}{2}\right),$$

proving the claim.

Case 2. We have by Lemma 4 and Corollary 3

$$h^{(2)}(\hat x)\hat\sigma^4 \underset{t\to\infty}{\sim} \frac{1}{t} \in RV(-1)$$

and

$$h^{(2)}(\hat x)\hat\sigma^3 L(t) \underset{t\to\infty}{\sim} \frac{L(t)}{\sqrt{t\psi(t)\varepsilon(t)}} \in RV\left(-\frac{1}{2}\right),$$

which concludes the proof of Lemma 7. $\qquad\square$

We now define some functions to be used in the sequel. A Taylor–Lagrange expansion of $K(x,t)$ in a neighborhood of $\hat x$ yields

$$K(x,t) = K(\hat x,t) - \frac{1}{2}h'(\hat x)(x-\hat x)^2 - \frac{1}{6}h^{(2)}(\hat x)(x-\hat x)^3 + \varepsilon(x,t), \qquad (20)$$

where, for some $\theta \in (0,1)$,

$$\varepsilon(x,t) = -\frac{1}{24}h^{(3)}(\hat x+\theta(x-\hat x))(x-\hat x)^4. \qquad (21)$$

Lemma 8 *We have*

$$\sup_{y \in [-L(t), L(t)]} \frac{|\xi(\hat{\sigma} y + \hat{x}, t)|}{h^{(2)}(\hat{x})\hat{\sigma}^3} \xrightarrow[t \to \infty]{} 0,$$

where $\xi(x, t) = \varepsilon(x, t) + q(x)$ and $\varepsilon(x, t)$ is defined in (21).

Proof For $y \in [-L(t), L(t)]$, by 21, it holds

$$\left| \frac{\varepsilon(\hat{\sigma} y + \hat{x}, t)}{h^{(2)}(\hat{x})\hat{\sigma}^3} \right| \leq \left| \frac{h^{(3)}(\hat{x} + \theta \hat{\sigma} y)(\hat{\sigma} y)^4}{h^{(2)}(\hat{x})\hat{\sigma}^3} \right| \leq \left| \frac{h^{(3)}(\hat{x} + \theta \hat{\sigma} y)\hat{\sigma} L^4(t)}{h^{(2)}(\hat{x})} \right|,$$

with $\theta \in (0, 1)$. Let $x = \theta \hat{\sigma} y$. It then holds $|x| \leq \hat{\sigma} L(t)$. Therefore by Lemma 6

$$\sup_{y \in [-L(t), L(t)]} \left| \frac{\varepsilon(\hat{\sigma} y + \hat{x}, t)|}{h^{(2)}(\hat{x})\hat{\sigma}^3} \right| \leq \sup_{|x| \leq \hat{\sigma} L(t)} \left| \frac{h^{(3)}(\hat{x} + x)}{h^{(2)}(\hat{x})} \hat{\sigma} L^4(t) \right| \xrightarrow[t \to \infty]{} 0.$$

It remains to prove that

$$\sup_{y \in [-L(t), L(t)]} \left| \frac{q(\hat{\sigma} y + \hat{x})}{h^{(2)}(\hat{x})\hat{\sigma}^3} \right| \xrightarrow[t \to \infty]{} 0. \tag{22}$$

Case 1. By (11) and by composition, $|h^{(2)}(\hat{x})| \in RV(\theta/\beta)$. Using Corollary 1 we obtain

$$|h^{(2)}(\hat{x})\hat{\sigma}^3| \underset{t \to \infty}{\sim} \frac{|h^{(2)}(\hat{x})|\psi^{3/2}(t)}{\beta^{3/2} t^{3/2}} \in RV\left(\frac{\theta}{\beta} + \frac{3}{2\beta} - \frac{3}{2} \right).$$

Since, by (14), $|q(\hat{x})| \in RV(\eta/\beta)$, for $\eta < \theta - 3\beta/2 + 3/2$ and putting $x = \hat{\sigma} y$, we obtain

$$\sup_{y \in [-L(t), L(t)]} \left| \frac{q(\hat{\sigma} y + \hat{x})}{h^{(2)}(\hat{x})\hat{\sigma}^3} \right| = \sup_{|x| \leq \hat{\sigma} L(t)} \left| \frac{q(\hat{x} + x)}{h^{(2)}(\hat{x})\hat{\sigma}^3} \right|$$

$$\underset{t \to \infty}{\sim} \frac{|q(\hat{x})|}{|h^{(2)}(\hat{x})\hat{\sigma}^3|} \in RV\left(\frac{\eta - \theta}{\beta} - \frac{3}{2\beta} + \frac{3}{2} \right),$$

which proves (22).

Case 2. By Lemma 4 and Corollary 3, we have

$$|h^{(2)}(\hat{x})\hat{\sigma}^3| \underset{t \to \infty}{\sim} \frac{1}{\sqrt{t\psi(t)\varepsilon(t)}} \in RV\left(-\frac{1}{2} \right).$$

As in Lemma 6, since by (15), $q(\psi(t)) \in RV(\eta)$, then we obtain, with $\eta < -1/2$

$$\sup_{y \in [-L(t), L(t)]} \left| \frac{q(\hat{\sigma} y + \hat{x})}{h^{(2)}(\hat{x})\hat{\sigma}^3} \right| = \sup_{|x| \le \hat{\sigma} L(t)} \left| \frac{q(\hat{x} + x)}{h^{(2)}(\hat{x})\hat{\sigma}^3} \right|$$

$$\le \sup_{|v| \le at} \left| \frac{q(\psi(t + v))}{h^{(2)}(\hat{x})\hat{\sigma}^3} \right|$$

$$\le (1 + a)^{\eta} q(\psi(t)) \sqrt{t \psi(t) \varepsilon(t)} (1 + \delta) \in RV\left(\eta + \frac{1}{2}\right),$$

for all $\delta > 0$, with $a < 1$, t large enough and $\eta + 1/2 < 0$. This proves (22). $\square$

Lemma 9 *For $\alpha \in \mathbb{N}$, denote*

$$\Psi(t, \alpha) := \int_0^\infty (x - \hat{x})^\alpha e^{tx} p(x) dx.$$

Then

$$\Psi(t, \alpha) \underset{t \to \infty}{=} \hat{\sigma}^{\alpha+1} e^{K(\hat{x}, t)} T_1(t, \alpha)(1 + o(1)),$$

where

$$T_1(t, \alpha) = \int_{-\frac{L^{1/3}(t)}{\sqrt{2}}}^{\frac{L^{1/3}(t)}{\sqrt{2}}} y^\alpha \exp\left(-\frac{y^2}{2}\right) dy - \frac{h^{(2)}(\hat{x})\hat{\sigma}^3}{6} \int_{-\frac{L^{1/3}(t)}{\sqrt{2}}}^{\frac{L^{1/3}(t)}{\sqrt{2}}} y^{3+\alpha} \exp\left(-\frac{y^2}{2}\right) dy.$$

$$(23)$$

Proof We define the interval I_t as follows

$$I_t := \left(-\frac{L^{\frac{1}{3}}(t)\hat{\sigma}}{\sqrt{2}}, \frac{L^{\frac{1}{3}}(t)\hat{\sigma}}{\sqrt{2}}\right).$$

For large enough τ, when $t \to \infty$ we can partition $\mathbb{R}_+$ into

$$\mathbb{R}_+ = \{x : 0 < x < \tau\} \cup \{x : x \in \hat{x} + I_t\} \cup \{x : x \ge \tau, x \notin \hat{x} + I_t\},$$

where for $x > \tau$, $q(x) < \log 2$. Thus we have

$$p(x) < 2e^{-g(x)}. \tag{24}$$

For fixed τ, $\{x : 0 < x < \tau\} \cap \{x : x \in \hat{x} + I_t\} = \emptyset$. Therefore $\tau < \hat{x} - \frac{L^{\frac{1}{3}}(t)\hat{\sigma}}{\sqrt{2}} \le \hat{x}$ for t large enough. Hence it holds

$$\Psi(t, \alpha) =: \Psi_1(t, \alpha) + \Psi_2(t, \alpha) + \Psi_3(t, \alpha), \tag{25}$$

where

$$\Psi_1(t, \alpha) = \int_0^\tau (x - \hat{x})^\alpha e^{tx} p(x)dx,$$

$$\Psi_2(t, \alpha) = \int_{x \in \hat{x} + I_t} (x - \hat{x})^\alpha e^{tx} p(x)dx,$$

$$\Psi_3(t, \alpha) = \int_{x \notin \hat{x} + I_t, x \geq \tau} (x - \hat{x})^\alpha e^{tx} p(x)dx.$$

We estimate $\Psi_1(t, \alpha)$, $\Psi_2(t, \alpha)$ and $\Psi_3(t, \alpha)$ in Step 1, Step 2 and Step 3.

Step 1: Since q is bounded, we consider

$$\log d = \sup_{x \in (0,\tau)} q(x)$$

and for t large enough, we have

$$|\Psi_1(t, \alpha)| \leq \int_0^\tau \left| x - \hat{x} \right|^\alpha e^{tx} p(x)dx \leq d \int_0^\tau \hat{x}^\alpha e^{tx}dx,$$

since when $0 < x < \tau < \hat{x}$ then $\left| x - \hat{x} \right| = \hat{x} - x < \hat{x}$ for t large enough and g is positive.

Since, for t large enough, we have

$$\int_0^\tau \hat{x}^\alpha e^{tx}dx = \hat{x}^\alpha \frac{e^{t\tau}}{t} - \frac{\hat{x}^\alpha}{t} \leq \hat{x}^\alpha \frac{e^{t\tau}}{t},$$

we obtain

$$|\Psi_1(t, \alpha)| \leq d\hat{x}^\alpha \frac{e^{t\tau}}{t}. \tag{26}$$

We now show that for $h \in \mathscr{R}$, it holds

$$\hat{x}^\alpha \frac{e^{t\tau}}{t} \underset{t \to \infty}{=} o(|\hat{\sigma}^{\alpha+1}| e^{K(\hat{x},t)} |h^{(2)}(\hat{x})\hat{\sigma}^3|), \tag{27}$$

with $K(x, t)$ defined as in (16). This is equivalent to

$$\frac{\hat{x}^\alpha e^{t\tau}}{t|\hat{\sigma}^{\alpha+4} h^{(2)}(\hat{x})|} \underset{t \to \infty}{=} o(e^{K(\hat{x},t)}),$$

which is implied by

$$- (\alpha + 4) \log |\hat{\sigma}| - \log t + \alpha \log \hat{x} + \tau t - \log |h^{(2)}(\hat{x})| \underset{t \to \infty}{=} o(K(\hat{x}, t)), \quad (28)$$

if $K(\hat{x}, t) \underset{t \to \infty}{\longrightarrow} \infty$.

Setting $u = h(v)$ in $\int_1^t \psi(u)du$, we have

$$\int_1^t \psi(u)du = t\psi(t) - \psi(1) - g(\psi(t)) + g(\psi(1)).$$

Since $K(\hat{x}, t) = t\psi(t) - g(\psi(t))$, we obtain

$$K(\hat{x}, t) = \int_1^t \psi(u)du + \psi(1) - g(\psi(1)) \underset{t \to \infty}{\sim} \int_1^t \psi(u)du. \quad (29)$$

Let us denote (19) by

$$K(\hat{x}, t) \underset{t \to \infty}{\sim} at\psi(t), \quad (30)$$

with

$$a = \begin{cases} \frac{\beta}{\beta+1} & \text{if } h \in RV(\beta) \\ 1 & \text{if } h \text{ is rapidly varying} \end{cases}.$$

We have to show that each term in (28) is $o(K(\hat{x}, t))$.

1. By Lemma 5, $\log \hat{\sigma} \underset{t \to \infty}{=} o(\int_1^t \psi(u)du)$. Hence $\log \hat{\sigma} \underset{t \to \infty}{=} o(K(\hat{x}, t))$.
2. By Corollaries 2 and 3, we have

$$\frac{t}{K(\hat{x}, t)} \underset{t \to \infty}{\sim} \frac{1}{a\psi(t)} \underset{t \to \infty}{\longrightarrow} 0.$$

Thus $t \underset{t \to \infty}{=} o(K(\hat{x}, t))$.
3. Since $\hat{x} = \psi(t) \underset{t \to \infty}{\longrightarrow} \infty$, it holds

$$\left| \frac{\log \hat{x}}{K(\hat{x}, t)} \right| \leq C \frac{\psi(t)}{K(\hat{x}, t)},$$

for some positive constant C and t large enough. Moreover by (30), we have

$$\frac{\psi(t)}{K(\hat{x}, t)} \underset{t \to \infty}{\sim} \frac{1}{at} \underset{t \to \infty}{\longrightarrow} 0.$$

Hence $\log \hat{x} \underset{t \to \infty}{=} o(K(\hat{x}, t))$.
4. Using (30), $\log |h^{(2)}(\hat{x})| \in RV(0)$ and $\log |h^{(2)}(\hat{x})| \underset{t \to \infty}{=} o(K(\hat{x}, t))$.

5. Since $\log t \underset{t\to\infty}{=} o(t)$ and $t \underset{t\to\infty}{=} o(K(\hat{x}, t))$, we obtain $\log t \underset{t\to\infty}{=} o(K(\hat{x}, t))$.

Since (28) holds and $K(\hat{x}, t) \underset{t\to\infty}{\longrightarrow} \infty$ by (29) and (30), we then get (27).

Eqs. (26) and (27) yield together

$$|\Psi_1(t, \alpha)| \underset{t\to\infty}{=} o(|\hat{\sigma}^{\alpha+1}|e^{K(\hat{x}, t)}|h^{(2)}(\hat{x})\hat{\sigma}^3|). \tag{31}$$

When α is even,

$$T_1(t, \alpha) = \int_{-\frac{t^{1/3}}{\sqrt{2}}}^{\frac{t^{1/3}}{\sqrt{2}}} y^\alpha \exp\left(-\frac{y^2}{2}\right) dy \underset{t\to\infty}{\sim} \sqrt{2\pi} M_\alpha, \tag{32}$$

where M_α is the moment of order α of a standard normal distribution. Thus by Lemma 7 we have

$$\frac{h^{(2)}(\hat{x})\hat{\sigma}^3}{T_1(t, \alpha)} \underset{t\to\infty}{\longrightarrow} 0. \tag{33}$$

When α is odd,

$$T_1(t, \alpha) = -\frac{h^{(2)}(\hat{x})\hat{\sigma}^3}{6} \int_{-\frac{t^{\frac{1}{3}}}{\sqrt{2}}}^{\frac{t^{\frac{1}{3}}}{\sqrt{2}}} y^{3+\alpha} \exp\left(-\frac{y^2}{2}\right) dy \underset{t\to\infty}{\sim} -\frac{h^{(2)}(\hat{x})\hat{\sigma}^3}{6} \sqrt{2\pi} M_{\alpha+3},$$

$$\tag{34}$$

where $M_{\alpha+3}$ is the moment of order $\alpha + 3$ of a standard normal distribution. Thus we have

$$\frac{h^{(2)}(\hat{x})\hat{\sigma}^3}{T_1(t, \alpha)} \underset{t\to\infty}{\sim} -\frac{6}{\sqrt{2\pi} M_{\alpha+3}}. \tag{35}$$

Combined with (31), (33) and (35) imply for $\alpha \in \mathbb{N}$

$$|\Psi_1(t, \alpha)| \underset{t\to\infty}{=} o(\hat{\sigma}^{\alpha+1} e^{K(\hat{x}, t)} T_1(t, \alpha)). \tag{36}$$

Step 2: By (6) and (20)

$$\Psi_2(t, \alpha) = \int_{x\in\hat{x}+I_t} (x - \hat{x})^\alpha e^{K(x, t)+q(x)} dx$$

$$= \int_{x\in\hat{x}+I_t} (x - \hat{x})^\alpha e^{K(\hat{x}, t) - \frac{1}{2}h'(\hat{x})(x-\hat{x})^2 - \frac{1}{6}h^{(2)}(\hat{x})(x-\hat{x})^3 + \xi(x, t)} dx,$$

where $\xi(x, t) = \varepsilon(x, t) + q(x)$. Making the substitution $y = (x - \hat{x})/\hat{\sigma}$, it holds

$$\Psi_2(t, \alpha) = \hat{\sigma}^{\alpha+1} e^{K(\hat{x},t)} \int_{-\frac{L^{\frac{1}{3}}(t)}{\sqrt{2}}}^{\frac{L^{\frac{1}{3}}(t)}{\sqrt{2}}} y^\alpha \exp\left(-\frac{y^2}{2} - \frac{\hat{\sigma}^3 y^3}{6} h^{(2)}(\hat{x}) + \xi(\hat{\sigma} y + \hat{x}, t)\right) dy,$$

(37)

since $h'(\hat{x}) = 1/\hat{\sigma}^2$.

On $\left\{y : y \in \left(-L^{\frac{1}{3}}(t)/\sqrt{2}, L^{\frac{1}{3}}(t)/\sqrt{2}\right)\right\}$, by Lemma 7, we have

$$\left|h^{(2)}(\hat{x})\hat{\sigma}^3 y^3\right| \le \left|h^{(2)}(\hat{x})\hat{\sigma}^3 L(t)\right|/2^{\frac{3}{2}} \xrightarrow[t\to\infty]{} 0.$$

Perform the first-order Taylor expansion

$$\exp\left(-\frac{h^{(2)}(\hat{x})\hat{\sigma}^3}{6} y^3 + \xi(\hat{\sigma} y + \hat{x}, t)\right) \underset{t\to\infty}{=} 1 - \frac{h^{(2)}(\hat{x})\hat{\sigma}^3}{6} y^3 + \xi(\hat{\sigma} y + \hat{x}, t) + o_1(t, y),$$

where

$$o_1(t, y) = o\left(-\frac{h^{(2)}(\hat{x})\hat{\sigma}^3}{6} y^3 + \xi(\hat{\sigma} y + \hat{x}, t)\right). \tag{38}$$

We obtain

$$\int_{-\frac{L^{\frac{1}{3}}(t)}{\sqrt{2}}}^{\frac{L^{\frac{1}{3}}(t)}{\sqrt{2}}} y^\alpha \exp\left(-\frac{y^2}{2} - \frac{\hat{\sigma}^3 y^3}{6} h^{(2)}(\hat{x}) + \xi(\hat{\sigma} y + \hat{x}, t)\right) dy =: T_1(t, \alpha) + T_2(t, \alpha),$$

where $T_1(t, \alpha)$ is defined in (23) and

$$T_2(t, \alpha) := \int_{-\frac{L^{\frac{1}{3}}(t)}{\sqrt{2}}}^{\frac{L^{\frac{1}{3}}(t)}{\sqrt{2}}} \left(\xi(\hat{\sigma} y + \hat{x}, t) + o_1(t, y)\right) y^\alpha e^{-\frac{y^2}{2}} dy. \tag{39}$$

Using (38) we have for t large enough

$$|T_2(t, \alpha)| \le \sup_{y\in[-\frac{L^{\frac{1}{3}}(t)}{\sqrt{2}}, \frac{L^{\frac{1}{3}}(t)}{\sqrt{2}}]} \left|\xi(\hat{\sigma} y + \hat{x}, t)\right| \int_{-\frac{L^{\frac{1}{3}}(t)}{\sqrt{2}}}^{\frac{L^{\frac{1}{3}}(t)}{\sqrt{2}}} |y|^\alpha e^{-\frac{y^2}{2}} dy$$

$$+ \int_{-\frac{L^{\frac{1}{3}}(t)}{\sqrt{2}}}^{\frac{L^{\frac{1}{3}}(t)}{\sqrt{2}}} \left(\left|o\left(\frac{h^{(2)}(\hat{x})\hat{\sigma}^3}{6} y^3\right)\right| + \left|o(\xi(\hat{\sigma} y + \hat{x}, t))\right|\right) |y|^\alpha e^{-\frac{y^2}{2}} dy,$$

where $\sup_{y\in[-L^{\frac{1}{3}}(t)/\sqrt{2},L^{\frac{1}{3}}(t)/\sqrt{2}]}\left|\xi(\hat{\sigma}y+\hat{x},t)\right| \leq \sup_{y\in[-L(t),L(t)]}\left|\xi(\hat{\sigma}y+\hat{x},t)\right|$ since $L^{\frac{1}{3}}(t)/\sqrt{2} \leq L(t)$ holds for t large enough. Thus

$$|T_2(t,\alpha)| \leq 2\sup_{y\in[-L(t),L(t)]}\left|\xi(\hat{\sigma}y+\hat{x},t)\right|\int_{-\frac{L^{\frac{1}{3}}(t)}{\sqrt{2}}}^{\frac{L^{\frac{1}{3}}(t)}{\sqrt{2}}}|y|^{\alpha}\,e^{-\frac{y^2}{2}}dy$$

$$+\left|o\left(\frac{h^{(2)}(\hat{x})\hat{\sigma}^3}{6}\right)\right|\int_{-\frac{L^{\frac{1}{3}}(t)}{\sqrt{2}}}^{\frac{L^{\frac{1}{3}}(t)}{\sqrt{2}}}|y|^{3+\alpha}\,e^{-\frac{y^2}{2}}dy$$

$$\underset{t\to\infty}{=}\left|o\left(\frac{h^{(2)}(\hat{x})\hat{\sigma}^3}{6}\right)\right|\left(\int_{-\frac{L^{\frac{1}{3}}(t)}{\sqrt{2}}}^{\frac{L^{\frac{1}{3}}(t)}{\sqrt{2}}}|y|^{\alpha}\,e^{-\frac{y^2}{2}}dy+\int_{-\frac{L^{\frac{1}{3}}(t)}{\sqrt{2}}}^{\frac{L^{\frac{1}{3}}(t)}{\sqrt{2}}}|y|^{3+\alpha}\,e^{-\frac{y^2}{2}}dy\right),$$

where the last equality holds from Lemma 8. Since the integrals in the last equality are both bounded, it holds

$$T_2(t,\alpha)\underset{t\to\infty}{=}o(h^{(2)}(\hat{x})\hat{\sigma}^3). \tag{40}$$

When α is even, using (32) and Lemma 7

$$\left|\frac{T_2(t,\alpha)}{T_1(t,\alpha)}\right| \leq \frac{|h^{(2)}(\hat{x})\hat{\sigma}^3|}{\sqrt{2\pi}\,M_{\alpha}}\underset{t\to\infty}{\longrightarrow}0. \tag{41}$$

When α is odd, using (34), we get

$$\frac{T_2(t,\alpha)}{T_1(t,\alpha)}\underset{t\to\infty}{=}-\frac{6}{\sqrt{2\pi}\,M_{\alpha+3}}o(1)\underset{t\to\infty}{\longrightarrow}0. \tag{42}$$

Now with $\alpha\in\mathbb{N}$, by (41) and (42)

$$T_2(t,\alpha)\underset{t\to\infty}{=}o(T_1(t,\alpha)),$$

which, combined with (37), yields

$$\Psi_2(t,\alpha)=c\hat{\sigma}^{\alpha+1}e^{K(\hat{x},t)}T_1(t,\alpha)(1+o(1)). \tag{43}$$

Step 3: The Three Chords Lemma implies, for $x\mapsto K(x,t)$ concave and $(x,y,z)\in\mathbb{R}_+^3$ such that $x<y<z$

$$\frac{K(y,t)-K(z,t)}{y-z}\leq\frac{K(x,t)-K(z,t)}{x-z}\leq\frac{K(x,t)-K(y,t)}{x-y}. \tag{44}$$

Since $x \mapsto K(x, t)$ is concave and attains its maximum in $\hat{x}$, we consider two cases: $x < \hat{x}$ and $x \geq \hat{x}$. After some calculus using (44) in each case, we get

$$K(x, t) - K(\hat{x}, t) \leq \frac{K(\hat{x} + sgn(x - \hat{x})\frac{L^{1/3}(t)\hat{\sigma}}{\sqrt{2}}) - K(\hat{x}, t)}{sgn(x - \hat{x})\frac{L^{1/3}(t)\hat{\sigma}}{\sqrt{2}}} (x - \hat{x}), \qquad (45)$$

where

$$sgn(x - \hat{x}) = \begin{cases} 1 & \text{if } x \geq \hat{x} \\ -1 & \text{if } x < \hat{x} \end{cases}.$$

Using Lemma 7, a third-order Taylor expansion in the numerator of (45) gives

$$K(\hat{x} + sgn(x - \hat{x})\frac{L^{1/3}(t)\hat{\sigma}}{\sqrt{2}}) - K(\hat{x}, t) \leq -\frac{1}{4}h'(\hat{x})L^{2/3}(t)\hat{\sigma}^2 = -\frac{1}{4}L^{2/3}(t),$$

which yields

$$K(x, t) - K(\hat{x}, t) \leq -\frac{\sqrt{2}}{4}\frac{L^{1/3}(t)}{\hat{\sigma}}|x - \hat{x}|.$$

Using (24), we obtain for large enough fixed τ

$$|\Psi_3(t, \alpha)| \leq 2 \int_{x \notin \hat{x}+I_t, x > \tau} |x - \hat{x}|^\alpha e^{K(x,t)} dx$$

$$\leq 2e^{K(\hat{x},t)} \int_{|x-\hat{x}| > \frac{L^{1/3}(t)\hat{\sigma}}{\sqrt{2}}, x > \tau} |x - \hat{x}|^\alpha \exp\left(-\frac{\sqrt{2}}{4}\frac{L^{1/3}(t)}{\hat{\sigma}}|x - \hat{x}|\right) dx$$

$$= 2e^{K(\hat{x},t)}\hat{\sigma}^{\alpha+1}\left[\int_{\frac{L^{1/3}(t)}{\sqrt{2}}}^{+\infty} y^\alpha e^{-\frac{\sqrt{2}}{4}L^{1/3}(t)y} dy \right.$$

$$\left. + \int_{\frac{\tau-\hat{x}}{\hat{\sigma}}}^{-\frac{L^{1/3}(t)}{\sqrt{2}}} (-y)^\alpha e^{\frac{\sqrt{2}}{4}L^{1/3}(t)y} dy\right]$$

$$:= 2e^{K(\hat{x},t)}\hat{\sigma}^{\alpha+1}(I_\alpha + J_\alpha).$$

It is easy but a bit tedious to show by recursion that

$$I_\alpha = \int_{\frac{L^{1/3}(t)}{\sqrt{2}}}^{+\infty} y^\alpha \exp\left(-\frac{\sqrt{2}}{4}L^{1/3}(t)y\right) dy$$

$$= \exp\left(-\frac{1}{4}L^{2/3}(t)\right) \sum_{i=0}^{\alpha} 2^{\frac{4i+3-\alpha}{2}} L^{\frac{\alpha-(2i+1)}{3}}(t) \frac{\alpha!}{(\alpha-i)!}$$

$$\underset{t\to\infty}{\sim} 2^{\frac{3-\alpha}{2}} \exp\left(-\frac{1}{4}L^{2/3}(t)\right) L^{\frac{\alpha-1}{3}}(t)$$

and

$$J_\alpha = \int_{\frac{\tau-\hat{x}}{\hat{\sigma}}}^{-\frac{L^{1/3}(t)}{\sqrt{2}}} (-y)^\alpha \exp\left(\frac{\sqrt{2}}{4}L^{1/3}(t)y\right) dy$$

$$= I_\alpha - \exp\left(\frac{\sqrt{2}}{4}L^{1/3}(t)\frac{\tau-\hat{x}}{\hat{\sigma}}\right) \sum_{i=0}^{\alpha} \left(\frac{\hat{x}-\tau}{\hat{\sigma}}\right)^{\alpha-i} L^{-\frac{i+1}{3}}(t) 2^{\frac{3i+3}{2}} \frac{\alpha!}{(\alpha-i)!}$$

$$= I_\alpha + M(t),$$

with $\hat{x}/\hat{\sigma} \in RV((1+1/\beta)/2)$ when $h \in RV(\beta)$ and $\hat{x}/\hat{\sigma} \in RV(1/2)$ when h is rapidly varying. Moreover, $\tau - \hat{x} < 0$, thus $M(t) \underset{t\to\infty}{\longrightarrow} 0$ and we have for some positive constant Q_1

$$|\Psi_3(t,\alpha)| \leq Q_1 e^{K(\hat{x},t)}\hat{\sigma}^{\alpha+1} \exp\left(-\frac{1}{4}L^{2/3}(t)\right) L^{\frac{\alpha-1}{3}}(t).$$

With (43), we obtain for some positive constant Q_2

$$\left|\frac{\Psi_3(t,\alpha)}{\Psi_2(t,\alpha)}\right| \leq \frac{Q_2 \exp(-\frac{1}{4}L^{2/3}(t))L^{\frac{\alpha-1}{3}}(t)}{|T_1(t,\alpha)|}.$$

In Step 1, we saw that $T_1(t,\alpha) \underset{t\to\infty}{\sim} \sqrt{2\pi}M_\alpha$, for α even and $T_1(t,\alpha) \underset{t\to\infty}{\sim} -\frac{h^{(2)}(\hat{x})\hat{\sigma}^3}{6}\sqrt{2\pi}M_{\alpha+3}$, for α odd. Hence for α even and t large enough

$$\left|\frac{\Psi_3(t,\alpha)}{\Psi_2(t,\alpha)}\right| \leq Q_3 \frac{\exp(-\frac{1}{4}L^{2/3}(t))L^{\frac{\alpha-1}{3}}(t)}{\sqrt{2\pi}M_\alpha} \underset{t\to\infty}{\longrightarrow} 0, \tag{46}$$

and for α odd and t large enough

$$\left| \frac{\Psi_3(t,\alpha)}{\Psi_2(t,\alpha)} \right| \le Q_4 \frac{\exp(-\frac{1}{4}L^{2/3}(t))L^{\frac{\alpha-1}{3}}(t)}{\frac{|h^{(2)}(\hat{x})\hat{\sigma}^3|}{6}\sqrt{2\pi}\, M_{\alpha+3}},$$

for positive constants Q_3 and Q_4.

As in Lemma 7, we have

$$|h^{(2)}(\hat{x})\hat{\sigma}^3| \in RV\left(\frac{\theta}{\beta} + \frac{3}{2\beta} - \frac{3}{2}\right) \quad \text{if } h \in RV(\beta)$$

and

$$|h^{(2)}(\hat{x})\hat{\sigma}^3| \in RV\left(-\frac{1}{2}\right) \quad \text{if } h \text{ is rapidly varying.}$$

Let us denote

$$|h^{(2)}(\hat{x})\hat{\sigma}^3| = t^\rho L_1(t),$$

for some slowly varying function L_1 and $\rho < 0$ defined as

$$\rho = \begin{cases} \frac{\theta}{\beta} + \frac{3}{2\beta} - \frac{3}{2} & \text{if } h \in RV(\beta) \\ -\frac{1}{2} & \text{if } h \text{ is rapidly varying} \end{cases}.$$

We have for some positive constant C

$$\left| \frac{\Psi_3(t,\alpha)}{\Psi_2(t,\alpha)} \right| \le C \exp\left(-\frac{1}{4}L^{2/3}(t) - \rho \log t - \log L_1(t)\right) L^{\frac{\alpha-1}{3}}(t) \xrightarrow[t\to\infty]{} 0,$$

since $-(\log t)^2/4 - \rho \log t - \log L_1(t) \underset{t\to\infty}{\sim} -(\log t)^2/4 \xrightarrow[t\to\infty]{} -\infty.$

Hence we obtain

$$\Psi_3(t,\alpha) \underset{t\to\infty}{=} o(\Psi_2(t,\alpha)). \tag{47}$$

The proof is completed by combining (25), (36), (43), and (47). $\qquad\square$

Proof (Proof of Theorem 1) By Lemma 9, if $\alpha = 0$, it holds

$$T_1(t,0) \xrightarrow[t\to\infty]{} \sqrt{2\pi},$$

since $L(t) \xrightarrow[t\to\infty]{} \infty$. Approximate the moment generating function of X

$$\Phi(t) = \Psi(t,0) \underset{t\to\infty}{=} \hat{\sigma} e^{K(\hat{x},t)} T_1(t,0)(1+o(1)) \underset{t\to\infty}{=} \sqrt{2\pi}\,\hat{\sigma} e^{K(\hat{x},t)}(1+o(1)). \tag{48}$$

If $\alpha = 1$, it holds

$$T_1(t, 1) \underset{t \to \infty}{=} -\frac{h^{(2)}(\hat{x})\hat{\sigma}^3}{6} M_4 \sqrt{2\pi}(1 + o(1)),$$

where $M_4 = 3$ denotes the fourth-order moment of the standard normal distribution. Consequently, we obtain

$$\Psi(t, 1) \underset{t \to \infty}{=} -\sqrt{2\pi}\hat{\sigma}^2 e^{K(\hat{x},t)} \frac{h^{(2)}(\hat{x})\hat{\sigma}^3}{2}(1+o(1)) \underset{t \to \infty}{=} -\Phi(t)\frac{h^{(2)}(\hat{x})\hat{\sigma}^4}{2}(1+o(1)), \tag{49}$$

which, together with the definition of $\Psi(t, \alpha)$, yields

$$\int_0^\infty x e^{tx} p(x)dx = \Psi(t, 1) + \hat{x}\Phi(t) \underset{t \to \infty}{=} \left(\hat{x} - \frac{h^{(2)}(\hat{x})\hat{\sigma}^4}{2}(1 + o(1))\right)\Phi(t).$$

Hence we get

$$m(t) = \frac{\int_0^\infty x e^{tx} p(x)dx}{\Phi(t)} = \hat{x} - \frac{h^{(2)}(\hat{x})\hat{\sigma}^4}{2}(1 + o(1)). \tag{50}$$

By Lemma 7, we obtain

$$m(t) \underset{t \to \infty}{\sim} \hat{x} = \psi(t). \tag{51}$$

If $\alpha = 2$, it follows:

$$T_1(t, 2) \underset{t \to \infty}{=} \sqrt{2\pi}(1 + o(1)).$$

Thus we have

$$\Psi(t, 2) \underset{t \to \infty}{=} \hat{\sigma}^2 \Phi(t)(1 + o(1)). \tag{52}$$

Using (49), (50) and (52), it follows:

$$\int_0^\infty (x - m(t))^2 e^{tx} p(x)dx = \int_0^\infty (x - \hat{x} + \hat{x} - m(t))^2 e^{tx} p(x)dx$$
$$= \Psi(t, 2) + 2(\hat{x} - m(t))\Psi(t, 1) + (\hat{x} - m(t))^2\Phi(t)$$
$$\underset{t \to \infty}{=} \hat{\sigma}^2\Phi(t)(1 + o(1)) - \hat{\sigma}^2\Phi(t)\frac{(h^{(2)}(\hat{x})\hat{\sigma}^3)^2}{4}(1 + o(1)) \underset{t \to \infty}{=} \hat{\sigma}^2\Phi(t)(1 + o(1)),$$

where the last equality holds since $|h^{(2)}(\hat{x})\hat{\sigma}^3| \underset{t \to \infty}{\longrightarrow} 0$ by Lemma 7.

Hence we obtain

$$s^2(t) = \frac{\int_0^\infty (x - m(t))^2 e^{tx} p(x) dx}{\Phi(t)} \underset{t\to\infty}{\sim} \hat\sigma^2 = \psi'(t). \tag{53}$$

If $\alpha = 3$, it holds

$$T_1(t, 3) = -\frac{h^{(2)}(\hat x)\hat\sigma^3}{6} \int_{-\frac{L^{\frac{1}{3}}(t)}{\sqrt 2}}^{\frac{L^{\frac{1}{3}}(t)}{\sqrt 2}} y^6 e^{-\frac{y^2}{2}} dy.$$

Thus we have

$$\Psi(t, 3) = -\sqrt{2\pi}\,\hat\sigma^4 e^{K(\hat x,t)} \frac{h^{(2)}(\hat x)\hat\sigma^3}{6} \int_{-\frac{L^{\frac{1}{3}}(t)}{\sqrt 2}}^{\frac{L^{\frac{1}{3}}(t)}{\sqrt 2}} \frac{1}{\sqrt{2\pi}} y^6 e^{-\frac{y^2}{2}} dy \tag{54}$$

$$\underset{t\to\infty}{=} -M_6 \frac{h^{(2)}(\hat x)\hat\sigma^6}{6} \Phi(t)(1 + o(1)),$$

where $M_6 = 15$ denotes the sixth-order moment of standard normal distribution. Using (49), (50), (52) and (54), we have

$$\int_0^\infty (x - m(t))^3 e^{tx} p(x) dx = \int_0^\infty (x - \hat x + \hat x - m(t))^3 e^{tx} p(x) dx$$

$$= \Psi(t, 3) + 3(\hat x - m(t))\Psi(t, 2) + 3(\hat x - m(t))^2 \Psi(t, 1) + (\hat x - m(t))^3 \Phi(t)$$

$$\underset{t\to\infty}{=} -h^{(2)}(\hat x)\hat\sigma^6 \Phi(t)(1 + o(1)) - h^{(2)}(\hat x)\hat\sigma^6 \Phi(t)\frac{(h^{(2)}(\hat x)\hat\sigma^3)^2}{4}(1 + o(1))$$

$$\underset{t\to\infty}{=} -h^{(2)}(\hat x)\hat\sigma^6 \Phi(t)(1 + o(1)),$$

where the last equality holds since $|h^{(2)}(\hat x)\hat\sigma^3| \underset{t\to\infty}{\longrightarrow} 0$ by Lemma 7. Hence we get

$$\mu_3(t) = \frac{\int_0^\infty (x - m(t))^3 e^{tx} p(x) dx}{\Phi(t)} \underset{t\to\infty}{\sim} -h^{(2)}(\hat x)\hat\sigma^6$$

$$= \frac{\psi^{(2)}(t)}{(\psi'(t))^3}(\psi'(t))^3 = \psi^{(2)}(t). \tag{55}$$

We now consider $\alpha = j > 3$ for even j. Using (50) and Lemma 9, we have

$$\int_0^\infty (x - m(t))^j e^{tx} p(x) dx = \int_0^\infty (x - \hat x + \hat x - m(t))^j e^{tx} p(x) dx \tag{56}$$

$$= \sum_{i=0}^j \binom{j}{i} \left(\frac{h^{(2)}(\hat x)\hat\sigma^4}{2}\right)^i \hat\sigma^{j-i+1} e^{K(\hat x,t)} T_1(t, j - i)(1 + o(1)),$$

with

$$
T_1(t, j-i) = \begin{cases} \displaystyle\int_{-\frac{L^{\frac{1}{3}}(t)}{\sqrt{2}}}^{\frac{L^{\frac{1}{3}}(t)}{\sqrt{2}}} y^{j-i} e^{-\frac{y^2}{2}} dy & \text{for even } i \\[2em] -\dfrac{h^{(2)}(\hat{x})\hat{\sigma}^3}{6} \displaystyle\int_{-\frac{L^{\frac{1}{3}}(t)}{\sqrt{2}}}^{\frac{L^{\frac{1}{3}}(t)}{\sqrt{2}}} y^{3+j-i} e^{-\frac{y^2}{2}} dy & \text{for odd } i \end{cases}
$$

$$
\underset{t\to\infty}{=} \begin{cases} \sqrt{2\pi}\, M_{j-i}(1 + o(1)) & \text{if } i \text{ is even} \\[1em] -\sqrt{2\pi}\, \dfrac{h^{(2)}(\hat{x})\hat{\sigma}^3}{6} M_{3+j-i} & \text{if } i \text{ is odd} \end{cases}.
$$

Using (48), we obtain

$$
\int_0^\infty (x - m(t))^j e^{tx} p(x)\,dx
$$

$$
\underset{t\to\infty}{=} \sum_{i=0}^j \binom{j}{i} \left(\frac{h^{(2)}(\hat{x})\hat{\sigma}^4}{2}\right)^i \Phi(t) \times
$$

$$
\left[\hat{\sigma}^{j-i} M_{j-i}(1+o(1)) \mathbb{I}_{even\ i} - \frac{h^{(2)}(\hat{x})\hat{\sigma}^4}{2} \sigma^{j-i-1} \frac{M_{3+j-i}}{3}(1+o(1)) \mathbb{I}_{odd\ i} \right]
$$

$$
\underset{t\to\infty}{=} \sum_{k=0}^{j/2} \binom{j}{2k} \left(\frac{h^{(2)}(\hat{x})\hat{\sigma}^4}{2}\right)^{2k} \Phi(t)\hat{\sigma}^{j-2k} M_{j-2k}(1+o(1))
$$

$$
- \sum_{k=0}^{j/2-1} \binom{j}{2k+1} \left(\frac{h^{(2)}(\hat{x})\hat{\sigma}^4}{2}\right)^{2(k+1)} \Phi(t)\hat{\sigma}^{j-2k-2} \frac{M_{3+j-2k-1}}{3}(1+o(1))
$$

$$
\underset{t\to\infty}{\sim} \hat{\sigma}^j \Phi(t) \times
$$

$$
\left(M_j + \sum_{k=1}^{j/2} \binom{j}{2k} (h^{(2)}(\hat{x})\hat{\sigma}^3)^{2k} \frac{M_{j-2k}}{2^{2k}} \right.
$$

$$
\left. - \sum_{k=0}^{j/2-1} \binom{j}{2k+1} (h^{(2)}(\hat{x})\hat{\sigma}^3)^{2(k+1)} \frac{M_{3+j-2k-1}}{3 \times 2^{2(k+1)}} \right)
$$

$$
\underset{t\to\infty}{=} M_j \hat{\sigma}^j \Phi(t)(1 + o(1)),
$$

since $|h^{(2)}(\hat{x})\hat{\sigma}^3| \underset{t\to\infty}{\longrightarrow} 0$ by Lemma 7. Hence we get for even j

$$\mu_j(t) = \frac{\int_0^\infty (x - m(t))^j e^{tx} p(x) dx}{\Phi(t)} \underset{t\to\infty}{\sim} M_j \hat{\sigma}^j \underset{t\to\infty}{\sim} M_j s^j(t), \qquad (57)$$

by (53).

To conclude, we consider $\alpha = j > 3$ for odd j. (56) holds true with

$$T_1(t, j - i) = \begin{cases} \int_{-\frac{L^{\frac{1}{3}}(t)}{\sqrt{2}}}^{\frac{L^{\frac{1}{3}}(t)}{\sqrt{2}}} y^{j-i} e^{-\frac{y^2}{2}} dy & \text{for odd } i \\[2em] -\frac{h^{(2)}(\hat{x})\hat{\sigma}^3}{6} \int_{-\frac{L^{\frac{1}{3}}(t)}{\sqrt{2}}}^{\frac{L^{\frac{1}{3}}(t)}{\sqrt{2}}} y^{3+j-i} e^{-\frac{y^2}{2}} dy & \text{for even } i \end{cases}$$

$$\underset{t\to\infty}{=} \begin{cases} \sqrt{2\pi} M_{j-i}(1 + o(1)) & \text{if } i \text{ is odd} \\[1em] -\sqrt{2\pi}\frac{h^{(2)}(\hat{x})\hat{\sigma}^3}{6} M_{3+j-i} & \text{if } i \text{ is even} \end{cases}.$$

Thus, with the same tools as above, some calculus and making use of (57),

$$\int_0^\infty (x - m(t))^j e^{tx} p(x) dx \underset{t\to\infty}{=} \frac{M_{j+3} - 3j M_{j-1}}{6} \times (-h^{(2)}(\hat{x})\hat{\sigma}^{j+3})\Phi(t).$$

Hence we get for odd j

$$\mu_j(t) = \frac{\int_0^\infty (x - m(t))^j e^{tx} p(x) dx}{\Phi(t)} \underset{t\to\infty}{\sim} \frac{M_{j+3} - 3j M_{j-1}}{6} \times (-h^{(2)}(\hat{x})\hat{\sigma}^{j+3})$$

$$\underset{t\to\infty}{\sim} \frac{M_{j+3} - 3j M_{j-1}}{6} \mu_3(t) s^{j-3}(t), \qquad (58)$$

by (53) and (55).

The proof is complete by considering (51), (53), (55), (57) and (58). $\qquad\square$

Proof (*Proof of Theorem* 2) It is proved incidentally in (48). $\qquad\square$

Proof (*Proof of Theorem* 3) Consider the moment generating function of the random variable

$$Y_t := \frac{\mathcal{X}_t - m(t)}{s(t)}.$$

It holds for any λ

$$\log E \exp \lambda Y_t = -\lambda \frac{m(t)}{s(t)} + \log \frac{\Phi\left(t + \frac{\lambda}{s(t)}\right)}{\Phi(t)}$$

$$= \frac{\lambda^2}{2} \frac{s^2\left(t + \theta \frac{\lambda}{s(t)}\right)}{s^2(t)} = \frac{\lambda^2}{2} \frac{\psi'\left(t + \theta \frac{\lambda(1+o(1))}{\sqrt{\psi'(t)}}\right)}{\psi'(t)} (1 + o(1))$$

as $t \to \infty$, for some $\theta \in (0, 1)$ depending on t, where we used Theorem 1. Now making use of Corollaries 2 and 3 it follows that

$$\lim_{t \to \infty} \log E \exp \lambda Y_t = \frac{\lambda^2}{2},$$

which proves the claim. $\qquad\square$

References

Balkema, A. A., Klüppelberg, C., & Resnick, S. I. (1993). Densities with Gaussian tails. *Proceedings of the London Mathematical Society, 66*, 568–588.

Barndorff-Nielsen, O. (1978). *Information and exponential families in statistical theory.* Chichester: Wiley.

Bingham, N. H., Goldie, C. M., & Teugels, J. L. (1987). *Regular variation.* Cambridge: Cambridge University Press.

Borovkov, A. A. (2008). Tauberian and Abelian theorems for rapidly decreasing distributions and their applications to stable laws. *Siberian Mathematical Journal, 49*, 796–805.

Broniatowski, M., & Fuchs, A. (1995). Tauberian theorems, Chernoff inequality, and the tail behavior of finite convolutions of distribution functions. *Advances in Mathematics, 116*, 12–33.

Broniatowski, M., & Mason, D. M. (1994). Extended large deviations. *Journal of Theoretical Probability, 7*, 647–666.

Feigin, P. D., & Yashchin, E. (1983). On a strong Tauberian result. *Zeitschrift für Wahrscheinlichkeitstheorie und Verwandte Gebiete, 65*, 35–48.

Feller, W. (1971). *An introduction to probability theory and its applications* (2nd ed., Vol. II). New York: Wiley.

Juszczak, D., & Nagaev, A. V. (2004). Local large deviation theorem for sums of i.i.d. random vectors when the Cramér condition holds in the whole space. *Probability and Mathematical Statistics, 24*, 297–320.

On Bahadur–Kiefer Type Processes for Sums and Renewals in Dependent Cases

Endre Csáki and Miklós Csörgő

Abstract We study the asymptotic behavior of Bahadur–Kiefer processes that are generated by summing partial sums of (weakly or strongly dependent) random variables and their renewals. Known results for i.i.d. case will be extended to dependent cases.

Keywords Partial sums · Renewals · Bahadur–Kiefer type processes · Wiener process · Fractional Brownian motion · Strong approximations

Mathematics Subject Classification (2010): Primary 60F17 · Secondary 60F15

1 Introduction

In this work we intend to deal with Bahadur–Kiefer type processes that are based on partial sums and their renewals of weakly, as well as strongly, dependent sequences of random variables. In order to initiate our approach, let $\{Y_0, Y_1, Y_2, \ldots\}$ be random variables which have the same marginal distribution and, to begin with, satisfy the following assumptions:

(i) $EY_0 = \mu > 0$;
(ii) $E(Y_0^2) < \infty$.

E. Csáki (✉)
Alfréd Rényi Institute of Mathematics, Hungarian Academy of Sciences,
P.O.Box 127, Budapest 1364, Hungary
e-mail: csaki.endre@renyi.mta.hu

M. Csörgő
School of Mathematics and Statistics, Carleton University, Ottawa,
ON K1S 5B6, Canada
e-mail: mcsorgo@math.carleton.ca

© Springer International Publishing Switzerland 2015
M. Hallin et al. (eds.), *Mathematical Statistics and Limit Theorems*,
DOI 10.1007/978-3-319-12442-1_6

In terms of the generic sequence $\{Y_j,\ j = 0, 1, 2, \ldots\}$, with $t \geq 0$, we define

$$S(t) := \sum_{i=1}^{[t]} Y_i, \tag{1.1}$$

$$N(t) := \inf\{s \geq 1 : S(s) > t\}, \tag{1.2}$$

$$Q(t) := S(t) + \mu N(\mu t) - 2\mu t, \tag{1.3}$$

whose respective appropriately normalized versions will be used in studying partial sums, their renewals, Bahadur–Kiefer type processes when the random variables in the sequence $Y_i,\ i = 0, 1, \ldots$ are weakly or strongly dependent.

The research area of what is known as Bahadur–Kiefer processes was initiated by Bahadur (1966) (cf. also Kiefer 1967) who established an almost sure representation of i.i.d. random variables-based sample quantiles in terms of their empiricals. Kiefer (1970) substantiated this work via studying the deviations between the sample quantile and its empirical processes. These three seminal papers have since been followed by many related further investigations (cf., e.g., Csörgő and Révész 1978, 1981, Chap. 5; Shorack 1982; Csörgő 1983; Deheuvels and Mason 1990, 1992; Deheuvels 1992a, b; Csörgő and Horváth 1993, Chaps. 3–6; Csörgő and Szyszkovicz 1998, and references in these works).

It follows from the results of Kiefer (1970), and also from Vervaat (1972a, b) as spelled out in Csáki et al. (2007), that the original i.i.d. based Bahadur–Kiefer process cannot converge weakly to any nondegenerate random element of the $D[0, 1]$ function space. On the other hand, Csörgő et al. (2007) showed the opposite conclusion to hold true for long-range dependence-based Bahadur–Kiefer processes. For an illustration and discussion of this conclusion, we refer to the Introduction and Corollary 1.2 of Csáki et al. (2013). For further results along these lines, we refer to Csörgő and Kulik (2008a, b).

The study of the almost sure asymptotic behavior of Bahadur–Kiefer type processes for sums and their renewals in the i.i.d. case was initiated by Horváth (1984), Deheuvels and Mason (1990), and augmented by further references and results as in Csörgő and Horváth (1993, Chap. 2).

Vervaat (1972a, b) initiated the study of limit theorems in general for processes with a positive drift and their inverses. For results on the asymptotic behavior of integrals of Bahadur–Kiefer type processes for sums and their renewals, the so-called Vervaat processes, we refer to Csáki et al. (2007) in the i.i.d. case, Csáki et al. (2010) in the weakly dependent case, and Csáki et al. (2013) in the strongly dependent case.

Back to the topics of this paper on Bahadur–Kiefer type processes for sums and their renewals, the forthcoming Sect. 2 is concerned with the weakly dependent case, and Sect. 3 concludes results in terms of long-range dependent sequences of random variables. Both of these sections contain the relevant proofs as well.

2 Weakly Dependent Case

In this section we deal with weakly dependent random variables-based Bahadur–Kiefer type processes. First, we summarize the main results in the case when Y_i are i.i.d. random variables with finite fourth moment.

Theorem A *Assume that $\{Y_i, i = 0, 1, \ldots\}$ are i.i.d. random variables with $EY_0 = \mu > 0$, $E(Y_0 - \mu)^2 = \sigma^2 > 0$, and $EY_0^4 < \infty$. Then we have*

$$Q(T) = \sigma \left(W(T) - W\left(T - \frac{\sigma}{\mu}W(T)\right)\right) + o_{\text{a.s.}}(T^{1/4}), \quad \text{as } T \to \infty, \quad (2.1)$$

$$\limsup_{T \to \infty} \frac{\sup_{0 \le t \le T} |Q(t/\mu)|}{(T \log \log T)^{1/4}(\log T)^{1/2}} = \frac{2^{1/4}\sigma^{3/2}}{\mu^{3/4}}, \quad \text{a.s.}, \quad (2.2)$$

$$\lim_{T \to \infty} \frac{\sup_{0 \le t \le T} |Q(t/\mu)|}{(\log T)^{1/2}(\sup_{0 \le t \le T} |\mu N(t) - t|)^{1/2}} = \frac{\sigma}{\mu^{1/2}}, \quad \text{a.s.}, \quad (2.3)$$

$$\lim_{T \to \infty} P(T^{-1/4}|Q(T/\mu)| \le y) = 2 \int_{-\infty}^{\infty} \Phi(y\mu^{3/4}\sigma^{-3/2}|x|^{-1/2})\varphi(x)\,dx - 1,$$
$$(2.4)$$

where Φ is the standard normal distribution function and φ is its density.

We note that (2.1) and (2.4) are due to Csörgő and Horváth (1993), (2.2) is due to Horváth (1984) and (2.3) is due to Deheuvels and Mason (1990). All these results can be found in Csörgő and Horváth (1993).

For the case of i.i.d. random variables when the fourth moment does not exist, we refer to Deheuvels and Steinebach (1992).

In this section, we assume that $S(t)$ can be approximated by a standard Wiener process as follows.

Assumption A On the same probability space there exist a sequence $\{Y_i, i = 0, 1, 2, \ldots\}$ of random variables, with the same marginal distribution, satisfying assumptions (i) and (ii), and a standard Wiener process $W(t)$, $t \ge 0$, such that

$$\sup_{0 \le t \le T} |S(t) - \mu t - \sigma W(t)| = O_{\text{a.s.}}(T^\beta) \qquad (2.5)$$

almost surely, as $T \to \infty$, with $\sigma > 0$, where $S(t)$ is defined by (1.1) and $\beta < 1/4$.

In the case of $1/4 \le \beta < 1/2$, there is a huge literature on strong approximation of the form (2.5) for weakly dependent random variables $\{Y_i\}$. The case $\beta < 1/4$ is treated in Berkes et al. (2014), where Komlós et al. (1975) type strong approximations as in (2.5) are proved under fairly general assumptions of dependence. For exact statements of, and conditions for, strong approximations that yield (2.5) to hold true for the partial sums as in Assumption A, we refer to Berkes et al. (2014).

Theorem 2.1 *Under Assumption A all the results (2.1)–(2.4) in Theorem A remain true.*

Proof In fact, we only have to prove (2.1), for the other results follow from the latter. It follows from Csörgő and Horváth (1993), Theorem 1.3 on p. 37, that under Assumption A we have

$$\limsup_{T \to \infty} \frac{\sup_{0 \le t \le T} \left| \frac{t}{\mu} - N(t) - \frac{\sigma}{\mu} W(t/\mu) \right|}{(T \log \log T)^{1/4} (\log T)^{1/2}} = 2^{1/4} \sigma^{3/2} \mu^{-7/4} \quad \text{a.s.}$$

and also

$$\sup_{0 \le t \le T} |\mu t - \mu S(N(\mu t))| = O_{\text{a.s.}}(T^{\beta})$$

as $T \to \infty$. Hence, as $T \to \infty$, we arrive at

$$Q(T) = S(T) + \mu N(\mu T) - 2\mu T = S(T) - \mu T - (S(N(\mu T)) - \mu N(\mu T)) + O_{\text{a.s.}}(T^{\beta})$$

$$= \sigma(W(t) - W(N(\mu T))) + O_{\text{a.s.}}(T^{\beta}) =$$

$$\sigma \left(W(T) - W \left(T - \frac{\sigma}{\mu} W(T) \right) \right) + o_{\text{a.s.}}(T^{1/4}),$$

i.e., having (2.1) as desired. $\square$

3 Strongly Dependent Case

In this section, we deal with long-range (strongly) dependent sequences, based on moving averages as defined by

$$\eta_j = \sum_{k=0}^{\infty} \psi_k \xi_{j-k}, \quad j = 0, 1, 2, \dots, \tag{3.1}$$

where $\{\xi_k, -\infty < k < \infty\}$ is a doubly infinite sequence of independent standard normal random variables, and the sequence of weights $\{\psi_k, k = 0, 1, 2, \dots\}$ is square summable. Then $\mathrm{E}(\eta_0) = 0$, $\mathrm{E}(\eta_0^2) = \sum_{k=0}^{\infty} \psi_k^2 =: \sigma^2$ and, on putting $\tilde{\eta}_j = \eta_j/\sigma$, $\{\tilde{\eta}_j, j = 0, 1, 2, \dots\}$ is a stationary Gaussian sequence with $\mathrm{E}(\tilde{\eta}_0) = 0$

and $E(\tilde{\eta}_0^2) = 1$. If $\psi_k \sim k^{-(1+\alpha)/2}\ell(k)$ with a slowly varying function, $\ell(k)$, at infinity, then $E(\eta_j \eta_{j+n}) \sim b_\alpha n^{-\alpha}\ell^2(n)$, where the constant b_α is defined by

$$b_\alpha = \int_0^\infty x^{-(1+\alpha)/2}(1+x)^{-(1+\alpha)/2}\,dx.$$

Now let $G(\cdot)$ be a real-valued Borel measurable function, and define the subordinated sequence $Y_j = G(\tilde{\eta}_j)$, $j = 0, 1, 2, \ldots$. We assume throughout that $J_1 := E(G(\tilde{\eta}_0)\tilde{\eta}_0) \neq 0$. We say in this case that the Hermite rank of the function $G(\cdot)$ is equal to 1 (cf. Introduction of Csáki et al. 2013).

For $1/2 < H < 1$ let $\{W_H(t),\, t \geq 0\}$ be a fractional Brownian motion (fbm), i.e., a mean-zero Gaussian process with covariance

$$EW_H(s)W_H(t) = \frac{1}{2}(s^{2H} + t^{2H} - |s-t|^{2H}). \tag{3.2}$$

Based on a strong approximation result of Wang et al. (2003), what follows next, was proved in Sect. 2 of Csáki et al. (2013).

Theorem B *Let η_j be defined by* (3.1) *with $\psi_k \sim k^{-(1+\alpha)/2}$, $0 < \alpha < 1$, and put $\tilde{\eta}_j = \eta_j/\sigma$ with $\sigma^2 := E(\eta_0^2) = \sum_{k=0}^\infty \psi_k^2$. Let $G(\cdot)$ be a function whose Hermite rank is 1, and put $Y_j = G(\tilde{\eta}_j)$, $j = 0, 1, 2, \ldots$. Furthermore, let $\{S(t),\, t \geq 0\}$ be as in* (1.1) *and assume condition* (ii). *Then, on an appropriate probability space for the sequence $\{Y_j = G(\tilde{\eta}_j),\, j = 0, 1, \ldots\}$, one can construct a fractional Brownian motion $W_{1-\alpha/2}(\cdot)$ such that, as $T \to \infty$, we have*

$$\sup_{0 \leq t \leq T}\left| S(t) - \mu t - \frac{J_1 \kappa_\alpha}{\sigma} W_{1-\alpha/2}(t)\right| = o_{a.s.}(T^{\gamma/2+\delta}), \tag{3.3}$$

where $\mu = E(Y_0)$,

$$\kappa_\alpha^2 = 2\frac{\int_0^\infty x^{-(\alpha+1)/2}(1+x)^{-(\alpha+1)/2}\,dx}{(1-\alpha)(2-\alpha)}, \tag{3.4}$$

$\gamma = 2 - 2\alpha$ for $\alpha < 1/2$, $\gamma = 1$ for $\alpha \geq 1/2$ and $\delta > 0$ is arbitrary.

Moreover, if we also assume condition (i), *then, as $T \to \infty$,*

$$\sup_{0 \leq t \leq T}\left| \mu N(\mu t) - \mu t + \frac{J_1 \kappa_\alpha}{\sigma} W_{1-\alpha/2}(t)\right| = o_{a.s.}(T^{\gamma/2+\delta} + T^{(1-\alpha/2)^2+\delta}), \tag{3.5}$$

with γ as right above, and arbitrary $\delta > 0$.

Now, for use in the sequel, we state iterated logarithm results for fractional Brownian motion and its increments, which follows from Ortega's extension in Ortega (1984) of Csörgő and Révész (1979), Csörgő and Révész (1981, Sect. 1.2).

Theorem C *For $T > 0$ let a_T be a nondecreasing function of T such that $0 < a_T \leq T$ and a_T / T is nonincreasing. Then*

$$\limsup_{T \to \infty} \frac{\sup_{0 \leq t \leq T - a_T} \sup_{0 \leq s \leq a_T} |W_{1-\alpha/2}(t+s) - W_{1-\alpha/2}(t)|}{a_T^{1-\alpha/2} (2(\log T/a_T + \log\log T))^{1/2}} = 1 \quad \text{a.s.} \quad (3.6)$$

If $\lim_{T \to \infty} (\log(T/a_T))/(\log\log T) = \infty$, then we have $\lim$ *instead of* $\limsup$ *in (3.6).*

First, we give an invariance principle for $Q(T)$ defined by (1.3) if $\gamma/2 < (1 - \alpha/2)^2$, which corresponds to the i.i.d. case when the fourth moment exists. Equivalently, we assume that

$$0 < \alpha < 2 - \sqrt{2}. \tag{3.7}$$

Note that in (3.8) below, the random time argument of $W_{1-\alpha/2}$ is strictly positive for large enough T with probability 1. So, without loss of generality, we may define $W_{1-\alpha/2}(T - u) = 0$ if $u > T$.

Theorem 3.1 *Under the conditions of Theorem B, including* (i) *and* (ii), *assuming* (3.7), *as $T \to \infty$, we have*

$$Q(T) = \frac{J_1 \kappa_\alpha}{\sigma} (W_{1-\alpha/2}(T) - W_{1-\alpha/2}(N(\mu T)) + o_{\text{a.s.}}(T^{\gamma/2+\delta})$$

$$= \frac{J_1 \kappa_\alpha}{\sigma} \left(W_{1-\alpha/2}(T) - W_{1-\alpha/2} \left(T - \frac{J_1 \kappa_\alpha}{\sigma \mu} W_{1-\alpha/2}(T) \right) \right) + o_{\text{a.s.}}(T^{\gamma/2+\delta}). \tag{3.8}$$

Proof Put $c = J_1 \kappa_\alpha / \sigma$. Then

$$Q(T) = S(T) - \mu T + \mu N(\mu T) - \mu T$$

$$= c W_{1-\alpha/2}(T) + o_{\text{a.s.}}(T^{\gamma/2+\delta}) + \mu(N(\mu T) - T).$$

But

$$\mu(T - N(\mu T)) = S(N(\mu T)) - \mu N(\mu T) + \mu T - S(N(\mu T))$$

$$= c W_{1-\alpha/2}(N(\mu T)) + o_{\text{a.s.}}((N(\mu T))^{\gamma/2+\delta}) + \mu T - S(N(\mu T)),$$

and using (3.5) and Theorem C, we have

$$c W_{1-\alpha/2}(N(\mu T)) = c W_{1-\alpha/2} \left(T - \frac{c}{\mu} W_{1-\alpha/2}(T) + o_{\text{a.s.}}(T^{\gamma/2+\delta} + T^{(1-\alpha/2)^2+\delta}) \right)$$

$$= c W_{1-\alpha/2} \left(T - \frac{c}{\mu} W_{1-\alpha/2}(T) \right) + o_{\text{a.s.}}(T^{(\gamma/2+\delta)(1-\alpha/2)} + T^{(1-\alpha/2)^3}).$$

On the other hand (cf. Csáki et al. 2013), $N(\mu T) = O_{\text{a.s.}}(T)$ and

$$\mu T - S(N(\mu T)) = o_{\text{a.s.}}(T^{\gamma/2+\delta}).$$

Since $(1-\alpha/2)^3 \le \gamma/2 < (1-\alpha/2)^2$, this dominates all the other remainder terms in the proof. Thus the proof of Theorem 3.1 is now complete. $\square$

The proof of Theorem 3.1 also yields the following result.

Proposition 1 *As $T \to \infty$,*

$$\mu T - \mu N(\mu T)) = \frac{J_1\kappa_\alpha}{\sigma} W_{1-\alpha/2}\left(T - \frac{J_1\kappa_\alpha}{\sigma\mu} W_{1-\alpha/2}(T)\right) + o_{\text{a.s.}}(T^{\gamma/2+\delta}).$$

Now we are to give a lim sup result for $Q(\cdot)$. For this we need a Strassen-type functional law of the iterated logarithm for fbm, due to Goodman and Kuelbs (1991).

Theorem D *Let*

$$\mathbf{K} = \{T_H g(t), \, 0 \le t \le 1, \, \int_{-\infty}^{1} g^2(u)\, du \le 1\},$$

where

$$T_H g(t) = \frac{1}{k_H}\int_0^t (t-u)^{H-1/2} g(u)\, du + \frac{1}{k_H}\int_{-\infty}^0 (t-u)^{H-1/2} - (-u)^{H-1/2}) g(u)\, du,$$

and

$$k_H^2 = \int_{-\infty}^0 ((1-s)^{H-1/2} - (-s)^{H-1/2})^2\, ds + \int_0^1 (1-s)^{2H-1}\, ds.$$

Then, almost surely, $\mathbf{K}$ is the set of limit points of the net of stochastic processes

$$\frac{W_H(nt)}{(2n^{2H}\log\log n)^{1/2}}, \quad 0 \le t \le 1, \tag{3.9}$$

as $n \to \infty$.

Theorem 3.2 *Under the conditions of Theorem 3.1, we have*

$$\limsup_{T\to\infty} \frac{|Q(T)|}{T^{(1-\alpha/2)^2}(\log\log T)^{1/2-\alpha/4}(\log T)^{1/2}} = \frac{2^{1-\alpha/4}(J_1\kappa_\alpha)^{2-\alpha/2}}{\sigma^{2-\alpha/2}\mu^{1-\alpha/2}} \quad \text{a.s.} \tag{3.10}$$

Proof It follows from Theorem C that

$$|W_{1-\alpha/2}(T)| \le (1+\delta)T^{1-\alpha/2}(2\log\log T)^{1/2}$$

with probability 1 for any $\delta > 0$ if T is large enough. Hence, applying Theorem C with $a_T = (1 + \delta)c/\mu T^{1-\alpha/2}(2\log\log T)^{1/2}$, $c = J_1\kappa_\alpha/\sigma$, we obtain

$$c \sup_{|s|\leq a_T} |W_{1-\alpha/2}(T) - W_{1-\alpha/2}(T - s)| \leq c(1+\delta)a_T^{1-\alpha/2}(2\log T)^{1/2},$$

almost surely for large enough T. Since $\delta > 0$ is arbitrary, we obtain the upper bound in (3.10).

To obtain the lower bound, we follow the proof in the i.i.d. case, given in Csörgő and Horváth (1993). On choosing

$$g(s) = \begin{cases} \frac{1}{k_H}((1 - s)^{H-1/2} - (-s)^{H-1/2}), & s \leq 0, \\ \frac{1}{k_H}(1 - s)^{H-1/2}, & 0 < s \leq 1, \end{cases}$$

in Theorem D, we have

$$f(t) = \frac{1}{k_H} \int_{-\infty}^{0} ((t-s)^{H-1/2} - (-s)^{H-1/2})g(s)\,ds + \frac{1}{k_H} \int_{0}^{t} (t-s)^{H-1/2}g(s)\,ds.$$

It can be seen that $\int_{-\infty}^{1} g^2(s)\,ds = 1$, and $\{f(t),\ 0 \leq t \leq 1\}$ is a continuous increasing function with $f(0) = 0$, $f(1) = 1$, and hence by Theorem D it is in $\mathbf{K}$. For $0 < \delta < 1$, on considering the function

$$g_\delta(s) = \begin{cases} g(s), & 0 \leq s \leq 1 - \delta, \\ 0, & 1 - \delta \leq s \leq 1, \end{cases}$$

we define

$$f_\delta(t) = \begin{cases} f(t), & 0 \leq t \leq 1 - \delta, \\ f(1 - \delta), & 1 - \delta \leq t \leq 1. \end{cases}$$

Then it can be seen that the latter function is in $\mathbf{K}$, and hence it is a limit function of the net of stochastic processes as in (3.9). It follows that there is a sequence T_k of random variables such that, in our context,

$$\lim_{k\to\infty} \sup_{0\leq t\leq 1} \left| \frac{W_{1-\alpha/2}(T_k t)}{T_k^{1-\alpha/2}(2\log\log T_k)^{1/2}} - f_\delta(t) \right| = 0.$$

Using Theorem C with $a_T = f(1 - \delta)c/\mu T^{1-\alpha/2}(2\log\log T)^{1/2}$, we get

$$\lim_{T\to\infty} \frac{\sup_{T(1-\delta)\leq t\leq T} c|W_{1-\alpha/2}(t + a_T) - W(t)|}{ca_T^{1-\alpha/2}(2\log T)^{1/2}} = 1 \quad \text{a.s.}$$

Since δ is arbitrary, and $\lim_{\delta\to 0} f_\delta(t) = f(t)$, the lower bound follows as in Csörgő and Horváth (1993, p. 28). This completes the proof of Theorem 3.2. $\qquad\square$

Next we give the limiting distribution of $Q(T)$.

Theorem 3.3 *Under the conditions of Theorem 3.1, we have*

$$\lim_{T\to\infty} P\left(Q(T)T^{-(1-\alpha/2)^2} \le y\right) = \int_{-\infty}^{\infty} \varphi(x)\Phi\left(\frac{y\sigma^{2-\alpha/2}\mu^{1-\alpha/2}}{|x|^{1-\alpha/2}(J_1\kappa_\alpha)^{2-\alpha/2}}\right)dx.$$

$$(3.11)$$

Proof According to Theorem 3.1 we have to determine the limiting distribution of

$$c\left(W_{1-\alpha/2}(T) - W_{1-\alpha/2}\left(T - \frac{c}{\mu}W_{1-\alpha/2}(T)\right)\right),$$

where $c = J_1\kappa_\alpha/\sigma$. Via the scaling property of fbm, i.e.,

$$\widetilde{W}(v) := T^{-1+\alpha/2}W_{1-\alpha/2}(Tv), \quad v \ge 0,$$

is also an fbm with parameter $1 - \alpha/2$. So we have to determine the limiting distribution of

$$c\left(\widetilde{W}(1) - \widetilde{W}(1 - c_1 T^{-\alpha/2}\widetilde{W}(1))\right),$$

as $T \to \infty$, where $c_1 = J_1\kappa_\alpha/(\sigma\mu)$.

For $u > 0$, the joint distribution of $\widetilde{W}(1)$, $\widetilde{W}(u)$ is bivariate normal with density

$$\frac{1}{2\pi\sigma_1\sigma_2\sqrt{1-r^2}}\exp\left(-\frac{1}{2(1-r^2)}\left(\frac{x^2}{\sigma_1^2} - 2r\frac{xy}{\sigma_1\sigma_2} + \frac{y^2}{\sigma_2^2}\right)\right),$$

where $\sigma_1^2 = E(W_{1-\alpha/2}^2(1)) = 1$, $\sigma_2^2 = E(W_{1-\alpha/2}^2(u)) = u^{2-\alpha}$ and

$$r = \frac{1 + u^{2-\alpha} - |1-u|^{2-\alpha}}{2\sigma_1\sigma_2}.$$

Now consider the conditional density

$$P(\widetilde{W}(u) \in dz | \widetilde{W}(1) = x) = \frac{1}{\sigma_2\sqrt{1-r^2}}\varphi\left(\frac{z - r\sigma_2 x}{\sigma_2\sqrt{1-r^2}}\right)dz,$$

where $u = 1 - c_1 x T^{-\alpha/2}$.

So the density function of $\widetilde{W}(1) - \widetilde{W}(u)$ is equal to

$$P(\widetilde{W}(1) - \widetilde{W}(u) \in dY) = \int_{-\infty}^{T^{\alpha/2}/c_1} \frac{1}{\sigma_2\sqrt{1-r^2}}\varphi(x)\varphi\left(\frac{x - Y - r\sigma_2 x}{\sigma_2\sqrt{1-r^2}}\right)dx\, dY$$

and hence its distribution function is

$$P(\widetilde{W}(1) - \widetilde{W}(u) \le Z) = \int_{-\infty}^{T^{\alpha/2}/c_1} \varphi(x)\Phi\left(\frac{Z - x + r\sigma_2 x}{\sigma_2\sqrt{1 - r^2}}\right) dx, \quad -\infty < Z < \infty.$$

It can be seen that, as $T \to \infty$,

$$\sigma_2\sqrt{1 - r^2} \sim \frac{|c_1 x|^{1-\alpha/2}}{T^{\alpha/2 - \alpha^2/4}},$$

$$\frac{x - xr\sigma_2}{\sigma_2\sqrt{1 - r^2}} = O(T^{-\alpha/2 + \alpha^2/4}).$$

Hence, as $T \to \infty$,

$$P(\widetilde{W}(1) - \widetilde{W}(u) \le Z) \sim \int_{-\infty}^{T^{\alpha/2}/c_1} \varphi(x)\Phi\left(\frac{ZT^{\alpha/2 - \alpha^2/4}}{|c_1 x|^{1-\alpha/2}}\right) dx.$$

Putting $Z = yT^{\alpha^2/4 - \alpha/2}/c$, and taking the limit $T \to \infty$, we finally obtain (3.11). $\qquad\square$

Acknowledgments We wish to thank two referees for their careful reading of, and constructive remarks on, our manuscript. Research supported by an NSERC Canada Discovery Grant at Carleton University, Ottawa and by the Hungarian National Foundation for Scientific Research, No. K108615.

References

Bahadur, R. R. (1966). A note on quantiles in large samples. *Annals of Mathematical Statistics, 37,* 577–580.

Berkes, I., Liu, W. D., & Wu, W. B. (2014). Komlós-Major-Tusnády approximation under dependence. *Annals of Probability, 42,* 794–817.

Csáki, E., Csörgő, M., & Kulik, R. (2010). On Vervaat processes for sums and renewals in weakly dependent cases. In I. Berkes, et al. (Eds.), *Dependence in probability, analysis and number theory. A Volume in Memory of Walter Philipp* (pp. 145–156). Heber City: Kendrick Press.

Csáki, E., Csörgő, M., & Kulik, R. (2013). Strong approximations for long memory sequences based partial sums, counting and their Vervaat processes. Submitted. arxiv:math.PR1302.3740.

Csáki, E., Csörgő, M., Rychlik, Z., & Steinebach, J. (2007). On Vervaat and Vervaat-error-type processes for partial sums and renewals. *Journal of Statistical Planning and Inference, 137,* 953–966.

Csörgő, M. (1983). *Quantile processes with statistical application. CBMS-NSF Regional Conference Series in Applied Mathematics* (Vol. 42). Philadelphia: SIAM.

Csörgő, M., & Horváth, L. (1993). *Weighted approximations in probability and statistics*. Chichester: Wiley.

Csörgő, M., & Kulik, R. (2008a). Reduction principles for quantile and Bahadur-Kiefer processes of long-range dependent sequences. *Probability Theory and Related Fields, 142,* 339–366.

Csörgő, M., & Kulik, R. (2008b). Weak Convergence of Vervaat and Vervaat error processes of long-range dependent sequences. *Journal of Theoretical Probability, 21,* 672–686.

Csörgő, M., & Révész, P. (1978). Strong approximations of the quantile process. *Annals of Statistics*, *6*, 882–894.

Csörgő, M., & Révész, P. (1979). How big are the increments of a Wiener process? *Annals of Probability*, *7*, 731–737.

Csörgő, M., & Révész, P. (1981). *Strong approximations in probability and statistics*. New York: Academic Press.

Csörgő, M., & Szyszkovicz, B. (1998). Sequential quantile and Bahadur-Kiefer processes. *Order statistics: Theory and methods. Handbook of Statistics*. (Vol. 16, pp. 631–688). Amsterdam: North-Holland.

Csörgő, M., Szyszkowicz, B., & Wang, L. H. (2006). Strong invariance principles for sequential Bahadur-Kiefer and Vervaat error processes of long-range dependent sequences. *Annals of Statistics*, *34*, 1013–1044. (Correction: *Annals of Statistics*, *35*, 2815–2817 (2007)).

Deheuvels, P. (1992a). Pointwise Bahadur-Kiefer-type theorems I. *Probability theory and applications. Mathematics and Its Applications* (Vol. 80, pp. 235–255). Dordrecht: Kluwer Academic Publishers.

Deheuvels, P. (1992b). Pointwise Bahadur-Kiefer-type theorems II. *Nonparametric statistics and related topics (Ottawa, ON)* (pp. 331–345). Amsterdam: North-Holland.

Deheuvels, P., & Mason, D. M. (1990). Bahadur-Kiefer-type processes. *Annals of Probability*, *18*, 669–697.

Deheuvels, P., & Mason, D. M. (1992) A functional LIL approach to pointwise Bahadur-Kiefer theorems. *Probability in Banach spaces, 8 (Brunswick, ME, 1991)*. Progress in Probability (Vol. 30, pp. 255–266). Boston: Birkhäuser.

Deheuvels, P., & Steinebach, J. (1992). On the limiting behavior of the Bahadur-Kiefer statistic for partial sums and renewal processes when the fourth moment does not exist. *Statistics & Probability Letters*, *13*, 179–188.

Goodman, V., & Kuelbs, J. (1991). Rates of clustering for some Gaussian self-similar processes. *Probability Theory and Related Fields*, *88*, 47–75.

Horváth, L. (1984). Strong approximations of renewal processes. *Stochastic Processes and Their Applications*, *18*, 127–138.

Kiefer, J. (1967). On Bahadur's representation of sample quantiles. *Annals of Mathematical Statistics*, *38*, 1323–1342.

Kiefer, J. (1970). Deviations between the sample quantile process and the sample df. *Nonparametric techniques in statistical inference* (pp. 299–319). London: Cambridge University Press.

Komlós, J., Major, P., & Tusnády, G. (1975). An approximation of partial sums of independent RV's and the sample DF I. *Zeitschrift für Wahrscheinlichkeitstheorie und Verwandte Gebiete*, *32*, 111–131.

Ortega, J. (1984). On the size of the increments of nonstationary Gaussian processes. *Stochastic Processes and their Applications*, *18*, 47–56.

Shorack, G. R. (1982). Kiefer's theorem via the Hungarian construction. *Zeitschrift für Wahrscheinlichkeitstheorie und Verwandte Gebiete*, *61*, 369–373.

Vervaat, W. (1972a). *Success Epochs in Bernoulli trials: With applications to number theory* (2nd ed. in 1977). *Mathematical Centre Tracts* (Vol. 42). Amsterdam: Matematisch Centrum.

Vervaat, W. (1972b). Functional central limit theorems for processes with positive drift and their inverses. *Zeitschrift für Wahrscheinlichkeitstheorie und Verwandte Gebiete*, *23*, 245–253.

Wang, Q., Lin, Y.-X., & Gulati, C. M. (2003). Strong approximation for long memory processes with applications. *Journal of Theoretical Probability*, *16*, 377–389.

Reduced-Bias Estimator of the Conditional Tail Expectation of Heavy-Tailed Distributions

El Hadji Deme, Stéphane Girard and Armelle Guillou

Abstract Several risk measures have been proposed in the literature. In this paper, we focus on the estimation of the Conditional Tail Expectation (CTE). Its asymptotic normality has been first established in the literature under the classical assumption that the second moment of the loss variable is finite, this condition being very restrictive in practical applications. Such a result has been extended by Necir et al., (Journal of Probability and Statistics 596839:17 2010) in the case of infinite second moment. In this framework, we propose a reduced-bias estimator of the CTE. We illustrate the efficiency of our approach on a small simulation study and a real data analysis.

1 Introduction

Different risk measures have been proposed in the literature and used to determine the amount of an asset to be kept in reserve in the financial framework. We refer to Goovaerts et al. (1984) for various examples and properties. One of the most popular examples in hydrology or climatology is undoubtedly the return period. A frequency analysis in hydrology focuses on the estimation of quantities (e.g., flows or annual rainfall) corresponding to a certain return period. It is closely related to the notion of a quantile. For a positive real-valued random variable X, the quantile of order $1 - \frac{1}{T}$

E.H. Deme
LERSTAD, Université Gaston Berger de Saint-Louis, Saint-Louis, Sénégal
e-mail: eltinoyaw@yahoo.fr

S. Girard
Team Mistis, Inria Grenoble Rhône-Alpes & Laboratoire Jean Kuntzmann,
Montbonnot, France
e-mail: Stephane.Girard@inria.fr

A. Guillou (✉)
Université de Strasbourg & CNRS, IRMA, UMR 7501, Strasbourg, France
e-mail: armelle.guillou@math.unistra.fr

© Springer International Publishing Switzerland 2015
M. Hallin et al. (eds.), *Mathematical Statistics and Limit Theorems*,
DOI 10.1007/978-3-319-12442-1_7

expresses the magnitude of the event which is exceeded with a probability equal to $\frac{1}{T}$. T is then called the return period. In an actuarial context, the Value at Risk (VaR) is defined as the $p-$quantile

$$Q(p) = \inf\{x \geq 0 : F(x) \geq p\}, \quad \text{for } p \in [0, 1],$$

with F the distribution function of the random variable X. A second important risk measure, based on the quantile notion, is the Conditional Tail Expectation (CTE) defined by

$$CTE_\alpha[X] = \mathbb{E}(X|X > Q(\alpha)), \quad \text{for } \alpha \in (0, 1).$$

The CTE satisfies all the desirable properties of a coherent risk measure (see Artzner et al. 1999) and it provides a more conservative measure of risk than the VaR for the same level of degree of confidence (see Landsman and Valdez 2003). For all these reasons, the CTE (sometimes referred to as expected shortfall) is preferable in many applications. It thus continues to receive an increased attention in the actuarial literature (see for instance Chaps. 2 and 7 in McNeil et al. 2005).

In the sequel we assume that F is continuous, which allows us to rewrite the $CTE_\alpha[X]$ as

$$\mathbb{C}_\alpha[X] = \frac{1}{1-\alpha} \int_\alpha^1 Q(s)ds.$$

Clearly, the CTE is unknown since it depends on F. Hence, it is desirable to define estimators for this quantity and to study their asymptotic properties. To this aim, suppose that we have at our disposal a sample $(X_1, \ldots, X_n)$ of independent and identically distributed random variables from F and denote by $X_{1,n} \leq \cdots \leq X_{n,n}$ the order statistics. The asymptotic behavior of $\mathbb{C}_{\alpha_n}[X]$ has been studied recently in Pan et al. (2013) and Zhu and Li (2012) when $\alpha_n \to 1$ as $n \to \infty$. On the contrary, in this paper we consider α fixed. A natural estimator of $\mathbb{C}_\alpha[X]$ can then be obtained by

$$\widehat{\mathbb{C}}_{n,\alpha}[X] = \frac{1}{1-\alpha} \int_\alpha^1 Q_n(s)ds, \tag{1}$$

where $Q_n(s)$ is the empirical quantile function, which is equal to the ith order statistic $X_{i,n}$, for all $s \in ((i-1)/n, i/n]$, and for all $i = 1, \ldots, n$. The asymptotic behavior of the estimator $\widehat{\mathbb{C}}_{n,\alpha}[X]$ has been studied by Brazauskas et al. (2008), when $\mathbb{E}[X^2] < \infty$. Unfortunately, this condition is quite restrictive. For instance, in the case of Pareto-type distributions, defined as $1 - F(x) = x^{-1/\gamma}\ell_F(x)$ where ℓ_F is a slowly varying function at infinity satisfying $\ell_F(\lambda x)/\ell_F(x) \to 1$ as $x \to \infty$, for all $\lambda > 0$, this condition of second moment implies that $\gamma \in (0, 1/2)$. When $\gamma \in (1/2, 1)$, we have $\mathbb{E}[X^2] = \infty$ but nevertheless the CTE is well defined and finite since $\mathbb{E}[X] < \infty$. Note that, in the case $\gamma = 1/2$, the finiteness of the second moment depends on the slowly varying function.

This framework will be the subject of this paper where we assume that

$$1 - F(x) = x^{-1/\gamma} \ell_F(x) \tag{2}$$

where $\gamma > 0$ is the extreme value index. We focus on the case where $\gamma \in (1/2, 1)$ and thus $\mathbb{E}[X^2] = \infty$, this range of values being excluded in the results of Brazauskas et al. (2008). The estimation of γ has been extensively studied in the literature and the most famous estimator is the Hill (1975) estimator defined as:

$$\widehat{\gamma}^H_{n,k} = \frac{1}{k} \sum_{j=1}^{k} j \left(\log X_{n-j+1,n} - \log X_{n-j,n} \right)$$

for an intermediate sequence $k = k(n)$, i.e., a sequence such that $k \to \infty$ and $k/n \to 0$ as $n \to \infty$. More generally, Csörgő et al. (1985) extended the Hill estimator into a kernel class of estimators

$$\widehat{\gamma}^K_{n,k} = \frac{1}{k} \sum_{j=1}^{k} K \left(\frac{j}{k+1} \right) Z_{j,k},$$

where K is a kernel integrating to one and $Z_{j,k} = j \left(\log X_{n-j+1,n} - \log X_{n-j,n} \right)$. Note that the Hill estimator corresponds to the particular case where $K(u) = \underline{K}(u) := \mathbb{I}_{\{0<u<1\}}$. Notice that $\mathbb{C}_\alpha[X]$ can be rewritten as

$$\mathbb{C}_\alpha[X] = \frac{1}{1-\alpha} \int_\alpha^{1-k/n} Q(s)ds + \frac{1}{1-\alpha} \int_0^{k/n} Q(1-s)ds.$$
$$=: \mathbb{C}_\alpha^{(1)}[X] + \mathbb{C}_\alpha^{(2)}[X].$$

In this spirit, Necir et al. (2010) introduced the following estimator of the CTE, which takes into account different asymptotic properties of moderate and high quantiles in the case of Pareto-type distributions:

$$\widetilde{\mathbb{C}}_{n,\alpha}[X] =: \widetilde{\mathbb{C}}_{n,\alpha}^{(1)}[X] + \widetilde{\mathbb{C}}_{n,\alpha}^{(2)}[X]$$
$$= \frac{1}{1-\alpha} \sum_{j=1}^{n-k} \left(\left(\frac{j}{n} - \alpha \right)_+ - \left(\frac{j-1}{n} - \alpha \right)_+ \right) X_{j,n}$$
$$+ \frac{k/n}{(1-\alpha)(1-\widehat{\gamma}^H_{n,k})} X_{n-k,n} \tag{3}$$

where $(s-\alpha)_+$ denotes the positive part of $(s-\alpha)$. The estimator $\widetilde{\mathbb{C}}_{n,\alpha}^{(1)}[X]$ is obtained similarly to (1) using the well-known properties of the empirical quantile function

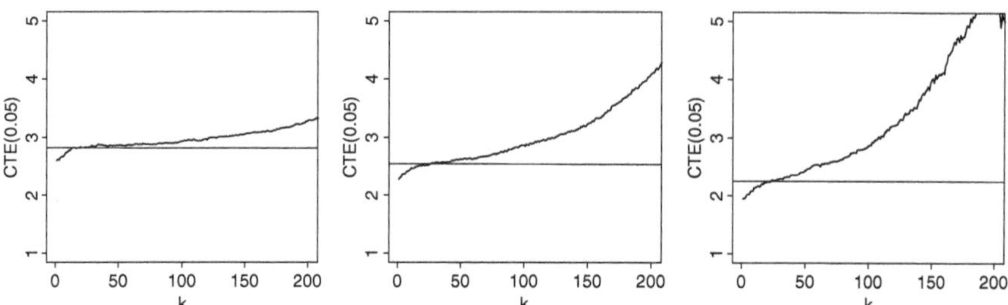

Fig. 1 Median of $\widetilde{\mathbb{C}}_{n,0.05}[X]$ as a function of k based on 500 samples of size 500 from a Burr distribution defined as $\overline{F}(x) = (1 + x^{-3\rho/2})^{1/\rho}$. From the *left* to the *right*: $\rho = -1.5$, $\rho = -1$, and $\rho = -0.75$. The *horizontal line* represents the true value of the $CTE_{0.05}[X]$

Q_n whereas $\widetilde{\mathbb{C}}_{n,\alpha}^{(2)}[X]$ is obtained using a Weissman estimator of Q: $\widehat{Q}(1 - s) := X_{n-k,n}\,(k/n)^{\widehat{\gamma}_{n,k}^{H}}\,s^{-\widehat{\gamma}_{n,k}^{H}}$, $s \to 0$ (see Weissman 1978).

This estimator may suffer from a high bias in finite sample situations, as illustrated on Fig. 1 on a Burr distribution with extreme value index $\gamma = 2/3$. Besides, it appears that the bias heavily depends on the intermediate sequence, making the choice of k difficult in practice.

The goal of this paper is twofold. First, we state an asymptotic normality result for $\widetilde{\mathbb{C}}_{n,\alpha}[X]$ exhibiting the bias term (Sect. 2) and thus generalizing the one of Necir et al. (2010). Second, the precise knowledge of the first order of the bias allows us to propose a reduced-bias approach. The efficiency of our method is illustrated on a small simulation study and a real dataset in Sect. 3. All the proofs are postponed to Sect. 4.

2 Main Results

As usual in the extreme value framework, to prove asymptotic normality results, we need a second-order condition on the function $\mathbb{U}(x) := Q(1 - 1/x)$, $x > 1$, such as the following:

Condition $(\mathscr{R}_{\mathbb{U}})$. *There exist a function $A(x) \to 0$ as $x \to \infty$ of constant sign for large values of x and a second-order parameter $\rho < 0$ such that, for every $x > 0$,*

$$\lim_{t \to \infty} \frac{\log \mathbb{U}(tx) - \log \mathbb{U}(t) - \gamma \log x}{A(t)} = \frac{x^\rho - 1}{\rho}.$$

Note that condition $(\mathscr{R}_{\mathbb{U}})$ implies that $|A|$ is regularly varying with index ρ (see, e.g., Geluk and de Haan 1987). It is satisfied for most of the classical distribution functions such as the Pareto, Burr, and Fréchet ones.

We begin by giving in Theorem 1 an expansion of $\widetilde{\mathbb{C}}_{n,\alpha}[X]$ in terms of functionals of a sequence of Brownian bridges $\mathbb{B}_n$, which leads to its asymptotic normality stated in Corollary 1. As it exhibits some bias, we propose a reduced-bias estimator whose expansion is formulated in Theorem 2 and its asymptotic normality is given in Corollary 2.

2.1 Asymptotic Results for the CTE Estimator

Theorem 1 *Assume that F satisfies $(\mathscr{R}_{\mathbb{U}})$ with $\gamma \in (1/2, 1)$. Then for any sequence of integer $k = k(n)$ satisfying $k \to \infty$, $k/n \to 0$, and $\sqrt{k}A(n/k) = O(1)$ as $n \to \infty$, we have*

$$\frac{n(1-\alpha)}{\sqrt{k}\mathbb{U}(n/k)} \left(\widetilde{\mathbb{C}}_{n,\alpha}[X] - \mathbb{C}_\alpha[X] \right) \overset{\mathscr{D}}{=} \sqrt{k}A\left(\frac{n}{k}\right) \mathscr{AB}(\gamma, \rho) + \mathbb{W}_{n,1} + \mathbb{W}_{n,2}$$

$$+ \mathbb{W}_{n,3} + o_{\mathbb{P}}(1)$$

where

$$\mathscr{AB}(\gamma, \rho) := \frac{\gamma\rho}{(1-\rho)(\gamma + \rho - 1)(1-\gamma)^2}$$

and

$$\begin{cases} \mathbb{W}_{n,1} := -\dfrac{\int_0^{1-k/n} \mathbb{B}_n(s)dQ(s)}{\sqrt{k/n}Q(1-k/n)} \\[3ex] \mathbb{W}_{n,2} := -\dfrac{\gamma}{1-\gamma}\sqrt{\dfrac{n}{k}}\,\mathbb{B}_n(1-k/n) \\[3ex] \mathbb{W}_{n,3} := \dfrac{\gamma}{(1-\gamma)^2}\sqrt{\dfrac{n}{k}}\int_0^1 s^{-1}\mathbb{B}_n(1-sk/n)d(s\underline{K}(s)) \end{cases}$$

with $\mathbb{B}_n$ a sequence of Brownian bridges.

Now, by computing the asymptotic variances of the different processes appearing in Theorem 1, we deduce the following corollary:

Corollary 1 *Under the assumptions of Theorem 1, if $\sqrt{k}A(n/k) \to \lambda \in \mathbb{R}$, we have*

$$\frac{n(1-\alpha)}{\sqrt{k}\mathbb{U}(n/k)} \left(\widetilde{\mathbb{C}}_{n,\alpha}[X] - \mathbb{C}_\alpha[X] \right) \overset{\mathscr{D}}{\longrightarrow} \mathscr{N}\left(\lambda\mathscr{AB}(\gamma, \rho), \mathscr{AV}(\gamma)\right)$$

where $\mathscr{AB}(\gamma, \rho)$ is as above and

$$\mathscr{AV}(\gamma) = \frac{\gamma^4}{(2\gamma - 1)(1-\gamma)^4}.$$

Since $\rho < 0$ and $\gamma \in (1/2, 1)$, we can easily check that $\mathscr{AB}(\gamma, \rho)$ is always positive and thus the sign of the function $A(.)$ determines the sign of the bias of $\widetilde{\mathbb{C}}_{n,\alpha}[X]$. Note that the asymptotic variance $\mathscr{AV}(\gamma)$ does not depend on α and that this result generalizes Theorem 3.1 in Necir et al. (2010) in case $\lambda \neq 0$. The goal of the next section is to propose a reduced-bias estimator of $\mathbb{C}_{\alpha}[X]$.

2.2 Reduced-Bias Method with the Least Squared Approach

From Theorem 1, it is clear that the estimator $\widetilde{\mathbb{C}}_{n,\alpha}[X]$ exhibits a bias due to the use in its construction of the Weissman's estimator which is known to have such a problem. To overcome this issue, we propose to use the exponential regression model introduced in Feuerverger and Hall (1999) and Beirlant et al. (1999) to construct a reduced-bias estimator.

More precisely, using $(\mathscr{R}_{\mathbb{U}})$, Feuerverger and Hall (1999) and Beirlant et al. (1999, 2002) proposed the following exponential regression model for the log-spacings of order statistics:

$$
Z_{j,k} \sim \left(\gamma + A(n/k) \left(\frac{j}{k+1} \right)^{-\rho} \right) + \varepsilon_{j,k}, \quad 1 \leq j \leq k, \tag{4}
$$

where $\varepsilon_{j,k}$ are zero-centered error terms. If we ignore the term $A(n/k)$ in (4), we retrieve the Hill-type estimator $\widehat{\gamma}_{n,k}^{H}$ by taking the mean of the left-hand side of (4). By using a least-squares approach, (4) can be further exploited to propose a reduced-bias estimator of γ in which ρ is substituted by a consistent estimator $\widehat{\rho} = \widehat{\rho}_{n,k}$ (see for instance Beirlant et al. (2002), Deme et al. (2013a), Fraga Alves (2003)) or by a canonical choice, such as $\rho = -1$ (see, e.g., Feuerverger and Hall (1999) or Beirlant et al. (1999)). The least squares estimators of γ and $A(n/k)$ are then given by

$$
\left\{
\begin{aligned}
\widehat{\gamma}_{n,k}^{LS}(\widehat{\rho}) &= \frac{1}{k} \sum_{j=1}^{k} Z_{j,k} - \frac{\widehat{A}_{n,k}^{LS}(\widehat{\rho})}{1 - \widehat{\rho}}, \\
\widehat{A}_{n,k}^{LS}(\widehat{\rho}) &= \frac{(1 - 2\widehat{\rho})(1 - \widehat{\rho})^2}{\widehat{\rho}^2} \frac{1}{k} \sum_{j=1}^{k} \left(\left(\frac{j}{k+1} \right)^{-\widehat{\rho}} - \frac{1}{1 - \widehat{\rho}} \right) Z_{j,k}.
\end{aligned}
\right.
$$

The main asymptotic properties of $\widehat{\gamma}_{n,k}^{LS}(\widehat{\rho})$ and $\widehat{A}_{n,k}^{LS}(\widehat{\rho})$ as functionals of a sequence of Brownian bridges $\mathbb{B}_n$ have been established in Deme et al. (2013b, Lemma 5). Note that $\widehat{\gamma}_{n,k}^{LS}(\widehat{\rho})$ can be viewed as a kernel estimator

$$
\widehat{\gamma}_{n,k}^{LS}(\widehat{\rho}) = \frac{1}{k} \sum_{j=1}^{k} K_{\widehat{\rho}} \left(\frac{j}{k+1} \right) Z_{j,k},
$$

where for $0 < u \leq 1$:

$$K_\rho(u) = \frac{1-\rho}{\rho} \underline{K}(u) + \left(1 - \frac{1-\rho}{\rho}\right) \underline{K}_\rho(u)$$

with $\underline{K}_\rho(u) = ((1-\rho)/\rho)(u^{-\rho} - 1)\mathbb{I}_{\{0<u<1\}}$, see Sect. 3 of Beirlant et al. (2002). Now, using the second-order refinements of assumption $(\mathscr{R}_U)$, we can construct the following asymptotically unbiased estimator of the quantile:

$$\widehat{Q}^{LS,\widehat{\rho}}(1-s) = (ns/k)^{-\widehat{\gamma}_{n,k}^{LS}(\widehat{\rho})} X_{n-k,n} \left(1 - \widehat{\rho}^{-1} \widehat{A}_{n,k}^{LS}(\widehat{\rho}) \left(1 - (ns/k)^{-\widehat{\rho}}\right)\right),$$

see, e.g., Matthys et al. (2004). Thus, in the spirit of (3), we arrive at the following asymptotically unbiased estimator of $\mathbb{C}_\alpha[X]$

$$\widetilde{\mathbb{C}}_{n,\alpha}^{LS,\widehat{\rho}}[X] := \frac{1}{1-\alpha} \sum_{j=1}^{n-k} \left(\left(\frac{j}{n} - \alpha\right)_+ - \left(\frac{j-1}{n} - \alpha\right)_+\right) X_{j,n}$$

$$+ \frac{k/n}{(1-\alpha)(1 - \widehat{\gamma}_{n,k}^{LS}(\widehat{\rho}))} \left(1 - \frac{\widehat{A}_{n,k}^{LS}(\widehat{\rho})}{\widehat{\gamma}_{n,k}^{LS}(\widehat{\rho}) + \widehat{\rho} - 1}\right) X_{n-k,n}.$$

Our next goal is to establish, under suitable assumptions, the asymptotic normality of $\widetilde{\mathbb{C}}_{n,\alpha}^{LS,\widehat{\rho}}[X]$. This is done in the following theorem.

Theorem 2 *Under the assumptions of Theorem 1, if $\widehat{\rho}$ is a consistent estimator of ρ, then we have*

$$\frac{n(1-\alpha)}{\sqrt{k}U(n/k)} \left(\widetilde{\mathbb{C}}_{n,\alpha}^{LS,\widehat{\rho}}[X] - \mathbb{C}_\alpha[X]\right) \overset{\mathscr{D}}{=} \mathbb{W}_{n,1} + \mathbb{W}_{n,2} + \mathbb{W}_{n,4} + \mathbb{W}_{n,5} + o_\mathbb{P}(1)$$

where $\mathbb{W}_{n,1}$, $\mathbb{W}_{n,2}$, and $\mathbb{W}_{n,3}$ are defined in Theorem 1, and

$$\begin{cases} \mathbb{W}_{n,4} := \dfrac{\rho\gamma^2}{(\gamma + \rho - 1)(1-\gamma)^2} \sqrt{\dfrac{n}{k}} \displaystyle\int_0^1 s^{-1}\mathbb{B}_n(1 - sk/n)d(sK_\rho(s)) \\[2em] \mathbb{W}_{n,5} := -\dfrac{(1-\gamma)(1-\rho)}{\gamma + \rho - 1}\mathbb{W}_{n,3}. \end{cases}$$

Now, by computing the asymptotic variances of the different processes appearing in Theorem 2, we deduce the following corollary.

Corollary 2 *Under the assumptions of Theorem 1, if $\widehat{\rho}$ is a consistent estimator of ρ, then we have*

$$\frac{n(1-\alpha)}{\sqrt{k}U(n/k)} \left(\widetilde{\mathbb{C}}_{n,\alpha}^{LS,\widehat{\rho}}[X] - \mathbb{C}_\alpha[X]\right) \overset{\mathscr{D}}{\longrightarrow} \mathscr{N}\left(0, \widetilde{\mathscr{AV}}(\gamma, \rho)\right)$$

with

$$\widetilde{\mathscr{A}\mathscr{V}}(\gamma, \rho) = \frac{\gamma^4(\gamma - \rho)^2}{(2\gamma - 1)(1 - \gamma)^4(\gamma + \rho - 1)^2}.$$

As expected, the asymptotic bias of our new estimator of the CTE is equal to zero whereas its asymptotic variance $\widetilde{\mathscr{A}\mathscr{V}}(\gamma, \rho)$ is larger than the one of the original estimators $\mathscr{A}\mathscr{V}(\gamma)$ exhibited in Corollary 1.

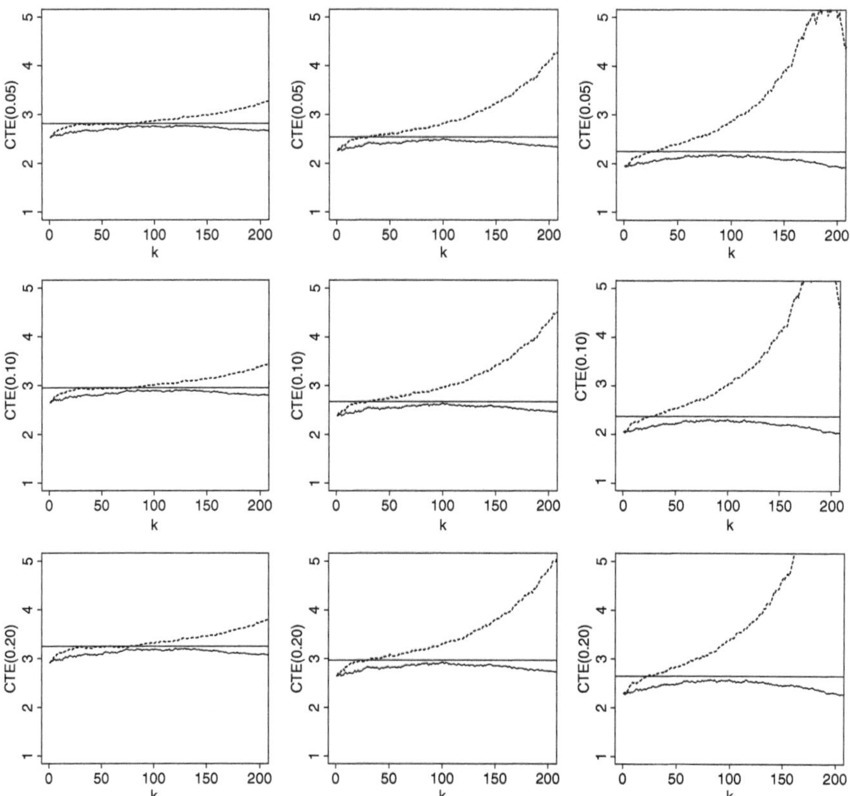

Fig. 2 Median of $\widetilde{\mathbb{C}}_{n,\alpha}[X]$ (*dotted line*) and $\widetilde{\mathbb{C}}_{n,\alpha}^{LS,-1}[X]$ (*full line*) as a function of k based on 500 samples of size 500 for $\alpha = 0.05$ (*top*), $\alpha = 0.10$ (*middle*), and $\alpha = 0.20$ (*bottom*) from a Burr distribution defined as $\overline{F}(x) = (1 + x^{-\frac{3\rho}{2}})^{1/\rho}$. From the *left* to the *right*: $\rho = -1.5$, $\rho = -1$, and $\rho = -0.75$. The *horizontal line* represents the true value of the $CTE_\alpha[X]$

3 Finite Sample Behavior

3.1 A Small Simulation Study

In this section, the biased estimator $\widetilde{\mathbb{C}}_{n,\alpha}[X]$ and the reduced-bias one $\widetilde{\mathbb{C}}_{n,\alpha}^{LS,-1}[X]$ are compared on a small simulation study. To this aim, 500 samples of size 500 are simulated from a Burr distribution defined as: $\overline{F}(x) = (1 + x^{-3\rho/2})^{1/\rho}$ which satisfied condition $(\mathscr{R}_{\mathbb{U}})$ with $A(t) = 2t^{\rho}/3$. The associated extreme value index is $\gamma = 2/3$ and ρ is the second-order parameter. Three values for $\alpha \in \{0.05, 0.10, 0.20\}$ are used and different values of $\rho \in \{-0.75, -1, -1.5\}$ are considered to assess its impact. The median and median squared error (MSE) of these estimators are estimated over 500 replications. The results are displayed in Figs. 2 and 3. It appears on Fig. 2 that the closer ρ is to 0, the more important is the bias of $\widetilde{\mathbb{C}}_{n,\alpha}[X]$ whatever the value

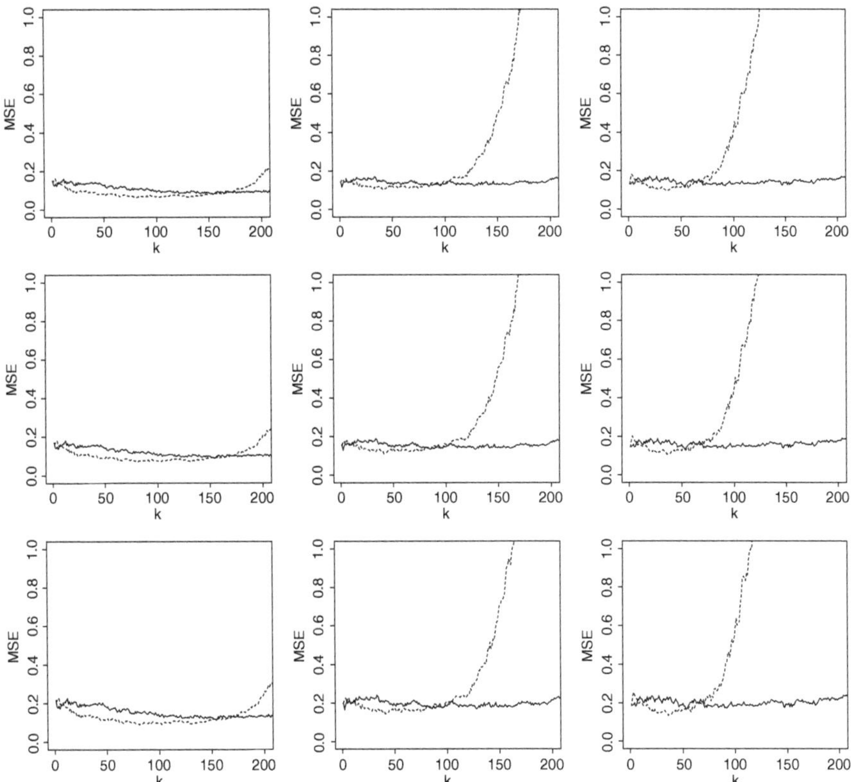

Fig. 3 MSE of $\widetilde{\mathbb{C}}_{n,\alpha}[X]$ (*dotted line*) and $\widetilde{\mathbb{C}}_{n,\alpha}^{LS,-1}[X]$ (*full line*) as a function of k based on 500 samples of size 500 for $\alpha = 0.05$ (*top*), $\alpha = 0.10$ (*middle*), and $\alpha = 0.20$ (*bottom*) from a Burr distribution defined as $\overline{F}(x) = (1 + x^{-\frac{3\rho}{2}})^{1/\rho}$. From the *left* to the *right*: $\rho = -1.5$, $\rho = -1$, and $\rho = -0.75$

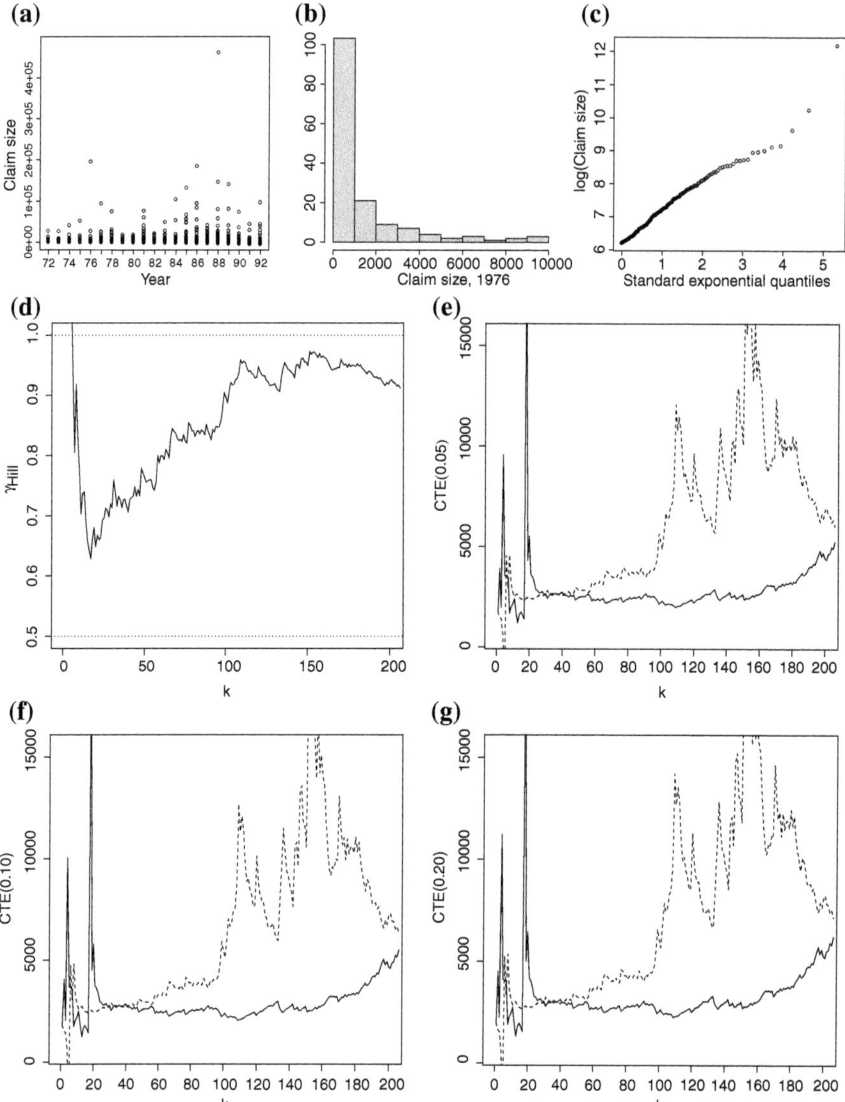

Fig. 4 Time plot for the Norwegian fire insurance data (**a**); histogram of the claim size for the year 1976 (**b**); quantile–quantile plot (**c**); hill estimator as a function of k for the year 1976 (**d**); biased estimator $\widetilde{\mathbb{C}}_{n,\alpha}[X]$ (*dotted line*) and reduced-bias one $\widetilde{\mathbb{C}}_{n,\alpha}^{LS,-1}[X]$ (*full line*) as a function of k for $\alpha = 0.05$ (**e**), $\alpha = 0.10$ (**f**), and $\alpha = 0.20$ (**g**)

of α is. The effect of the bias correction on the MSE is illustrated on Fig. 3. We can observe that the MSE of the reduced-bias estimator $\widetilde{\mathbb{C}}_{n,\alpha}^{LS,-1}[X]$ is almost constant with respect to k, especially when the bias of $\widetilde{\mathbb{C}}_{n,\alpha}[X]$ is strong, i.e. when ρ is close to 0.

3.2 Real Data Analysis

Our real dataset concerns a Norwegian fire insurance portfolio from 1972 until 1992. Together with the year of occurrence, the data contain the value ($\times1\,000$ Krone) of the claims. A priority of 500 units was in force. These data were of some concern in that the number of claims had risen systematically with a maximum in 1988 as illustrated in Fig. 4a. We concentrate here on the year 1976 where the average claim size per year reached a peak as was the case in 1988. The sample size is $n = 207$. Figure 4b shows the histogram corresponding to this year 1976. The assumption of a Pareto-type tail (2) can be visually checked thanks to a quantile–quantile plot (Fig. 4c). It appears that the logarithm of the claim sizes (vertical axis) is approximately a linear function of the standard exponential quantiles (horizontal axis). From Fig. 4d we can observe the difficulty to find a stable part in the plot of the Hill estimator $\widehat{\gamma}_{n,k}^{H}$ as a function of k, due to the bias of this estimator. We can apply our methodology to this real dataset as the extreme value index (or at least its estimator) is in the interval $(1/2, 1)$ whatever the value of k is. Figure 4e–g shows the biased estimator $\widetilde{\mathbb{C}}_{n,\alpha}[X]$ (dashed line) and the reduced-bias one $\widetilde{\mathbb{C}}_{n,\alpha}^{LS,-1}[X]$ (full line) for three different values of α: 0.05, 0.10, and 0.20. The reduced-bias estimator $\widetilde{\mathbb{C}}_{n,\alpha}^{LS,-1}[X]$ is almost constant for a large range of values of k which makes the choice of k easier in practice.

4 Proofs

Let $Y_1, \ldots, Y_n$ be independent and identically distributed random variables from the unit Pareto distribution G, defined as $G(y) = 1 - y^{-1}$, $y \geq 1$. For each $n \geq 1$, let $Y_{1,n} \leq \cdots \leq Y_{n,n}$ be the order statistics pertaining to $Y_1, \ldots, Y_n$. Clearly, $X_{j,n} \overset{\mathscr{D}}{=} \mathbb{U}(Y_{j,n})$, $j = 1, \ldots, n$. We shall assume that we are on the probability space $(\Omega, \mathbb{A}, \mathbb{P})$ of Theorem 2.1 of Csörgő et al. (1986) carrying a sequence of independent and identically distributed uniform $(0, 1)$ random variables $\xi_1, \xi_2, \ldots$ and a sequence of Brownian bridges $\mathbb{B}_n(s)$, $0 \leq s \leq 1, n = 1, 2 \ldots$ such that for all $0 \leq \nu < 1/2$ and $\lambda > 0$

$$\sup_{\lambda/n \leq t \leq 1-\lambda/n} \frac{|\beta_n(t) - \mathbb{B}_n(t)|}{(t(1-t))^{1/2-\nu}} = O_{\mathbb{P}}(n^{-\nu}),$$

where β_n is the uniform quantile process

$$\beta_n(t) = \sqrt{n}\,(t - \mathbb{V}_n(t))$$

with $\mathbb{V}_n$ denoting the empirical uniform quantile function defined to be $\mathbb{V}_n(t) = \xi_{j,n}$, $\frac{j-1}{n} < t \leq \frac{j}{n}$, $j = 1, \ldots, n$ and $\mathbb{V}_n(0) = 0$.

The following lemma gives an asymptotic expansion for the second random term appearing in (3).

Lemma 1 *Under the assumptions of Theorem 1, we have*

$$\frac{n(1-\alpha)}{\sqrt{k}\,\mathbb{U}(n/k)} \left(\widetilde{\mathbb{C}}_{n,\alpha}^{(2)}[X] - \mathbb{C}_{\alpha}^{(2)}[X] \right) \overset{\mathscr{D}}{=} \sqrt{k}\,A\left(\frac{n}{k}\right) \mathscr{A}\,\mathscr{B}(\gamma,\rho) + \mathbb{W}_{n,2}$$
$$+ \mathbb{W}_{n,3} + o_{\mathbb{P}}(1).$$

Proof of Lemma 1. Note that $\widetilde{\mathbb{C}}_{n,\alpha}^{(2)}[X]$ can be rewritten as follows

$$(1-\alpha)\widetilde{\mathbb{C}}_{n,\alpha}^{(2)}[X] \overset{\mathscr{D}}{=} \frac{k/n}{1-\widehat{\gamma}_{n,k}^{H}}\,\mathbb{U}\left(Y_{n-k,n}\right).$$

As a consequence, the following expansion holds:

$$\frac{n(1-\alpha)}{\sqrt{k}\,\mathbb{U}(n/k)} \left(\widetilde{\mathbb{C}}_{n,\alpha}^{(2)}[X] - \mathbb{C}_{\alpha}^{(2)}[X] \right) = \sum_{j=1}^{4} T_{n,j},$$

where

$$T_{n,1} := \frac{\sqrt{k}}{1-\widehat{\gamma}_{n,k}^{H}} \left[\frac{\mathbb{U}\left(Y_{n-k,n}\right)}{\mathbb{U}(n/k)} - \left(\frac{k}{n}Y_{n-k,n}\right)^{\gamma} \right],$$

$$T_{n,2} := \frac{\sqrt{k}}{1-\widehat{\gamma}_{n,k}^{H}} \left[\left(\frac{k}{n}Y_{n-k,n}\right)^{\gamma} - 1 \right],$$

$$T_{n,3} := \frac{1}{(1-\widehat{\gamma}_{n,k}^{H})(1-\gamma)}\,\sqrt{k}\left(\widehat{\gamma}_{n,k}^{H} - \gamma\right),$$

$$T_{n,4} := \frac{n}{\sqrt{k}\,\mathbb{U}(n/k)} \left[\frac{k/n}{1-\gamma}\mathbb{U}(n/k) - (1-\alpha)\mathbb{C}_{\alpha}^{(2)}[X] \right].$$

We study each term separately.

Term $T_{n,1}$. According to de Haan and Ferreira (2006, Theorem 2.3.9), for any $\delta > 0$, we have

$$\frac{\mathbb{U}\left(Y_{n-k,n}\right)}{\mathbb{U}(n/k)} - \left(\frac{k}{n}Y_{n-k,n}\right)^{\gamma}$$
$$= A_0\left(\frac{n}{k}\right) \left\{ \left(\frac{k}{n}Y_{n-k,n}\right)^{\gamma} \frac{\left(\frac{k}{n}Y_{n-k,n}\right)^{\rho} - 1}{\rho} + o_{\mathbb{P}}(1)\left(\frac{k}{n}Y_{n-k,n}\right)^{\gamma+\rho\pm\delta} \right\},$$

where $A_0(t) \sim A(t)$ as $t \to \infty$.

Thus, since $k Y_{n-k,n}/n = 1 + o_{\mathbb{P}}(1)$ and $\widehat{\gamma}_{n,k}^H \xrightarrow{\mathbb{P}} \gamma$, it readily follows that

$$T_{n,1} = o_{\mathbb{P}}(1). \tag{5}$$

Term $T_{n,2}$. The equality $Y_{n-k,n} \overset{\mathscr{D}}{=} (1 - \xi_{n-k,n})^{-1}$ yields

$$\sqrt{k}\left[\left(\frac{k}{n}Y_{n-k,n}\right)^{\gamma} - 1\right] \overset{\mathscr{D}}{=} \sqrt{k}\left(\left(\frac{n}{k}(1 - \xi_{n-k,n})\right)^{-\gamma} - 1\right)$$

$$= -\gamma\sqrt{k}\left(\frac{n}{k}(1 - \xi_{n-k,n}) - 1\right)$$

$$\times (1 + o_{\mathbb{P}}(1)) \text{ by a Taylor expansion}$$

$$= -\gamma\sqrt{\frac{n}{k}}\beta_n\left(1 - \frac{k}{n}\right)(1 + o_{\mathbb{P}}(1))$$

$$= -\gamma\sqrt{\frac{n}{k}}\left(\mathbb{B}_n\left(1 - \frac{k}{n}\right) + O_{\mathbb{P}}(n^{-\nu})\left(\frac{k}{n}\right)^{1/2-\nu}\right)$$

$$\times (1 + o_{\mathbb{P}}(1)),$$

for $0 \leq \nu < 1/2$, by the Csörgő et al. (1986) result cited above. Thus, using again that $\widehat{\gamma}_{n,k}^H \xrightarrow{\mathbb{P}} \gamma$, it follows that

$$T_{n,2} \overset{\mathscr{D}}{=} -\frac{\gamma}{1-\gamma}\sqrt{\frac{n}{k}}\mathbb{B}_n\left(1 - \frac{k}{n}\right)(1 + o_{\mathbb{P}}(1)) = \mathbb{W}_{n,2} + o_{\mathbb{P}}(1). \tag{6}$$

Term $T_{n,3}$. According to Theorem 1 in Deme et al. (2013b) and by the consistency in probability of $\widehat{\gamma}_{n,k}^H$, we have

$$T_{n,3} \overset{\mathscr{D}}{=} \frac{1}{(1-\gamma)^2}\left\{\frac{\sqrt{k}A(n/k)}{1-\rho} + \gamma\sqrt{\frac{n}{k}}\int_0^1 s^{-1}\mathbb{B}_n\left(1 - s\frac{k}{n}\right)d\left(s\underline{K}(s)\right)\right\} + o_{\mathbb{P}}(1)$$

$$= \frac{1}{(1-\rho)(1-\gamma)^2}\sqrt{k}A(n/k) + \mathbb{W}_{n,3} + o_{\mathbb{P}}(1). \tag{7}$$

Term $T_{n,4}$. A change of variables and an integration by parts yield

$$T_{n,4} = \sqrt{k}\left\{\frac{1}{1-\gamma} - \int_1^\infty x^{-2}\frac{\mathbb{U}(nx/k)}{\mathbb{U}(n/k)}dx\right\}$$

$$= -\sqrt{k}\int_1^\infty x^{-2}\left(\frac{\mathbb{U}(nx/k)}{\mathbb{U}(n/k)} - x^\gamma\right)dx.$$

Thus, Theorem 2.3.9 in de Haan and Ferreira (2006) entails that, for $\gamma \in (1/2, 1)$,

$$T_{n,4} = -\sqrt{k} A_0 \left(\frac{n}{k}\right) \int_1^\infty x^{\gamma-2} \frac{x^\rho - 1}{\rho} dx \, (1 + o(1))$$

$$= \sqrt{k} A \left(\frac{n}{k}\right) \frac{1}{(1 - \gamma)(\gamma + \rho - 1)} (1 + o(1)). \tag{8}$$

Combining (5)–(8), Lemma 1 follows. $\qquad\square$

Proof of Theorem 1. Combining Lemma 1 with statement (4.3) in Necir et al. (2010), we get

$$\frac{n(1 - \alpha)}{\sqrt{k}\mathbb{U}(n/k)} \left(\widetilde{\mathbb{C}}_{n,\alpha}[X] - \mathbb{C}_\alpha[X]\right) \stackrel{\mathscr{D}}{=} \sqrt{k} A \left(\frac{n}{k}\right) \mathscr{A}\mathscr{B}(\gamma, \rho) + \mathbb{W}_{n,1}$$

$$+ \mathbb{W}_{n,2} + \mathbb{W}_{n,3} + o_{\mathbb{P}}(1).$$

Theorem 1 is thus established. $\qquad\square$

Proof of Corollary 1. From Theorem 1, we only have to compute the asymptotic variance of the limiting process. The computations are tedious but quite direct. We only give below the main arguments, i.e.,

$$\mathbb{E}\mathbb{W}_{n,1}^2 = \frac{\int_0^{1-k/n}(1 - t)\left(\int_0^t s\,dQ(s)\right) dQ(t)}{k/n\,Q^2(1 - k/n)}$$

$$+ \frac{\int_0^{1-k/n} t\left(\int_t^{1-k/n}(1 - s)dQ(s)\right) dQ(t)}{k/n\,Q^2(1 - k/n)}$$

$$= \frac{\int_{k/n}^1 u\left(\int_u^1 dQ(1 - v)\right) dQ(1 - u)}{k/n\,Q^2(1 - k/n)}$$

$$- \frac{\int_{k/n}^1 u\left(\int_u^1 v\,dQ(1 - v)\right) dQ(1 - u)}{k/n\,Q^2(1 - k/n)}$$

$$+ \frac{\int_{k/n}^1 \left(\int_{k/n}^u v\,dQ(1 - v)\right) dQ(1 - u)}{k/n\,Q^2(1 - k/n)}$$

$$- \frac{\int_{k/n}^1 u\left(\int_{k/n}^u v\,dQ(1 - v)\right) dQ(1 - u)}{k/n\,Q^2(1 - k/n)}$$

$$=: Q_{1,n} + Q_{2,n} + Q_{3,n} + Q_{4,n}.$$

Recall now that $Q(1 - s) = s^{-\gamma}\ell(s)$ with ℓ a slowly varying function at 0. By integration by parts and using Lemma 6 in Deme et al. (2013b), it follows that

$$Q_{1,n} = \frac{1}{2}\left[1 + \frac{\int_{k/n}^1 Q^2(1-u)du}{k/n\,Q^2(1-k/n)}\right] \longrightarrow \frac{\gamma}{2\gamma - 1}.$$

Now remark that $d\left(\int_u^1 v\,dQ(1-v)\right) = -u\,dQ(1-u)$ which implies that

$$Q_{2,n} = -\frac{1}{2}\frac{k}{n}\left[\frac{\int_{k/n}^1 v\,dQ(1-v)}{k/n\,Q(1-k/n)}\right]^2 = o(1) \tag{9}$$

this last result is coming from the fact that, according to Proposition 1.3.6 in Bingham et al. (1987): for all $\varepsilon > 0, x^{-\varepsilon}\ell(x) \longrightarrow \infty$ as $x \to 0$. Thus, choosing $0 < \varepsilon < \gamma - \frac{1}{2}$ entails

$$0 \le s\left(\frac{\int_s^1 t\,d(Q(1-t))}{s\,Q(1-s)}\right)^2 = s\left(1 + \frac{\int_s^1 t^{-\gamma}\ell(t)dt}{s^{1-\gamma}\ell(s)}\right)^2$$

$$\le s\left(1 + Cs^{\gamma-1-\varepsilon}\right)^2 = O\left(s^{1+2[\gamma-1-\varepsilon]}\right) = o(1)$$

where C is a suitable constant. Consequently, $Q_{2,n} \longrightarrow 0$. The two others terms, $Q_{3,n}$ and $Q_{4,n}$, can be treated similarly, leading to

$$Q_{3,n} = Q_{1,n} \longrightarrow \frac{\gamma}{2\gamma - 1}$$
$$Q_{4,n} = Q_{2,n} \longrightarrow 0.$$

Finally,

$$\mathbb{E}W_{n,1}^2 \longrightarrow \frac{2\gamma}{2\gamma - 1},$$

and direct computations now lead to

$$\mathbb{E}W_{n,2}^2 \longrightarrow \frac{\gamma^2}{(1-\gamma)^2}$$

$$\mathbb{E}W_{n,3}^2 \longrightarrow \frac{\gamma^2}{(1-\gamma)^4} \quad \text{by Corollary 1 in Deme et al. (2013b)}$$

$$\mathbb{E}(W_{n,1}W_{n,2}) \longrightarrow \frac{\gamma}{1-\gamma} \quad \text{by (9)}$$

$$\mathbb{E}(W_{n,1}W_{n,3}) = 0$$

$$\mathbb{E}(W_{n,2}W_{n,3}) = 0.$$

Combining all these results, Corollary 1 follows. $\square$

Proof of Theorem 2. We use the following decomposition

$$\frac{n(1-\alpha)}{\sqrt{k}\,\mathbb{U}(n/k)}\left(\widetilde{\mathbb{C}}_{n,\alpha}^{LS,\widehat{\rho}}[X] - \mathbb{C}_\alpha[X]\right) = \sum_{i=1}^{7} S_{n,i}$$

where

$$S_{n,1} = \frac{n(1-\alpha)}{\sqrt{k}\,\mathbb{U}(n/k)}\left(\widetilde{\mathbb{C}}_{n,\alpha}^{(1)}[X] - \mathbb{C}_\alpha^{(1)}[X]\right)$$

$$S_{n,2} = \frac{1}{1-\widehat{\gamma}_{n,k}^{LS}(\widehat{\rho})}\left(1 - \frac{\widehat{A}_{n,k}^{LS}(\widehat{\rho})}{\widehat{\gamma}_{n,k}^{LS}(\widehat{\rho}) + \widehat{\rho} - 1}\right)\sqrt{k}\left[\frac{\mathbb{U}\left(Y_{n-k,n}\right)}{\mathbb{U}(n/k)} - \left(\frac{k}{n}Y_{n-k,n}\right)^\gamma\right]$$

$$S_{n,3} = \frac{1}{1-\widehat{\gamma}_{n,k}^{LS}(\widehat{\rho})}\left(1 - \frac{\widehat{A}_{n,k}^{LS}(\widehat{\rho})}{\widehat{\gamma}_{n,k}^{LS}(\widehat{\rho}) + \widehat{\rho} - 1}\right)\sqrt{k}\left[\left(\frac{k}{n}Y_{n-k,n}\right)^\gamma - 1\right]$$

$$S_{n,4} = \frac{1}{(1-\widehat{\gamma}_{n,k}^{LS}(\widehat{\rho}))(1-\gamma)}\sqrt{k}\left(\widehat{\gamma}_{n,k}^{LS}(\widehat{\rho}) - \gamma\right)$$

$$S_{n,5} = \sqrt{k}\,A(n/k)\left[\frac{1}{(1-\gamma)(\gamma+\rho-1)} - \frac{1}{\left(1-\widehat{\gamma}_{n,k}^{LS}(\widehat{\rho})\right)\left(\widehat{\gamma}_{n,k}^{LS}(\widehat{\rho}) + \widehat{\rho} - 1\right)}\right]$$

$$S_{n,6} = -\frac{1}{\left(1-\widehat{\gamma}_{n,k}^{LS}(\widehat{\rho})\right)\left(\widehat{\gamma}_{n,k}^{LS}(\widehat{\rho}) + \widehat{\rho} - 1\right)}\sqrt{k}\left(\widehat{A}_{n,k}^{LS}(\widehat{\rho}) - A(n/k)\right)$$

$$S_{n,7} = \frac{n}{\sqrt{k}\,\mathbb{U}(n/k)}\left[\frac{k/n}{1-\gamma}\left(1 - \frac{A(n/k)}{\gamma+\rho-1}\right)\mathbb{U}(n/k) - (1-\alpha)\mathbb{C}_\alpha^{(2)}[X]\right].$$

Now, we are going to study separately the terms $S_{n,1}, \ldots, S_{n,7}$.

Term $S_{n,1}$. Statement (4.3) in Necir et al. (2010) leads to

$$S_{n,1} = \mathbb{W}_{n,1} + o_{\mathbb{P}}(1). \tag{10}$$

Term $S_{n,2}$. Note that

$$S_{n,2} = \frac{1-\widehat{\gamma}_{n,k}^{H}}{1-\widehat{\gamma}_{n,k}^{LS}(\widehat{\rho})}\left(1 - \frac{\widehat{A}_{n,k}^{LS}(\widehat{\rho})}{\widehat{\gamma}_{n,k}^{LS}(\widehat{\rho}) + \widehat{\rho} - 1}\right)T_{n,1}$$

where $T_{n,1}$ is defined in the proof of Lemma 1. Thus, combining Lemma 5 in Deme et al. (2013b) with the consistency of $\widehat{\rho}$ and (5), we obtain that

$$S_{n,2} = o_{\mathbb{P}}(1). \tag{11}$$

Term $S_{n,3}$. Similarly, we observe that $S_{n,3} = T_{n,2}(1 + o_{\mathbb{P}}(1))$ where $T_{n,2}$ is defined in the proof of Lemma 1. Thus, according to (6), we have

$$S_{n,3} \overset{\mathscr{D}}{=} \mathbb{W}_{n,2} + o_{\mathbb{P}}(1). \tag{12}$$

Term $S_{n,4}$. Combining Lemma 5 in Deme et al. (2013b) with the consistency of $\widehat{\gamma}_{n,k}^{LS}(\widehat{\rho})$, we infer that

$$S_{n,4} \overset{\mathscr{D}}{=} \frac{\gamma + \rho - 1}{\rho \gamma} \mathbb{W}_{n,4} + o_{\mathbb{P}}(1). \tag{13}$$

Term $S_{n,5}$. Under the assumption that $\sqrt{k}A(n/k) = O(1)$ and by the consistency of $\widehat{\rho}$ and $\widehat{\gamma}_{n,k}^{LS}(\widehat{\rho})$ we have

$$S_{n,5} = o_{\mathbb{P}}(1). \tag{14}$$

Term $S_{n,6}$. Using Lemma 5 in Deme et al. (2013b), we get

$$
\begin{aligned}
S_{n,6} \overset{\mathscr{D}}{=} &-\frac{\gamma(1-\rho)}{(1-\gamma)(\gamma+\rho-1)} \sqrt{\frac{n}{k}} \int_0^1 s^{-1} \mathbb{B}_n\left(1 - \frac{sk}{n}\right) \\
&\times d(s(\underline{K}(s) - K_\rho(s))) + o_{\mathbb{P}}(1) \\
=&-\frac{(1-\rho)(1-\gamma)}{\gamma+\rho-1}\left(\mathbb{W}_{n,3} - \frac{\gamma+\rho-1}{\rho\gamma}\mathbb{W}_{n,4}\right) + o_{\mathbb{P}}(1) \\
=& \; \mathbb{W}_{n,5} + \frac{(1-\rho)(1-\gamma)}{\gamma\rho}\mathbb{W}_{n,4} + o_{\mathbb{P}}(1).
\end{aligned}
\tag{15}
$$

Term $S_{n,7}$. Remark that

$$S_{n,7} = -\frac{\sqrt{k}A(n/k)}{(1-\gamma)(\gamma+\rho-1)} + T_{n,4},$$

where $T_{n,4}$ is defined in the proof of Lemma 1. Thus using (8) and the assumption that $\sqrt{k}A(n/k) = O(1)$, we deduce that

$$S_{n,7} = o_{\mathbb{P}}(1). \tag{16}$$

Combining (10)–(16), Theorem 2 follows. $\qquad\square$

Proof of Corollary 2. From Theorem 2, we only have to compute the asymptotic variance of the limiting process. As in Corollary 1, the computations are quite direct and the desired asymptotic variance can be obtained by noticing that

$$\mathbb{E}W^2_{n,5} \longrightarrow \frac{\gamma^2(1-\rho)^2}{(1-\gamma)^2(\gamma+\rho-1)^2}$$

$$\mathbb{E}(\mathbb{W}_{n,1}\mathbb{W}_{n,4}) = 0$$

$$\mathbb{E}(\mathbb{W}_{n,1}\mathbb{W}_{n,5}) = 0$$

$$\mathbb{E}W^2_{n,4} = \frac{\gamma^4(1-\rho)^2}{(1-\gamma)^4(\gamma+\rho-1)^2}$$

$$\mathbb{E}(\mathbb{W}_{n,2}\mathbb{W}_{n,5}) = 0$$

$$\mathbb{E}(\mathbb{W}_{n,2}\mathbb{W}_{n,4}) = 0$$

$$\mathbb{E}(\mathbb{W}_{n,4}\mathbb{W}_{n,5}) = -\frac{\rho\gamma^3(1-\rho)}{(1-\gamma)^3(\gamma+\rho-1)^2}.$$

$\square$

Acknowledgments The authors thank the referee for the comments concerning the previous version. The first author acknowledges support from AIRES-Sud (AIRES-Sud is a program from the French Ministry of Foreign and European Affairs, implemented by the "Institut de Recherche pour le Développement", IRD-DSF) and from the "Ministère de la Recherche Scientifique" of Sénégal.

References

Artzner, P., Delbaen, F., Eber, J.-M., & Heath, D. (1999). Coherent measures of risk. *Mathematical Finance, 9*, 203–228.

Beirlant, J., Dierckx, G., Goegebeur, M., & Matthys, G. (1999). Tail index estimation and an exponential regression model. *Extremes, 2*, 177–200.

Beirlant, J., Dierckx, G., Guillou, A., & Starica, C. (2002). On exponential representations of log-spacings of extreme order statistics. *Extremes, 5*, 157–180.

Bingham, N. H., Goldie, C. M., & Teugels, J. L. (1987). Regular variation, Cambridge.

Brazauskas, V., Jones, B., Puri, M., & Zitikis, R. (2008). Estimating conditional tail expectation with actuarial applications in view. *Journal of Statistical Planning and Inference, 138*, 3590–3604.

Csörgő, M., Csörgő, S., Horváth, L., & Mason, D. M. (1986). Weighted empirical and quantile processes. *Annals of Probability, 14*, 31–85.

Csörgő, S., Deheuvels, P., & Mason, D. M. (1985). Kernel estimates of the tail index of a distribution. *Annals of Statistics, 13*, 1050–1077.

de Haan, L., & Ferreira, A. (2006). *Extreme value theory: An introduction.* Springer.

Deme, E., Gardes, L., & Girard, S. (2013a). On the estimation of the second order parameter for heavy-tailed distributions. *REVSTAT—Statistical Journal, 11*(3), 277–299.

Deme, E., Girard, S., & Guillou, A. (2013b). Reduced-bias estimator of the proportional hazard premium for heavy-tailed distributions. *Insurance Mathematic & Economics, 52*, 550–559.

Feuerverger, A., & Hall, P. (1999). Estimating a tail exponent by modelling departure from a Pareto distribution. *Annals of Statistics, 27*, 760–781.

Fraga Alves, M. I., Gomes, M. I., & de Haan, L. (2003). A new class of semi-parametric estimators of the second order parameter. *Portugaliae Mathematica, 60*(2), 193–213.

Geluk, J. L., & de Haan, L. (1987). Regular variation, extensions and Tauberian theorems, CWI tract 40, Center for Mathematics and Computer Science, P.O. Box 4079, 1009 AB Amsterdam, The Netherlands.

Goovaerts, M. J., de Vylder, F., & Haezendonck, J. (1984). *Insurance premiums, theory and applications*. Amsterdam: North Holland.

Hill, B. M. (1975). A simple approach to inference about the tail of a distribution. *Annals of Statistics, 3*, 1136–1174.

Landsman, Z., & Valdez, E. (2003). Tail conditional expectations for elliptical distributions. *North American Actuarial Journal, 7*, 55–71.

Matthys, G., Delafosse, E., Guillou, A., & Beirlant, J. (2004). Estimating catastrophic quantile levels for heavy-tailed distributions. *Insurance Mathematic & Economics, 34*, 517–537.

McNeil, A.J., Frey, R., & Embrechts, P. (2005) Quantitative risk management: concepts, techniques, and tools, Princeton University Press.

Necir, A., Rassoul, A., & Zitikis, R. (2010) Estimating the conditional tail expectation in the case of heavy-tailed losses, Journal of Probability and Statistics, ID 596839, 17.

Pan, X., Leng, X., & Hu, T. (2013). Second-order version of Karamata's theorem with applications. *Statistics and Probability Letters, 83*, 1397–1403.

Weissman, I. (1978). Estimation of parameters and larges quantiles based on the k largest observations. *Journal of American Statistical Association, 73*, 812–815.

Zhu, L., & Li, H. (2012). Asymptotic analysis of multivariate tail conditional expectations. *North American Actuarial Journal, 16*, 350–363.

On Sequential Empirical Distribution Functions for Random Variables with Mean Zero

Erich Haeusler and Stefan Horni

Abstract The classical sequential empirical distribution function incorporates all subsamples of a sample of independent and identically distributed random variables and is therefore well suited to construct tests for detecting a distributional change occurring somewhere in the sample. If the independent and identically distributed variables are replaced by the residuals of appropriate time series models tests for a distributional change in the unobservable errors (or innovations) of these models are obtained; see Bai (Annals of Statistics, 22:2051–2061, 1994) for the discussion of ARMA models. These errors are often assumed to have mean zero, an information which is not taken into account by the classical sequential empirical distribution function. Based upon ideas from empirical likelihood, see Owen (London/Boca Raton: Chapman & Hall/CRC, 2001), we consider a modified sequential empirical distribution function for random variables with mean zero which does exploit this information.

1 Introduction

The classical sequential empirical distribution function incorporates all subsamples $\varepsilon_1, \ldots, \varepsilon_k$ for $k = 1, \ldots, n$ of a sample $\varepsilon_1, \ldots, \varepsilon_n$ of size n of independent and identically distributed (iid) random variables and is therefore well suited to construct tests for detecting a distributional change occurring somewhere in the whole sample $\varepsilon_1, \ldots, \varepsilon_n$. If the iid variables are replaced by the residuals of appropriate time series models tests for a distributional change in the unobservable errors (or innovations) of these models are obtained; see Bai (1994) for the discussion of ARMA models. These errors are often assumed to have mean zero, an information which is not taken

---

Dedicated to Paul Deheuvels on the occasion of his 65th anniversary.

E. Haeusler (✉) · S. Horni
Mathematical Institute, University of Giessen, Arndtstrasse 2, 35392 Giessen, Germany
e-mail: erich.haeusler@math.uni-giessen.de

S. Horni
e-mail: stefan.horni@math.uni-giessen.de

© Springer International Publishing Switzerland 2015

M. Hallin et al. (eds.), *Mathematical Statistics and Limit Theorems*,
DOI 10.1007/978-3-319-12442-1_8

into account by the classical sequential empirical distribution function. Based upon ideas from empirical likelihood, see Owen (2001), we consider a modified sequential empirical distribution function for random variables with mean zero which does exploit this information. It will be shown that for autoregressive processes of order one this modified sequential empirical distribution function leads to changepoint tests of higher power than the classical sequential empirical distribution function. The approach applies to more general time series models like ARMA models and ARCH(1) models with similar results, but we will restrict ourselves here to the AR(1) case to avoid the technicalities involved in the discussion of more complex models.

2 Independent and Identically Distributed Random Variables

For sample size $n \in \mathbb{N}$ let $\varepsilon_1, \ldots, \varepsilon_n$ be iid random variables with common continuous distribution function F (which is independent of n). The *sequential empirical distribution function* pertaining to the sample $\varepsilon_1, \ldots, \varepsilon_n$ is defined by

$$F_n^{seq}(s, x) = \frac{1}{[ns]} \sum_{i=1}^{[ns]} 1_{\{\varepsilon_i \leq x\}}, \quad s \in [1/n, 1], \ x \in [-\infty, \infty],$$

with $F_n^{seq}(s, x) = 0$ for $s \in [0, 1/n)$, where $[x]$ is the integer part of $x \in \mathbb{R}$. By Bickel and Wichura (1971),

$$\left(\sqrt{n}s \left(F_n^{seq}(s, x) - F(x)\right)\right)_{(s,x) \in [0,1] \times [-\infty, \infty]} \xrightarrow{\mathscr{L}} K \quad \text{as } n \to \infty \qquad (1)$$

for a centered Gaussian process $K = (K(s, x))_{(s,x) \in [0,1] \times [-\infty, \infty]}$ with continuous paths and the covariance function $\text{cov}(K(s, x), K(\tilde{s}, \tilde{x})) = (s \wedge \tilde{s})(F(x \wedge \tilde{x}) - F(x) F(\tilde{x}))$. Here and elsewhere throughout the paper, convergence in distribution may be considered as weak convergence of stochastic processes with paths in the Skorohod space $D([0, 1] \times [-\infty, \infty])$ in the sense of Neuhaus (1971) (identify the time intervals $[-\infty, \infty]$ and $[0, 1]$ by a smooth, strictly increasing one-to-one transformation) or as weak convergence of stochastic processes in the space $B([0, 1] \times [-\infty, \infty])$ of bounded measurable functions on $[0, 1] \times [-\infty, \infty]$ in the sense of Hoffmann–Jørgensen; see, e.g., van der Vaart and Wellner (1996).

Sequential empirical distribution functions are an appropriate tool to construct tests in the following changepoint problem: Assume that there exists some $\tau \in (0, 1)$ such that $\varepsilon_1, \ldots, \varepsilon_{[n\tau]}$ are iid F_1 for some distribution function F_1 and $\varepsilon_{[n\tau]+1}, \ldots, \varepsilon_n$ are iid F_2 for some distribution function F_2. Furthermore, assume that F_1, F_2, and τ are all unknown. Then the case $F_1 = F_2$ corresponds to the case that $\varepsilon_1, \ldots, \varepsilon_n$ are iid for some unknown distribution function $F (= F_1 = F_2)$, whereas the case

$F_1 \neq F_2$ corresponds to a sudden change occuring in the distribution of the ε_i's at the (unknown) changepoint $[n\tau]$ (or $[n\tau]+1$). For testing the hypothesis $H_0 : F_1 = F_2$ against the alternative $H_1 : F_1 \neq F_2$ we can use F_n^{seq} and its counterpart

$$G_n^{seq}(s, x) = \frac{1}{n - [ns]} \sum_{i=[ns]+1}^{n} 1_{\{\varepsilon_i \leq x\}}, \quad s \in [0, 1), \ x \in [-\infty, \infty],$$

with $G_n^{seq}(1, x) = 0$, which is the sequential empirical distribution function of $\varepsilon_1, \ldots, \varepsilon_n$ with reversed order of time. The Kolmogorov–Smirnov-type statistic $D_n^\infty = \sup_{(s,x) \in [0,1] \times [-\infty,\infty]} |D_n(s, x)|$ based on the stochastic process

$$D_n(s, x) = \frac{[ns](n - [ns])}{n^2} \left(F_n^{seq}(s, x) - G_n^{seq}(s, x) \right), \quad s \in [0, 1], \ x \in [-\infty, \infty],$$

is clearly suitable to distinguish between H_0 and H_1: Under $H_0 : F_1 = F_2 (= F)$ we have $F_n^{seq}(s, \cdot) \approx F \approx G_n^{seq}(s, \cdot)$ for all s and all large n by the Glivenko–Cantelli theorem so that D_n^∞ will be small with high probability. Therefore, it is reasonable to reject H_0 for large values of D_n^∞. Under $H_0 : F_1 = F_2 (= F)$ the functional central limit theorem corresponding to (1) is

$$\sqrt{n} D_n \xrightarrow{\mathscr{L}} Y \quad \text{as } n \to \infty \tag{2}$$

for a centered Gaussian process $Y = (Y(s, x))_{(s,x) \in [0,1] \times [-\infty,\infty]}$ with continuous paths and covariance function $\mathrm{cov}(Y(s, x), Y(\tilde{s}, \tilde{x})) = (s \wedge \tilde{s} - s\tilde{s})(F(x \wedge \tilde{x}) - F(x)F(\tilde{x}))$. The continuous mapping theorem implies

$$\sqrt{n} D_n^\infty \xrightarrow{\mathscr{L}} \sup_{(s,x) \in [0,1] \times [-\infty,\infty]} |Y(s, x)| \quad \text{as } n \to \infty.$$

Note that the process Y may be written as $Y(s, x) = B(s, F(x))$ for a centered Gaussian process $B = (B(s, t))_{(s,t) \in [0,1]^2}$ with continuous paths and covariance function $\mathrm{cov}\left(B(s, t), B(\tilde{s}, \tilde{t})\right) = (s \wedge \tilde{s} - s\tilde{s})(t \wedge \tilde{t} - t\tilde{t})$. Consequently, for continuous F we get

$$\sup_{(s,x) \in [0,1] \times [-\infty,\infty]} |Y(s, x)| \stackrel{\mathscr{L}}{=} \sup_{(s,x) \in [0,1] \times [-\infty,\infty]} |B(s, F(x))| = \sup_{(s,t) \in [0,1]^2} |B(s, t)|.$$

This makes an asymptotic test for H_0 against H_1 based on D_n^∞ asymptotically distribution free under H_0 for continuous F. Obviously, the asymptotic null distributions of other test statistics for testing H_0 against H_1 can be derived from (2) as well, but we will restrict our considerations here upon D_n^∞.

3 Autoregressive Processes of Order One

Now we consider a stable autoregressive process $(X_i)_{i \geq 0}$ of order one, i.e., the X_i satisfy the recursive equation

$$X_i = \rho X_{i-1} + \varepsilon_i \quad \text{for all } i \in \mathbb{N},$$

some real autoregression parameter ρ with $|\rho| < 1$ and a sequence ε_i, $i \in \mathbb{N}$, of iid square integrable errors with $E(\varepsilon_i) = 0$. The distribution function F of the ε_i's is always assumed to be continuous. The starting value X_0 is also square integrable with $E(X_0) = 0$ and is independent of the sequence $(\varepsilon_i)_{i \in \mathbb{N}}$. If interpreted as a statistical model, the random variables $X_0, X_1, \ldots, X_n$ are observable at sample size $n \in \mathbb{N}$, whereas the autoregression parameter ρ is unknown and the errors $\varepsilon_1, \ldots, \varepsilon_n$ are unobservable. The statement that $\varepsilon_1, \ldots, \varepsilon_{[n\tau]}$ are iid F_1 and $\varepsilon_{[n\tau]+1}, \ldots, \varepsilon_n$ are iid F_2 for some $\tau \in (0, 1)$ and possibly different distribution functions F_1 and F_2 now reflects a possible change in the error distributions of the sample $X_0, X_1, \ldots, X_n$. Because $\varepsilon_1, \ldots, \varepsilon_n$ are not observable, tests for $H_0 : F_1 = F_2$ against $H_1 : F_1 \neq F_2$ cannot be based on D_n and D_n^∞ from Sect. 2. Instead, for each sample size $n \in \mathbb{N}$ we will use the residuals $\widehat{\varepsilon}_{ni} = X_i - \widehat{\rho}_n X_{i-1}$, $i = 1, \ldots, n$, for some estimator $\widehat{\rho}_n$ of ρ based upon the sample $X_0, X_1, \ldots, X_n$. The only property of $\widehat{\rho}_n$ that will be required and therefore always assumed to hold in the sequel is $\sqrt{n}$-consistency under $H_0 : F_1 = F_2$, i.e., boundedness in probability of the sequence $\sqrt{n}\,(\widehat{\rho}_n - \rho)$, $n \in \mathbb{N}$, if ρ is the true value of the autoregression parameter. This is satisfied, for example, for the usual least squares estimator, which was therefore used in our simulation study to obtain the results that will be presented in Sect. 5. The versions of F_n^{seq}, G_n^{seq}, D_n, and D_n^∞ as functions of the residuals are now defined by

$$\widehat{F}_n^{seq}(s, x) = \frac{1}{[ns]} \sum_{i=1}^{[ns]} 1_{\{\widehat{\varepsilon}_{ni} \leq x\}}, \quad s \in [1/n, 1], \ x \in [-\infty, \infty],$$

with $\widehat{F}_n^{seq}(s, x) = 0$ for $s \in [0, 1/n)$,

$$\widehat{G}_n^{seq}(s, x) = \frac{1}{n - [ns]} \sum_{i=[ns]+1}^{n} 1_{\{\widehat{\varepsilon}_{ni} \leq x\}}, \quad s \in [0, 1), \ x \in [-\infty, \infty],$$

with $\widehat{G}_n^{seq}(1, x) = 0$,

$$\widehat{D}_n(s, x) = \frac{[ns]\,(n - [ns])}{n^2} \left(\widehat{F}_n^{seq}(s, x) - \widehat{G}_n^{seq}(s, x) \right), \quad s \in [0, 1], \ x \in [-\infty, \infty],$$

and $\widehat{D}_n^\infty = \sup_{(s,x) \in [0, 1] \times [-\infty, \infty]} \left| \widehat{D}_n(s, x) \right|$. For a stationary AR(1)-process $(X_i)_{i \geq 0}$ it follows from the results for ARMA models in Sect. 2 of Bai (1994) that under the hypothesis $H_0 : F_1 = F_2(= F)$

$$\sup_{(s,x)\in[0,1]\times[-\infty,\infty]} \left|\widehat{D}_n\,(s,x) - D_n\,(s,x)\right| = o_P\left(1/\sqrt{n}\right) \quad \text{as } n \to \infty, \qquad (3)$$

provided that F has a uniformly continuous strictly positive density. Consequently, under these assumptions we have

$$\sqrt{n}\widehat{D}_n^{\infty} \overset{\mathscr{L}}{\to} \sup_{(s,t)\in[0,1]^2} |B\,(s,t)| \quad \text{as } n \to \infty,$$

so that the same asymptotic test for H_0 against H_1 as in case of independent random variables works. Note that this test does not make use of the fact that the errors ε_i have mean zero. In the next section, we will study a modified sequential empirical distribution function which takes this information into account.

4 Random Variables with Mean Zero

Consider for the time being a sequence ε_i, $i \in \mathbb{N}$, of iid random variables with common continuous distribution function F. It is well known that the classical empirical distribution function

$$F_n\,(x) = \frac{1}{n}\sum_{i=1}^{n} 1_{\{\varepsilon_i \leq x\}}, \quad x \in [-\infty, \infty],$$

is the nonparametric maximum likelihood estimator for F based on the sample $\varepsilon_1, \ldots, \varepsilon_n$ of sample size $n \in \mathbb{N}$. The mean $\overline{\varepsilon}_n = \frac{1}{n}\sum_{i=1}^{n}\varepsilon_i$ of F_n is always a.s. different from zero, even if F has mean zero.

If F has mean zero, it can be estimated from a sample $\varepsilon_1, \ldots, \varepsilon_n$ by its nonparametric maximum likelihood estimator $F_{n,0}$ in the set of all distribution functions with mean zero. The estimator $F_{n,0}$ is the maximizer of the empirical likelihood

$$L\left(\widetilde{F}\right) = \prod_{i=1}^{n} \widetilde{F}\,(\varepsilon_i) - \widetilde{F}\,(\varepsilon_i - 0)$$

of all distribution functions $\widetilde{F}$ under the contraint $\int_{-\infty}^{\infty} x\,d\widetilde{F}\,(dx) = 0$. Here $\widetilde{F}\,(x - 0)$ denotes the left-hand limit of $\widetilde{F}$ at $x \in \mathbb{R}$. The maximizer can be obtained by the method of Lagrange multipliers and is given as follows (for computational details of the maximization procedure see (Owen 1990, 2001; Qin and Lawless 1994): If

$$\min_{1\leq i\leq n}\varepsilon_i < 0 < \max_{1\leq i\leq n}\varepsilon_i, \qquad (4)$$

then there exists a unique $t_n = t_n (\varepsilon_1, \ldots, \varepsilon_n)$ with

$$\left(\frac{1}{n} - 1\right) \frac{1}{\max_{1 \leq i \leq n} \varepsilon_i} < t_n < \left(\frac{1}{n} - 1\right) \frac{1}{\min_{1 \leq i \leq n} \varepsilon_i}$$

(which is the Lagrange multiplier in the maximization problem) and

$$\sum_{i=1}^{n} \frac{\varepsilon_i}{1 + t_n \varepsilon_i} = 0, \tag{5}$$

and $F_{n,0}$ is defined by

$$F_{n,0}(x) = \sum_{i=1}^{n} \frac{1}{n} \frac{1}{1 + t_n \varepsilon_i} 1_{\{\varepsilon_i \leq x\}}, \quad x \in [-\infty, \infty].$$

Note that (5) is tantamount to $\int_{-\infty}^{\infty} x \, d F_{n,0}(dx) = 0$ so that $F_{n,0}$ is a distribution function with mean zero. If condition (4) is violated, then all the data $\varepsilon_1, \ldots, \varepsilon_n$ are either positive or negative, and it is obviously impossible to construct a reasonable estimator of F with mean zero. However, from $E(\varepsilon_1) = 0$ and $E(\varepsilon_1^2) > 0$ it easily follows that

$$P\left(\min_{1 \leq i \leq n} \varepsilon_i < 0 < \max_{1 \leq i \leq n} \varepsilon_i\right) \to 1 \quad \text{as } n \to \infty, \tag{6}$$

which means that with probability converging to one the estimator $F_{n,0}$ is well defined. For square integrable ε_i a functional central limit theorem for $F_{n,0}$ is proven in Zhang (1997) which in conjunction with Example 2 in Sect. 5.3 of Bickel et al. (1993) shows that $F_{n,0}$ is asymptotically efficient for estimating distribution functions with mean zero.

In view of the definition of $F_{n,0}$ a sequential version $F_{n,0}^{seq}$ of $F_{n,0}$ can be defined as follows: For every $n \in \mathbb{N}$ with $n \geq 2$ and $k = 2, \ldots, n$ on the event $\{\min_{1 \leq i \leq k} \varepsilon_i < 0 < \max_{1 \leq i \leq k} \varepsilon_i\}$ let t_{nk} be the unique number with

$$\left(\frac{1}{k} - 1\right) \frac{1}{\max_{1 \leq i \leq k} \varepsilon_i} < t_{nk} < \left(\frac{1}{k} - 1\right) \frac{1}{\min_{1 \leq i \leq k} \varepsilon_i} \tag{7}$$

and

$$\sum_{i=1}^{k} \frac{\varepsilon_i}{1 + t_{nk} \varepsilon_i} = 0. \tag{8}$$

Then for $s \in [0, 1]$, $x \in [-\infty, \infty]$ we set

$$
F_{n,0}^{seq}(s, x) =
\begin{cases}
\displaystyle\sum_{i=1}^{[ns]} \frac{1}{[ns]} \frac{1}{1 + t_{n[ns]}\varepsilon_i} 1_{\{\varepsilon_i \le x\}}, & \text{if } \min_{1 \le i \le [ns]} \varepsilon_i < 0 < \max_{1 \le i \le [ns]} \varepsilon_i \\[2em]
F_n^{seq}(s, x) & \text{, otherwise}.
\end{cases}
$$

For $a \in (0, 1)$ and $n \in \mathbb{N}$ set $A_{n,a} = \left\{ \min_{1 \le i \le [n^a]} \varepsilon_i < 0 < \max_{1 \le i \le [n^a]} \varepsilon_i \right\}$ and note that on $A_{n,a}$, $F_{n,0}^{seq}(s, x)$ is defined by the first case in the definition so that $F_{n,0}^{seq}(s, \cdot)$ is a distribution function with mean zero for all $s \in \left[n^{a-1}, 1 \right]$. Moreover, from (6), for any $a \in (0, 1)$, we have $P\left(A_{n,a} \right) \to 1$ as $n \to \infty$. Thus with probability tending to one as $n \to \infty$ the empirical distribution functions appearing in the sequential empirical distribution function $F_{n,0}^{seq}$ have mean zero.

For our first result about $F_{n,0}^{seq}$ we introduce the function

$$
U(x) = E\left(\varepsilon_1 1_{\{\varepsilon_1 \le x\}} \right), \quad x \in [-\infty, \infty], \tag{9}
$$

and the partial sum processes

$$
S_n(s) = \frac{1}{[ns]} \sum_{i=1}^{[ns]} \varepsilon_i, \quad s \in [1/n, 1],
$$

with $S_n(s) = 0$ for $s \in [0, 1/n)$ and $n \in \mathbb{N}$. We also set $\sigma^2 = E\left(\varepsilon_1^2 \right)$ and require a slightly stronger moment condition than square integrability of the ε_i, namely

$$
E\left(\varepsilon_1^2 \log\log\left(3 + |\varepsilon_1| \right) \right) < \infty. \tag{10}
$$

Proposition 1 *Under (10) we have, as $n \to \infty$,*

$$
\sup_{(s,x)\in[0,1]\times[-\infty,\infty]} [ns] \left| F_{n,0}^{seq}(s, x) - \left(F_n^{seq}(s, x) - \frac{1}{\sigma^2} S_n(s) U(x) \right) \right| = o_P\left(\sqrt{n} \right).
$$

From Proposition 1 we can derive the following functional central limit theorem for $F_{n,0}^{seq}$ which corresponds to (1):

Theorem 1 *Under (10) we have*

$$
\left(\sqrt{n} s \left(F_{n,0}^{seq}(s, x) - F(x) \right) \right)_{(s,x)\in[0,1]\times[-\infty,\infty]} \xrightarrow{\mathscr{L}} Z \quad \text{as } n \to \infty
$$

for a centered Gaussian process $Z = (Z(s, x))_{(s,x) \in [0,1] \times [-\infty, \infty]}$ with continuous paths and covariance function

$$\mathrm{cov}\left(Z(s, x), Z(\tilde{s}, \tilde{x})\right) = (s \wedge \tilde{s})\left(F(x \wedge \tilde{x}) - F(x)F(\tilde{x}) - \frac{1}{\sigma^2}U(x)U(\tilde{x})\right).$$

To define the version of D_n for testing $H_0 : F_1 = F_2$ against $H_1 : F_1 \neq F_2$ if all distribution functions have mean zero we first have to define the corresponding version of G_n^{seq}: For every $n \in \mathbb{N}$ with $n \geq 2$ and $k = 0, \ldots, n - 2$ on the event $\{\min_{k+1 \leq i \leq n} \varepsilon_i < 0 < \max_{k+1 \leq i \leq n} \varepsilon_i\}$ let u_{nk} denote the unique number with

$$\left(\frac{1}{n-k} - 1\right)\frac{1}{\max_{k+1 \leq i \leq n} \varepsilon_i} < u_{nk} < \left(\frac{1}{n-k} - 1\right)\frac{1}{\min_{k+1 \leq i \leq n} \varepsilon_i}$$

and

$$\sum_{i=k+1}^{n} \frac{\varepsilon_i}{1 + u_{nk}\varepsilon_i} = 0.$$

Then for $s \in [0, 1]$, $x \in [-\infty, \infty]$ we set

$$G_{n,0}^{seq}(s, x) = \begin{cases} \displaystyle\sum_{i=[ns]+1}^{n} \frac{1}{n - [ns]}\frac{1}{1 + u_{n[ns]}\varepsilon_i} 1_{\{\varepsilon_i \leq x\}} \,, & \text{if } \min_{[ns]+1 \leq i \leq n} \varepsilon_i < 0 < \max_{[ns]+1 \leq i \leq n} \varepsilon_i \\[4mm] G_n^{seq}(s, x) & , \text{ otherwise}. \end{cases}$$

For $a \in (0, 1)$ and $n \in \mathbb{N}$ set $B_{n,a} = \{\min_{n-[n^a] \leq i \leq n} \varepsilon_i < 0 < \max_{n-[n^a] \leq i \leq n} \varepsilon_i\}$ and note that on $B_{n,a}$, $G_{n,0}^{seq}(s, x)$ is defined by the first case in the definition so that $G_{n,0}^{seq}(s, \cdot)$ is a distribution function with mean zero for all $s \in \left[0, 1 - n^{a-1}\right]$. Moreover, for any $a \in (0, 1)$ we have $P\left(B_{n,a}\right) \to 1$ as $n \to \infty$. Thus, as above, with probability tending to one as $n \to \infty$ the empirical distribution functions appearing in $G_{n,0}^{seq}$ have mean zero. Now we can define the stochastic processes

$$D_{n,0}(s, x) = \frac{[ns](n - [ns])}{n^2}\left(F_{n,0}^{seq}(s, x) - G_{n,0}^{seq}(s, x)\right), \quad s \in [0, 1], \, x \in [-\infty, \infty].$$

We also need the partial sum processes $T_n(s) = \frac{1}{n-[ns]}\sum_{i=[ns]+1}^{n} \varepsilon_i$, $s \in [0, 1)$, with $T_n(1) = 0$ and the stochastic processes

$$W_n(s, x) = D_n(s, x) - \frac{1}{\sigma^2}\frac{[ns](n - [ns])}{n^2}(S_n(s) - T_n(s))U(x)$$

for $s \in [0, 1]$, $x \in [-\infty, \infty]$ and $n \in \mathbb{N}$. Our main result about $D_{n,0}$ is

Proposition 2 *Under* (10) *and* $H_0 : F_1 = F_2$ *we have*

$$\sup_{(s,x)\in[0,1]\times[-\infty,\infty]} \left| D_{n,0}(s,x) - W_n(s,x) \right| = o_P\left(1/\sqrt{n}\right) \quad as\ n \to \infty.$$

Proposition 2 implies

Theorem 2 *Under* (10) *and* $H_0 : F_1 = F_2$ *we have* $\sqrt{n}D_{n,0} \overset{\mathscr{L}}{\to} W$ *as* $n \to \infty$, *where* $W = (W(s,x))_{(s,x)\in[0,1]\times[-\infty,\infty]}$ *is a centered Gaussian process with continuous paths and covariance function*

$$\mathrm{cov}\left(W(s,x), W(\tilde{s}, \tilde{x})\right) = (s \wedge \tilde{s} - s\tilde{s})\left(F(x \wedge \tilde{x}) - F(x)F(\tilde{x}) - \frac{1}{\sigma^2}U(x)U(\tilde{x})\right).$$

Now we continue our study of the AR(1)-process from Sect. 3. Replacing the ε_i in the definitions of $F_{n,0}^{seq}$ and $G_{n,0}^{seq}$ everywhere by the residuals $\widehat{\varepsilon}_{ni}$, we obtain the residual sequential empirical distribution functions $\widehat{F}_{n,0}^{seq}$ and $\widehat{G}_{n,0}^{seq}$. For the events $\widehat{A}_{n,a} = \left\{\min_{1\leq i\leq[n^a]}\widehat{\varepsilon}_{ni} < 0 < \max_{1\leq i\leq[n^a]}\widehat{\varepsilon}_{ni}\right\}$ and $\widehat{B}_{n,a} = \left\{\min_{n-[n^a]\leq i\leq n}\widehat{\varepsilon}_{ni} < 0 < \max_{n-[n^a]\leq i\leq n}\widehat{\varepsilon}_{ni}\right\}$, defined for $n \in \mathbb{N}$ and $a \in (0,1)$, it can be shown that $P\left(\widehat{A}_{n,a}\right) \to 1$ and $P\left(\widehat{B}_{n,a}\right) \to 1$ as $n \to \infty$, for all $a \in (0,1)$. Consequently, as in the case of the ε_i, with probability tending to one as $n \to \infty$ the empirical distribution functions appearing in $\widehat{F}_{n,0}^{seq}$ and $\widehat{G}_{n,0}^{seq}$ have mean zero. Replacing $F_{n,0}^{seq}$ and $G_{n,0}^{seq}$ by $\widehat{F}_{n,0}^{seq}$ and $\widehat{G}_{n,0}^{seq}$ in the definition of the process $D_{n,0}$, we obtain its version $\widehat{D}_{n,0}$ for the residuals. Our main result in this paper is

Theorem 3 *Let the* AR(1)-*process* $(X_i)_{i\geq 0}$ *be stationary, let* (10) *be satisfied and let the distribution function* F *have a uniformly continuous strictly positive density. Then under* $H_0 : F_1 = F_2(= F)$, *where* W *is the same process as in Theorem 2, we have* $\sqrt{n}\widehat{D}_{n,0} \overset{\mathscr{L}}{\to} W$ *as* $n \to \infty$.

5 Bootstrap and Simulations

Theorem 3 and the continuous mapping theorem imply, as $n \to \infty$,

$$\sqrt{n}\widehat{D}_{n,0}^{\infty} = \sup_{(s,x)\in[0,1]\times[-\infty,\infty]} \sqrt{n}\left|\widehat{D}_{n,0}(s,x)\right| \overset{\mathscr{L}}{\to} \sup_{(s,x)\in[0,1]\times[-\infty,\infty]} |W(s,x)| \tag{11}$$

under $H_0 : F_1 = F_2(= F)$. Unlike in the case of the process Y of (2), the distribution of the limiting random variable in (11) heavily depends on F, as is apparent from the covariance function of W given in Theorem 2. Therefore, it is not possible to construct tests from (11) that are asymptotically distribution free under the null hypotheses $H_0 : F_1 = F_2$. A remedy is the bootstrap. For this, given a sample $X_0, X_1, \ldots, X_n$

from the AR(1)-process, let $\varepsilon_{n1}^*, \ldots, \varepsilon_{nn}^*$ be bootstrap random variables which are iid under the conditional probability $P_n^* = P\left(\cdot|\widehat{\varepsilon}_{n1}, \ldots, \widehat{\varepsilon}_{nn}\right)$ with common distribution function $\widehat{F}_{n,0}^{seq}(1, \cdot)$. Replacing the ε_i in $D_{n,0}$ by the bootstrap variables ε_{ni}^*, we obtain a bootstrap version $\widehat{D}_{n,0}^*$ of $D_{n,0}$ which is consistent according to the following theorem.

Theorem 4 *Let the* AR(1)-*process* $(X_i)_{i \geq 0}$ *be stationary, let* F *have a uniformly continuous strictly positive density, and assume* $E\left(\varepsilon_1^2 \log\left(2 + |\varepsilon_1|\right)^2\right) < \infty$. *Then under* $H_0 : F_1 = F_2 (= F)$, *where* W *is the same process as in Theorem 2,*

$$\sqrt{n}\widehat{D}_{n,0}^* \xrightarrow{\mathcal{L}} W \quad under\ P_n^*\ in\ probability\ as\ n \to \infty.$$

Thus the bootstrap is consistent and yields bootstrap critical values for tests of H_0 against H_1 based on $\widehat{D}_{n,0}^\infty$ in the usual way: For $\widehat{D}_{n,0}^{*\infty} = \sup_{(s,x) \in [0,1] \times [-\infty,\infty]} \left|\widehat{D}_{n,0}^*(s,x)\right|$ and

$$k_{n,\alpha}^* = \inf\left\{x \in [-\infty, \infty] : P_n^*\left(\sqrt{n}\widehat{D}_{n,0}^{*\infty} \leq x\right) \geq 1 - \alpha\right\},$$

defined for $n \in \mathbb{N}$ and $\alpha \in (0, 1)$, we have $P\left(\sqrt{n}\widehat{D}_{n,0}^\infty \geq k_{n,\alpha}^*\right) \to \alpha$ as $n \to \infty$ so that the test which rejects $H_0 : F_1 = F_2$ if and only if $\sqrt{n}\widehat{D}_{n,0}^\infty \geq k_{n,\alpha}^*$ holds is a test of asymptotic level $\alpha \in (0, 1)$.

In Table 1 we report some simulated rejection probabilities under H_0 of the bootstrap test using $\widehat{D}_{n,0}^\infty$ for some finite sample sizes n, levels α and autoregression parameters ρ if both F_1 and F_2 are the standard normal distribution function. The nominal asymptotic level is kept quite well already for small sample sizes in this case. In Tables 2, 3 and 4 we compare the power of this bootstrap test and the power of the test based on $\widehat{D}_n^\infty$ if the critical values from Table 1 in Picard (1985) are used. In Table 4, $t_4/\sqrt{2}$ is the t-distribution with 4 degrees of freedom, rescaled to variance one. The test based on $\widehat{D}_{n,0}^\infty$ has always larger and sometimes considerably larger power than the test based on $\widehat{D}_n^\infty$ which does not exploit the fact that the errors in the AR(1)-model have mean zero. Consequently, using the sequential empirical distribution functions $\widehat{F}_{n,0}^{seq}$ and $\widehat{G}_{n,0}^{seq}$, which do exploit mean zero of the errors, instead of $\widehat{F}_n^{seq}$ and $\widehat{G}_n^{seq}$ seems to be preferable in the testing problem under consideration.

Table 1 Simulated levels for $\widehat{D}_{n,0}^\infty$ under $H_0 : F_1 = F_2 = N(0, 1)$

$\rho = 0.3$				$\rho = 0.5$			
$n \backslash \alpha$	0.1	0.05	0.01	$n \backslash \alpha$	0.1	0.05	0.01
10	0.076	0.022	0.002	10	0.076	0.032	0.003
30	0.102	0.043	0.008	30	0.084	0.035	0.003
50	0.105	0.054	0.010	50	0.099	0.042	0.006
100	0.109	0.057	0.008	100	0.124	0.067	0.015

Table 2 Simulated power under H_1 for $\tau = 0.5$, $F_1 = N(0, 1)$, $F_2 = N(0, 4)$ with $\rho = 0.5$

$\widehat{D}_n^\infty$				$\widehat{D}_{n,0}^\infty$			
$n\backslash\alpha$	0.1	0.05	0.01	$n\backslash\alpha$	0.1	0.05	0.01
10	0.014	0.005	0.000	10	0.082	0.039	0.003
30	0.122	0.040	0.007	30	0.185	0.091	0.023
50	0.226	0.143	0.037	50	0.407	0.266	0.080
100	0.524	0.337	0.127	100	0.804	0.670	0.397

Table 3 Simulated power under H_1 for $\tau = 0.1$, $F_1 = N(0, 1)$, $F_2 = N(0, 4)$ with $\rho = 0.5$

$\widehat{D}_n^\infty$				$\widehat{D}_{n,0}^\infty$			
$n\backslash\alpha$	0.1	0.05	0.01	$n\backslash\alpha$	0.1	0.05	0.01
10	0.012	0.002	0.000	10	0.067	0.026	0.005
30	0.050	0.011	0.000	30	0.107	0.043	0.005
50	0.089	0.043	0.013	50	0.124	0.068	0.014
100	0.126	0.061	0.019	100	0.184	0.091	0.025

Table 4 Simulated power under H_1 for $\tau = 0.5$, $F_1 = N(0, 1)$, $F_2 = t_4/\sqrt{2}$ with $\rho = 0.5$

$\widehat{D}_n^\infty$				$\widehat{D}_{n,0}^\infty$			
$n\backslash\alpha$	0.1	0.05	0.01	$n\backslash\alpha$	0.1	0.05	0.01
10	0.009	0.006	0.000	10	0.074	0.026	0.006
30	0.090	0.030	0.002	30	0.106	0.046	0.006
50	0.092	0.043	0.005	50	0.140	0.061	0.012
100	0.155	0.078	0.020	100	0.200	0.117	0.026

6 Proofs

Let ε_i, $i \in \mathbb{N}$, be an iid sequence of random variables with continuous distribution function F, $0 < \sigma^2 = E\left(\varepsilon_1^2\right) < \infty$ and $E(\varepsilon_1) = 0$. We will prepare for the proof of Proposition 1 by a sequence of lemmas.

Note that $\sum_{i=1}^{k} \varepsilon_i^2 > 0$ a.s., for all $k \in \mathbb{N}$ by continuity of F so that the random variables $\xi_{n,a}^{(1)}$ and $\xi_{n,a}^{(2)}$ appearing in the following lemma are a.s. well defined.

Lemma 1 *For all $a \in (0, 1)$ we have, as $n \to \infty$,*

$$\xi_{n,a}^{(1)} = \max_{[n^a] \le k \le n} k \frac{\left|\sum_{i=1}^{k} \varepsilon_i\right|}{\sum_{i=1}^{k} \varepsilon_i^2} = O_P\left(\sqrt{n}\right) \tag{12}$$

and

$$\xi_{n,a}^{(2)} = \max_{[n^a] \le k \le n} \sqrt{\frac{k}{\log \log k}} \frac{\left|\sum_{i=1}^{k} \varepsilon_i\right|}{\sum_{i=1}^{k} \varepsilon_i^2} = O_P(1). \tag{13}$$

Proof For the proof of (12) note that for any random variables ξ_n and ξ, as $n \to \infty$,

$$\xi_n \to \xi \quad \text{a.s.} \quad \text{if and only if} \quad \sup_{k \geq n} |\xi_k - \xi| \to 0 \quad \text{in probability}. \tag{14}$$

Consequently, the strong law of large numbers implies

$$\max_{[n^a] \leq k \leq n} \left| \frac{1}{k} \sum_{i=1}^{k} \varepsilon_i^2 - \sigma^2 \right| = o_P(1) \quad \text{as } n \to \infty. \tag{15}$$

By Kolmogorov's maximal inequality,

$$\max_{[n^a] \leq k \leq n} \left| \sum_{i=1}^{k} \varepsilon_i \right| = O_P\left(\sqrt{n}\right) \quad \text{as } n \to \infty. \tag{16}$$

Now (12) follows from (15), (16) and the inequality, valid for all $K \in (0, \infty)$ and all large n,

$$P \left(\max_{[n^a] \leq k \leq n} k \frac{\left| \sum_{i=1}^{k} \varepsilon_i \right|}{\sum_{i=1}^{k} \varepsilon_i^2} \geq K \sqrt{n} \right)$$

$$\leq P \left(\max_{[n^a] \leq k \leq n} \left| \sum_{i=1}^{k} \varepsilon_i \right| \geq \frac{1}{2} K \sigma^2 \sqrt{n} \right) + P \left(\max_{[n^a] \leq k \leq n} \left| \frac{1}{k} \sum_{i=1}^{k} \varepsilon_i^2 - \sigma^2 \right| \geq \frac{1}{2} \sigma^2 \right).$$

Writing

$$\sqrt{\frac{k}{\log \log k}} \frac{\left| \sum_{i=1}^{k} \varepsilon_i \right|}{\sum_{i=1}^{k} \varepsilon_i^2} = \frac{\left| \sum_{i=1}^{k} \varepsilon_i \right|}{\sqrt{k \log \log k}} \frac{1}{\frac{1}{k} \sum_{i=1}^{k} \varepsilon_i^2},$$

we see that (13) follows immediately from the law of the iterated logarithm and the strong law of large numbers. $\qquad\qquad\square$

From now on we will always assume that the moment condition (10) is satisfied.

Lemma 2 *For all $a \in (0, 1)$ we have*

$$\xi_{n,a}^{(3)} = \max_{[n^a] \leq k \leq n} \sqrt{\frac{\log \log k}{k}} \max_{1 \leq i \leq k} |\varepsilon_i| = o_P(1) \quad \text{as } n \to \infty. \tag{17}$$

Proof A standard application of the first Borel–Cantelli lemma shows that (10) implies $\sqrt{(\log \log n)/n} \max_{1 \leq i \leq n} |\varepsilon_i| \to 0$ a.s. as $n \to \infty$, and (17) follows from (14). $\qquad\qquad\square$

For all $a \in (0, 1)$ and $n \in \mathbb{N}$ the random variables t_{nk}, $k = [n^a], \ldots, n$, are well defined on $A_{n,a}$ by (7) and (8). Observe that (7) implies $1 + t_{nk}\varepsilon_i > 0$ on $A_{n,a}$ for

$k = \left[n^a\right], \ldots, n$ and $i = 1, \ldots, k$. This fact will be of importance several times in the sequel.

Lemma 3 *For all $a \in (0, 1)$ and $n \in \mathbb{N}$ on $A_{n,a}$ we have*

$$\max_{[n^a] \leq k \leq n} k \, |t_{nk}| \leq \left(1 + \max_{[n^a] \leq k \leq n} |t_{nk}| \max_{1 \leq i \leq k} |\varepsilon_i|\right) \xi_{n,a}^{(1)} \tag{18}$$

and

$$\left(\max_{[n^a] \leq k \leq n} \sqrt{\frac{k}{\log \log k}} \, |t_{nk}|\right) \left(1 - \xi_{n,a}^{(2)} \xi_{n,a}^{(3)}\right) \leq \xi_{n,a}^{(2)}. \tag{19}$$

Proof On $A_{n,a}$ we have by (8), for $k = \left[n^a\right], \ldots, n$,

$$0 = \sum_{i=1}^{k} \frac{\varepsilon_i}{1 + t_{nk}\varepsilon_i} = \sum_{i=1}^{k} \varepsilon_i - t_{nk} \sum_{i=1}^{k} \frac{\varepsilon_i^2}{1 + t_{nk}\varepsilon_i}$$

so that

$$\left|\sum_{i=1}^{k} \varepsilon_i\right| = |t_{nk}| \sum_{i=1}^{k} \frac{\varepsilon_i^2}{1 + t_{nk}\varepsilon_i} \geq \frac{|t_{nk}|}{1 + \max_{[n^a] \leq m \leq n} |t_{nm}| \max_{1 \leq i \leq m} |\varepsilon_i|} \sum_{i=1}^{k} \varepsilon_i^2,$$

whence

$$|t_{nk}| \leq \left(1 + \max_{[n^a] \leq m \leq n} |t_{nm}| \max_{1 \leq i \leq m} |\varepsilon_i|\right) \frac{\left|\sum_{i=1}^{k} \varepsilon_i\right|}{\sum_{i=1}^{k} \varepsilon_i^2}.$$

Multiplying this inequality by k and taking the maximum over $k = \left[n^a\right], \ldots, n$ gives (18), and multiplying by $\sqrt{\frac{k}{\log \log k}}$ yields

$$\sqrt{\frac{k}{\log \log k}} \, |t_{nk}| \leq \left(1 + \max_{[n^a] \leq m \leq n} \sqrt{\frac{m}{\log \log m}} \, |t_{nm}| \sqrt{\frac{\log \log m}{m}} \max_{1 \leq i \leq m} |\varepsilon_i|\right) \xi_{n,a}^{(2)}$$

$$\leq \left(1 + \left(\max_{[n^a] \leq m \leq n} \sqrt{\frac{m}{\log \log m}} \, |t_{nm}|\right) \xi_{n,a}^{(3)}\right) \xi_{n,a}^{(2)},$$

which after taking the maximum over k again and rearranging terms leads to (19). $\qquad\square$

Recall that the random variables t_{nk}, $k = \left[n^a\right], \ldots, n$, are well defined on $A_{n,a}$. Because of $P\left(A_{n,a}\right) \to 1$ an $n \to \infty$, in all subsequent statements concerning convergence in probability or in distribution, the definition of t_{nk} on the event $\Omega \backslash A_{n,a}$ of the underlying probability space $(\Omega, \mathscr{F}, P)$ plays no role. Therefore, in the following proofs we can always argue entirely on $A_{n,a}$ and ignore the complement of $A_{n,a}$ completely.

Corollary 1 *For all $a \in (0, 1)$ we have*

$$\max_{[n^a] \leq k \leq n} \sqrt{\frac{k}{\log \log k}} \, |t_{nk}| = O_P(1) \quad \text{as } n \to \infty. \tag{20}$$

Proof From (13) and (17) we obtain

$$1 - \xi_{n,a}^{(2)} \xi_{n,a}^{(3)} = 1 + o_P(1) \quad \text{as } n \to \infty, \tag{21}$$

and (20) follows from (19), (13) and (21). $\qquad\qquad\qquad\qquad\qquad\qquad\square$

Lemma 4 *For all $a \in (0, 1)$ we have, as $n \to \infty$,*

$$\eta_{n,a}^{(1)} = \max_{[n^a] \leq k \leq n} |t_{nk}| \max_{1 \leq i \leq k} |\varepsilon_i| = o_P(1) \tag{22}$$

$$\eta_{n,a}^{(2)} = \max_{[n^a] \leq k \leq n} k \, |t_{nk}| = O_P\left(\sqrt{n}\right) \tag{23}$$

$$\eta_{n,a}^{(3)} = \max_{[n^a] \leq k \leq n} \max_{1 \leq i \leq k} \frac{1}{1 + t_{nk}\varepsilon_i} = O_P(1) \tag{24}$$

$$\eta_{n,a}^{(4)} = \max_{[n^a] \leq k \leq n} \sup_{x \in [-\infty,\infty]} \left| \frac{1}{k} \sum_{i=1}^{k} \varepsilon_i 1_{\{\varepsilon_i \leq x\}} - U(x) \right| = o_P(1) \tag{25}$$

$$\eta_{n,a}^{(5)} = \max_{[n^a] \leq k \leq n} \left| \frac{1}{\sigma^2} \sum_{i=1}^{k} \varepsilon_i - k t_{nk} \right| = o_P\left(\sqrt{n}\right). \tag{26}$$

Proof Assertion (22) follows from

$$\eta_{n,a}^{(1)} \leq \left(\max_{[n^a] \leq k \leq n} \sqrt{\frac{k}{\log \log k}} \, |t_{nk}| \right) \left(\max_{[n^a] \leq k \leq n} \sqrt{\frac{\log \log k}{k}} \max_{1 \leq i \leq k} |\varepsilon_i| \right)$$

in conjunction with (20) and (17). Assertion (23) is immediate from (18), (22) and (12), and (24) is immediate from (22). As to (25), Lemma 2.2 of Zhang (1997) implies

$$\sup_{x \in [-\infty,\infty]} \left| \frac{1}{n} \sum_{i=1}^{n} \varepsilon_i 1_{\{\varepsilon_i \leq x\}} - U(x) \right| \to 0 \quad \text{a.s. as } n \to \infty,$$

which gives (25) in view of (14). It remains to verify (26). Using $\frac{1}{1+x} = 1 - x + \frac{x^2}{1+x}$ on $A_{n,a}$ we obtain from (8), for $k = [n^a], \ldots, n$,

$$0 = \sum_{i=1}^{k} \frac{\varepsilon_i}{1 + t_{nk}\varepsilon_i} = \sum_{i=1}^{k} \varepsilon_i - t_{nk} \sum_{i=1}^{k} \varepsilon_i^2 + t_{nk}^2 \sum_{i=1}^{k} \frac{\varepsilon_i^3}{1 + t_{nk}\varepsilon_i}$$

which implies

$$\frac{1}{\sigma^2} \sum_{i=1}^{k} \varepsilon_i - k t_{nk} = \frac{1}{\sigma^2} t_{nk} \sum_{i=1}^{k} \left(\varepsilon_i^2 - \sigma^2 \right) - \frac{1}{\sigma^2} t_{nk}^2 \sum_{i=1}^{k} \frac{\varepsilon_i^3}{1 + t_{nk}\varepsilon_i}.$$

Now, as $n \to \infty$,

$$\max_{[n^a] \leq k \leq n} \left| t_{nk} \sum_{i=1}^{k} \left(\varepsilon_i^2 - \sigma^2 \right) \right| \leq \eta_{n,a}^{(2)} \max_{[n^a] \leq k \leq n} \frac{1}{k} \left| \sum_{i=1}^{k} \left(\varepsilon_i^2 - \sigma^2 \right) \right| = o_P \left(\sqrt{n} \right)$$

by (23) and (15), and

$$\max_{[n^a] \leq k \leq n} t_{nk}^2 \left| \sum_{i=1}^{k} \frac{\varepsilon_i^3}{1 + t_{nk}\varepsilon_i} \right| \leq \eta_{n,a}^{(1)} \eta_{n,a}^{(2)} \eta_{n,a}^{(3)} \max_{[n^a] \leq k \leq n} \frac{1}{k} \sum_{i=1}^{k} \varepsilon_i^2 = o_P \left(\sqrt{n} \right)$$

by (22), (23), and (24) and the strong law of large numbers. $\qquad\square$

Now we are prepared to give the
Proof of Propositon 1. For all $a \in (0, 1)$ on $A_{n,a}$ and for all $s \in [n^{a-1}, 1]$ and $x \in [-\infty, \infty]$ we have, using $\frac{1}{1+x} - 1 = \frac{x^2}{1+x} - x$,

$$\zeta_n(s, x) = [ns] \left| F_{n,0}^{seq}(s, x) - \left(F_n^{seq}(s, x) - \frac{1}{\sigma^2} S_n(s) U(x) \right) \right|$$

$$= \left| \sum_{i=1}^{[ns]} \left(\frac{1}{1 + t_{n[ns]}\varepsilon_i} - 1 \right) 1_{\{\varepsilon_i \leq x\}} + \frac{1}{\sigma^2} \sum_{i=1}^{[ns]} \varepsilon_i U(x) \right|$$

$$= \left| t_{n[ns]}^2 \sum_{i=1}^{[ns]} \frac{\varepsilon_i^2}{1 + t_{n[ns]}\varepsilon_i} 1_{\{\varepsilon_i \leq x\}} - t_{n[ns]} \sum_{i=1}^{[ns]} \varepsilon_i 1_{\{\varepsilon_i \leq x\}} \right.$$

$$\left. + \left(\frac{1}{\sigma^2} \sum_{i=1}^{[ns]} \varepsilon_i - [ns] t_{n[ns]} \right) U(x) + [ns] t_{n[ns]} U(x) \right|$$

$$\leq \eta_{n,a}^{(1)} \eta_{n,a}^{(2)} \eta_{n,a}^{(3)} \max_{1 \leq k \leq n} \frac{1}{k} \sum_{i=1}^{k} |\varepsilon_i| + \eta_{n,a}^{(2)} \eta_{n,a}^{(4)} + \eta_{n,a}^{(5)} E(|\varepsilon_1|),$$

because for the function U from (9) clearly $|U(x)| \le E(|\varepsilon_1|)$ for all $x \in [-\infty, \infty]$. Hence by Lemma 4 and the strong law of large numbers

$$\sup_{(s,x)\in[n^{a-1},1]\times[-\infty,\infty]} \zeta_n(s,x) = o_P\left(\sqrt{n}\right) \quad \text{as } n \to \infty.$$

For all $s \in \left[0, n^{a-1}\right]$ and $x \in [-\infty, \infty]$ we have

$$\zeta_n(s,x) \le [ns] + \frac{1}{\sigma^2}\left|\sum_{i=1}^{[ns]}\varepsilon_i\right| |U(x)| \le n^a + \frac{1}{\sigma^2}E(|\varepsilon_1|)\max_{1\le k\le[n^a]}\left|\sum_{i=1}^{k}\varepsilon_i\right|.$$

Since $\max_{1\le k\le[n^a]}\left|\sum_{i=1}^{k}\varepsilon_i\right| = O_P\left(n^{a/2}\right)$ as $n \to \infty$ by Kolmogorov's maximal inequality, the right-hand side of the last inequality is $o_P\left(\sqrt{n}\right)$ as $n \to \infty$ for all $a \in \left(0, \frac{1}{2}\right)$, which concludes the proof. $\qquad\square$

Based on Proposition 1 we can give the

Proof of Theorem 1. As a consequence of Proposition 1 the two sequences

$$\sqrt{n}s\left(F_{n,0}^{seq}(s,x) - F(x)\right), \ s \in [0,1], \ x \in [-\infty, \infty],$$

and

$$Z_n(s,x) = \frac{[ns]}{\sqrt{n}}\left(F_n^{seq}(s,x) - F(x) - \frac{1}{\sigma^2}S_n(s)U(x)\right), \ s \in [0,1], \ x \in [-\infty, \infty],$$

of stochastic processes have the same asymptotic distribution. Therefore, we can show convergence in distribution of the sequence Z_n, $n \in \mathbb{N}$, toward Z.

The classical methodology for proving convergence in distribution of stochastic processes is to prove convergence of the finite dimensional distributions (fidis) and uniform tightness. Now, because $Z_n(s,x)$ is a sum of independent random variables with mean zero, a routine application of the multivariate central limit theorem shows that the fidis of Z_n converge in distribution to the fidis of Z. In case of limit processes with continuous paths a suitable condition for uniform tightness of a sequence $V_n = (V_n(s,x))_{(s,x)\in[0,1]\times[-\infty,\infty]}$ of processes is

$$\lim_{\delta\downarrow0}\limsup_{n\to\infty} P\left(\omega(V_n,\delta) \ge \varepsilon\right) = 0 \quad \text{for all } \varepsilon > 0, \tag{27}$$

where the modulus of continuity $\omega(f,\delta)$ for a function $f : [0,1] \times [-\infty, \infty] \to \mathbb{R}$ and any $\delta > 0$ is defined by

$$\omega(f,\delta) = \sup_{\substack{(s,x),(\tilde{s},\tilde{x})\in[0,1]\times[-\infty,\infty] \\ |s-\tilde{s}|<\delta,\ m(x,\tilde{x})<\delta}} |f(s,x) - f(\tilde{s},\tilde{x})|.$$

Here m is a metric on $[-\infty, \infty]$ which is defined by $m(x, \tilde{x}) = |h(x) - h(\tilde{x})|$ for some smooth one-to-one function $h : [-\infty, \infty] \rightarrow [0, 1]$ with $h(-\infty) = 0$ and $h(\infty) = 1$ so that the time interval $[-\infty, \infty]$ is identified with the time interval $[0, 1]$. For (27) as a criterion for uniform tightness within the theory of convergence in distribution in Skorohod spaces of functions with multidimensional arguments the reader is referred to Theorem 2 in Wichura (1969) (see also the discussion after Theorem 1 in Neuhaus 1971), whereas Sects. 1.3 and 1.5 in van der Vaart and Wellner (1996) cover the theory of convergence in distribution in the sense of Hoffmann–Jørgensen. Because of $\omega(f + g, \delta) \leq \omega(f, \delta) + \omega(g, \delta)$ for all functions f and g and $\delta > 0$, it is trivial that (27) is satisfied for the sum of two processes whenever it is satisfied for the summands individually. Therefore, to verify that (27) holds for Z_n it is sufficient to verify it for the processes

$$\frac{[ns]}{\sqrt{n}} \left(F_n^{seq}(s, x) - F(x) \right), \quad s \in [0, 1], \ x \in [-\infty, \infty], \tag{28}$$

and

$$\frac{[ns]}{\sqrt{n}} \frac{1}{\sigma^2} S_n(s) U(x), \quad s \in [0, 1], \ x \in [-\infty, \infty], \tag{29}$$

individually. But uniform tightness for the processes in (28) holds because of (1), and uniform tightness of the processes in (29) follows from uniform tightness of the partial sum processes S_n (in their one-dimensional time parameter $s \in [0, 1]$) and uniform continuity of the function $U : [-\infty, \infty] \rightarrow \mathbb{R}$ (in its one-dimensional argument x). Consequently, the sequence $Z_n, n \in \mathbb{N}$, is uniformly tight, and the proof of Theorem 1 is complete. $\qquad\square$

Proof of Proposition 2. For every $n \in \mathbb{N}$ the random vectors $(\varepsilon_1, \varepsilon_2, \ldots, \varepsilon_n)$ and $(\varepsilon_n, \ldots, \varepsilon_2, \varepsilon_1)$ have the same distribution which implies that the stochastic processes

$$[ns] \left(F_{n,0}^{seq}(s, x) - \left(F_n^{seq}(s, x) - \frac{1}{\sigma^2} S_n(s) U(x) \right) \right), \quad s \in [0, 1] \times [-\infty, \infty],$$

and

$$(n - [ns]) \left(G_{n,0}^{seq}(s, x) - \left(G_n^{seq}(s, x) - \frac{1}{\sigma^2} T_n(s) U(x) \right) \right), \quad s \subset [0, 1] \times [-\infty, \infty],$$

have the same distribution as well. Consequently, by Proposition 1,

$$\sup_{(s,x)\in[0,1]\times[-\infty,\infty]} (n - [ns]) \left| G_{n,0}^{seq}(s, x) - \left(G_n^{seq}(s, x) - \frac{1}{\sigma^2} T_n(s) U(x) \right) \right| = o_P(\sqrt{n}),$$

and because for all $s \in [0, 1]$ and $x \in [-\infty, \infty]$ we have

$$\left| D_{n,0} (s, x) - W_n (s, x) \right| \le \frac{[ns]}{n} \left(F_{n,0}^{seq} (s, x) - \left(F_n^{seq} (s, x) - \frac{1}{\sigma^2} S_n (s) U (x) \right) \right)$$

$$+ \frac{(n - [ns])}{n} \left(G_{n,0}^{seq} (s, x) - \left(G_n^{seq} (s, x) - \frac{1}{\sigma^2} T_n (s) U (x) \right) \right),$$

the proposition follows. $\qquad\square$

Based on Proposition 2 we can give the

Proof of Theorem 2. As a consequence of Proposition 2 the two sequences of processes $\sqrt{n} D_{n,0}$, $n \in \mathbb{N}$, and $\sqrt{n} W_n$, $n \in \mathbb{N}$, have the same asymptotic distribution so that we can show that the sequence $\sqrt{n} W_n$, $n \in \mathbb{N}$, converges in distribution to W. As in the proof of Theorem 1, we need to verify convergence of the fidis and uniform tightness. But because $\sqrt{n} W_n (s, x)$ is a sum of independent random variables with mean zero, fidi convergence is obtained by a routine application of the multivariate central limit theorem. Uniform tightness of $\sqrt{n} W_n$, $n \in \mathbb{N}$, follows from the fact that the sequences $\sqrt{n} D_n$, $n \in \mathbb{N}$, and

$$\sqrt{n} \frac{[ns] (n - [ns])}{n^2} (S_n (s) - T_n (s)) U (x), \quad s \in [0, 1], \ x \in [-\infty, \infty], \quad n \in \mathbb{N},$$

are uniformly tight individually (recall (2) and the facts that S_n and T_n are partial sum processes and that U is uniformly continuous). This concludes the proof of Theorem 2. $\qquad\square$

Now we turn to the study of the AR(1)-model under the assumptions of Theorem 3. The key to the proof of Theorem 3 is a detailed study of the closeness of the residuals $\widehat{\varepsilon}_{ni}$ and the errors ε_i. The basic result is

Lemma 5 $\displaystyle \max_{1 \le i \le n} |\widehat{\varepsilon}_{ni} - \varepsilon_i| = o_P (1) \ \ as \ n \to \infty.$

Proof For all $n \in \mathbb{N}$ and $i = 1, \ldots, n$ by definition of the residuals

$$\widehat{\varepsilon}_{ni} - \varepsilon_i = X_i - \widehat{\rho}_n X_{i-1} - (X_i - \rho X_{i-1}) = (\rho - \widehat{\rho}_n) X_{i-1} \qquad (30)$$

so that $\max_{1 \le i \le n} |\widehat{\varepsilon}_{ni} - \varepsilon_i| \le |\widehat{\rho}_n - \rho| \max_{1 \le i \le n} |X_{i-1}|$. Now $|\widehat{\rho}_n - \rho| = O_P (1/\sqrt{n})$ by our basic assumption about $\widehat{\rho}_n$ and $\max_{1 \le i \le n} |X_{i-1}| = o_P (\sqrt{n})$ as $n \to \infty$ by stationarity of $(X_i)_{i \ge 0}$ and $E (X_1^2) < \infty$, which yields the assertion. $\qquad\square$

Lemma 5 implies that $\widehat{D}_{n,0}$ is well defined with probability converging to one as $n \to \infty$:

Corollary 2 *For all* $a \in (0, 1)$ *we have* $P(\widehat{A}_{n,a}) \to 1$ *and* $P(\widehat{B}_{n,a}) \to 1$ *as* $n \to \infty$.

Proof Because of $E(\varepsilon_1) = 0$ and $E(\varepsilon_1^2) > 0$ there exists a $\delta > 0$ with $P(\varepsilon_1 \leq \delta) < 1$ and $P(\varepsilon_1 \geq -\delta) < 1$. For all $a \in (0, 1)$ and all large $n \in \mathbb{N}$ we have

$$P\left(\Omega \setminus \widehat{A}_{n,a}\right) = P\left(\left\{0 \leq \min_{1 \leq i \leq [n^a]} \widehat{\varepsilon}_{ni}\right\} \cup \left\{\max_{1 \leq i \leq [n^a]} \widehat{\varepsilon}_{ni} \leq 0\right\}\right)$$

$$\leq P\left(\min_{1 \leq i \leq [n^a]} \varepsilon_i \geq -\delta\right) + P\left(\max_{1 \leq i \leq [n^a]} \varepsilon_i \leq \delta\right) + P\left(\max_{1 \leq i \leq [n^a]} |\widehat{\varepsilon}_{ni} - \varepsilon_i| \geq \delta\right)$$

$$= P(\varepsilon_1 \geq -\delta)^{[n^a]} + P(\varepsilon_1 \leq \delta)^{[n^a]} + P\left(\max_{1 \leq i \leq [n^a]} |\widehat{\varepsilon}_{ni} - \varepsilon_i| \geq \delta\right).$$

The right-hand side of this inequality converges to zero as $n \to \infty$ because of $[n^a] \to \infty$ and Lemma 5, whence $P(\widehat{A}_{n,a}) \to 1$. The proof of $P(\widehat{B}_{n,a}) \to 1$ is similar. $\qquad\square$

Proof of Theorem 3. As the proof of Theorem 2 shows Theorem 3 follows from

$$\sup_{(s,x) \in [0,1] \times [-\infty,\infty]} \left|\widehat{D}_{n,0}(s, x) - W_n(s, x)\right| = o_P(1/\sqrt{n}) \quad \text{as } n \to \infty. \tag{31}$$

By the triangle inequality, (31) follows from (3) and, as $n \to \infty$,

$$\sup_{(s,x) \in [0,1] \times [-\infty,\infty]} \left|\widehat{D}_{n,0}(s, x) - \left(\widehat{D}_n(s, x) - \frac{1}{\sigma^2} \frac{[ns](n - [ns])}{n^2}\right.\right.$$
$$\left.\left. \times (S_n(s) - T_n(s)) U(x)\right)\right| = o_P(1/\sqrt{n})$$

which in turn follows from

$$\sup_{(s,x) \in [0,1] \times [-\infty,\infty]} [ns] \left|\widehat{F}_{n,0}^{seq}(s, x) - \left(\widehat{F}_n^{seq}(s, x) - \frac{1}{\sigma^2} S_n(s) U(x)\right)\right| = o_P(\sqrt{n}) \tag{32}$$

and

$$\sup_{(s,x) \in [0,1] \times [-\infty,\infty]} (n - [ns]) \left|\widehat{G}_{n,0}^{seq}(s, x) - \left(\widehat{G}_n^{seq}(s, x) - \frac{1}{\sigma^2} T_n(s) U(x)\right)\right| = o_P(\sqrt{n}).$$

Note that the application of (3) from Bai (1994) requires the assumption that F has a uniformly continuous strictly positive density in our Theorem 3. Clearly, (32) is a version of Proposition 1 with the independent ε_i in $F_{n,0}^{seq}$ and F_n^{seq} replaced by the residuals (but not in S_n). The second statement is the counterpart for $\widehat{G}_{n,0}^{seq}$ and $\widehat{G}_n^{seq}$. We will present here the main steps in the approach to (32). The approach to the other statement is similar.

Recall that the random variables $\widehat{t}_{nk}$ are defined by (7) and (8) with t_{nk} replaced by $\widehat{t}_{nk}$ and ε_i by $\widehat{\varepsilon}_{ni}$. Therefore, by copying the proof of Proposition 1, for all $a \in (0, 1)$

on $\widehat{A}_{n,a}$ and for all $s \in \left[n^{a-1}, 1\right]$ and $x \in [-\infty, \infty]$ we get

$$\widehat{\zeta}_n(s, x) = [ns] \left| \widehat{F}^{seq}_{n,0}(s, x) - \left(\widehat{F}^{seq}_n(s, x) - \frac{1}{\sigma^2} S_n(s) U(x) \right) \right|$$

$$= \left| \sum_{i=1}^{[ns]} \left(\frac{1}{1 + \widehat{t}_{n[ns]}\widehat{\varepsilon}_{ni}} - 1 \right) 1_{\{\widehat{\varepsilon}_{ni} \leq x\}} + \frac{1}{\sigma^2} \sum_{i=1}^{[ns]} \varepsilon_i U(x) \right|$$

$$= \left| \widehat{t}^2_{n[ns]} \sum_{i=1}^{[ns]} \frac{\widehat{\varepsilon}^2_{ni}}{1 + \widehat{t}_{n[ns]}\widehat{\varepsilon}_{ni}} 1_{\{\widehat{\varepsilon}_{ni} \leq x\}} - \widehat{t}_{n[ns]} \sum_{i=1}^{[ns]} \widehat{\varepsilon}_{ni} 1_{\{\widehat{\varepsilon}_{ni} \leq x\}} \right.$$

$$\left. + \left(\frac{1}{\sigma^2} \sum_{i=1}^{[ns]} \varepsilon_i - [ns]\widehat{t}_{n[ns]} \right) U(x) + [ns]\widehat{t}_{n[ns]} U(x) \right|$$

$$\leq \widehat{\eta}^{(1)}_{n,a} \widehat{\eta}^{(2)}_{n,a} \widehat{\eta}^{(3)}_{n,a} \max_{1 \leq k \leq n} \frac{1}{k} \sum_{i=1}^{k} |\widehat{\varepsilon}_{ni}| + \widehat{\eta}^{(2)}_{n,a} \widehat{\eta}^{(4)}_{n,a} + \widehat{\eta}^{(5)}_{n,a} E(|\varepsilon_1|)$$

with

$$\widehat{\eta}^{(1)}_{n,a} = \max_{[n^a] \leq k \leq n} \left|\widehat{t}_{nk}\right| \max_{1 \leq i \leq k} |\widehat{\varepsilon}_{ni}|, \quad \widehat{\eta}^{(2)}_{n,a} = \max_{[n^a] \leq k \leq n} k\left|\widehat{t}_{nk}\right|,$$

$$\widehat{\eta}^{(3)}_{n,a} = \max_{[n^a] \leq k \leq n} \max_{1 \leq i \leq k} \frac{1}{1 + \widehat{t}_{nk}\widehat{\varepsilon}_{ni}}, \quad \widehat{\eta}^{(4)}_{n,a} = \max_{[n^a] \leq k \leq n} \sup_{x \in [-\infty,\infty]} \left| \frac{1}{k} \sum_{i=1}^{k} \widehat{\varepsilon}_{ni} 1_{\{\widehat{\varepsilon}_{ni} \leq x\}} - U(x) \right|,$$

$$\widehat{\eta}^{(5)}_{n,a} = \max_{[n^a] \leq k \leq n} \left| \frac{1}{\sigma^2} \sum_{i=1}^{k} \varepsilon_i - k\widehat{t}_{nk} \right|.$$

To complete the proof of Theorem 3 it remains to show that for $j = 1, \ldots, 5$ the random variables $\widehat{\eta}^{(j)}_{n,a}$ are exactly of the same order for $n \to \infty$ as their counterparts $\eta^{(j)}_{n,a}$ in Lemma 4, and $\max_{1 \leq k \leq n} \frac{1}{k} \sum_{i=1}^{k} |\widehat{\varepsilon}_{ni}| = O_P(1)$. The latter is implied by

$$\max_{1 \leq k \leq n} \frac{1}{k} \sum_{i=1}^{k} |\widehat{\varepsilon}_{ni}| \leq \max_{1 \leq i \leq n} |\widehat{\varepsilon}_{ni} - \varepsilon_i| + \max_{1 \leq k \leq n} \frac{1}{k} \sum_{i=1}^{k} |\varepsilon_i|,$$

Lemma 5 and the strong law of large numbers. For the proof of the required properties of the $\widehat{\eta}^{(j)}_{n,a}$ note that Lemma 3 follows entirely from the fact that the t_{nk} are defined by (7) and (8). Consequently, because the $\widehat{t}_{nk}$ are defined by (7) and (8) with t_{nk} replaced by $\widehat{t}_{nk}$ and ε_i by $\widehat{\varepsilon}_{ni}$, by the proof of Lemma 3 we have the versions

$$\max_{[n^a] \leq k \leq n} k\left|\widehat{t}_{nk}\right| \leq \left(1 + \max_{[n^a] \leq k \leq n} \left|\widehat{t}_{nk}\right| \max_{1 \leq i \leq k} |\widehat{\varepsilon}_{ni}| \right) \widehat{\xi}^{(1)}_{n,a}$$

and

$$\left(\max_{[n^a]\le k\le n} \sqrt{\frac{k}{\log\log k}}\,|\widehat{t}_{nk}| \right) \left(1 - \widehat{\xi}_{n,a}^{(2)}\,\widehat{\xi}_{n,a}^{(3)} \right) \le \widehat{\xi}_{n,a}^{(2)}$$

of (18) and (19) with

$$\widehat{\xi}_{n,a}^{(1)} = \max_{[n^a]\le k\le n} k\,\frac{\left|\sum_{i=1}^{k}\widehat{\varepsilon}_{ni}\right|}{\sum_{i=1}^{k}\widehat{\varepsilon}_{ni}^{2}}, \quad \widehat{\xi}_{n,a}^{(2)} = \max_{[n^a]\le k\le n} \sqrt{\frac{k}{\log\log k}}\,\frac{\left|\sum_{i=1}^{k}\widehat{\varepsilon}_{ni}\right|}{\sum_{i=1}^{k}\widehat{\varepsilon}_{ni}^{2}} \quad \text{and}$$

$$\widehat{\xi}_{n,a}^{(3)} = \max_{[n^a]\le k\le n} \sqrt{\frac{\log\log k}{k}}\,\max_{1\le i\le k}|\widehat{\varepsilon}_{ni}|.$$

Therefore, we can copy the proofs of (22)–(24) to obtain $\widehat{\eta}_{n,a}^{(1)} = o_P(1)$, $\widehat{\eta}_{n,a}^{(2)} = O_P(\sqrt{n})$ and $\widehat{\eta}_{n,a}^{(3)} = O_P(1)$ as $n \to \infty$ provided that we can show $\widehat{\xi}_{n,a}^{(1)} = O_P(\sqrt{n})$, $\widehat{\xi}_{n,a}^{(2)} = O_P(1)$ and $\widehat{\xi}_{n,a}^{(3)} = o_P(1)$.

We will provide here the details of the argument that establishes $\widehat{\xi}_{n,a}^{(1)} = O_P(\sqrt{n})$ as $n \to \infty$. For all $a \in (0, 1)$ and $n \in \mathbb{N}$ we have

$$\max_{[n^a]\le k\le n} \left|\sum_{i=1}^{k}\widehat{\varepsilon}_{ni}\right| \le n \max_{[n^a]\le k\le n} \frac{1}{k}\left|\sum_{i=1}^{k}(\widehat{\varepsilon}_{ni} - \varepsilon_i)\right| + \max_{[n^a]\le k\le n}\left|\sum_{i=1}^{k}\varepsilon_i\right|.$$

The second summand on the right-hand side is $O_P(\sqrt{n})$ as $n \to \infty$ by (16). By (30) the first summand equals $n\,|\widehat{\rho}_n - \rho|\,\max_{[n^a]\le k\le n}\frac{1}{k}\left|\sum_{i=1}^{k}X_{i-1}\right| = O_P(\sqrt{n})$ as $n \to \infty$ by our basic assumption on $\widehat{\rho}_n$ and the ergodic theorem applied to the stationary sequence X_i, $i \ge 0$. Hence $\max_{[n^a]\le k\le n}\left|\sum_{i=1}^{k}\widehat{\varepsilon}_{ni}\right| = O_P(\sqrt{n})$ as $n \to \infty$. Moreover,

$$\max_{[n^a]\le k\le n} \frac{1}{k}\sum_{i=1}^{k}|\widehat{\varepsilon}_{ni}^{2} - \varepsilon_i^{2}| = \max_{[n^a]\le k\le n} \frac{1}{k}\sum_{i=1}^{k}|\widehat{\varepsilon}_{ni} - \varepsilon_i|\,|\widehat{\varepsilon}_{ni} + \varepsilon_i|$$

$$\le \left(\max_{1\le i\le n}|\widehat{\varepsilon}_{ni} - \varepsilon_i| \right)\left(\max_{[n^a]\le k\le n}\frac{1}{k}\sum_{i=1}^{k}|\widehat{\varepsilon}_{ni} - \varepsilon_i| + 2\max_{[n^a]\le k\le n}\frac{1}{k}\sum_{i=1}^{k}|\varepsilon_i| \right) = o_P(1)$$

as $n \to \infty$ by Lemma 5 and the strong law of large numbers. This result combined with (15) yields $\max_{[n^a]\le k\le n}\frac{1}{k}\sum_{i=1}^{k}|\widehat{\varepsilon}_{ni}^{2} - \sigma^2| = o_P(1)$ as $n \to \infty$. Thus we have proved the versions of (15) and (16) for the residuals $\widehat{\varepsilon}_{ni}$ that are needed to copy the argument that leads to (12) in the proof of Lemma 1. This establishes $\widehat{\xi}_{n,a}^{(1)} = O_P(\sqrt{n})$ as $n \to \infty$, as desired.

The proofs of $\widehat{\xi}_{n,a}^{(2)} = O_P(1)$, $\widehat{\xi}_{n,a}^{(3)} = o_P(1)$, $\widehat{\eta}_{n,a}^{(4)} = o_P(1)$ and $\widehat{\eta}_{n,a}^{(5)} = o_P(\sqrt{n})$ as $n \to \infty$ all require similar deductions of the asymptotic properties of the residuals $\widehat{\varepsilon}_{ni}$ from the corresponding properties of the errors ε_i and will not be detailed here. $\qquad\square$

For the *Proof of Theorem 4* note that the process $\widehat{D}_{n,0}^*$ is based on the bootstrap variables $\varepsilon_{n1}^*, \ldots, \varepsilon_{nn}^*$ which are iid for every $n \in \mathbb{N}$ under the conditional probability P_n^*. Therefore $\widehat{D}_{n,0}^*$ is a bootstrap version of the process $D_{n,0}$ considered in Theorem 2 (and not of the process appearing in Theorem 3), and the proof of Theorem 4 is a repetition of the proof of Theorem 2 for the ε_{ni}^* and P_n^* instead of the ε_i and P. The slightly stronger moment condition than (10) is a consequence of the fact that the bootstrap variables ε_{ni}^* form a triangular array so that the bootstrap version of (13) cannot be derived from the law of the iterated logarithm but has to be established through an application of the Hájek–Rényi-inequality, which leads to a $\log k$ in this bootstrap version instead of the $\log \log k$ in (13). This $\log k$ has to be compensated by a $\log k$ instead of $\log \log k$ in the bootstrap version of Lemma 2, for which we need the stronger moment condition. Technical details can be found in Horni (2014). $\quad\square$

Acknowledgments The first author would like to express his gratitude to Paul Deheuvels and the organizers of the conference on *Mathematical Statistics and Limit Theorems* for inviting him. Suggestions by a referee which improved the presentation are also gratefully acknowledged.

References

Bai, J. (1994). Weak convergence of the sequential empirical processes of residuals in ARMA models. *Annals of Statistics, 22*, 2051–2061.

Bickel, P. J., & Wichura, M. J. (1971). Convergence criteria for multiparameter stochastic processes and some applications. *Annals of Mathematical Statistics, 42*, 1656–1670.

Bickel, P. J., Klaassen, C. A. J., Ritov, J., & Wellner, J. A. (1993). *Efficient and adaptive estimation for semiparametric models*. Baltimore: The John Hopkins University Press.

Horni, S. (2014). Changepointtests für die Fehlerverteilung in ARMA-Zeitreihen. Dissertation, University of Giessen.

Neuhaus, G. (1971). On weak convergence of stochastic processes with multidimensional time parameter. *Annals of Mathematical Statistics, 42*, 1285–1295.

Owen, A. B. (1990). Empirical likelihood confidence regions. *Annals of Statistics, 18*, 90–120.

Owen, A. B. (2001). *Empirical likelihood*. London/Boca Raton: Chapman & Hall/CRC.

Picard, D. (1985). Testing and estimating change-point in time series. *Advances in Applied Probability, 17*, 841–867.

Qin, J., & Lawless, J. (1994). Empirical likelihood and general estimating equations. *Annals of Statistics, 22*, 300–325.

van der Vaart, A. W., & Wellner, J. A. (1996). *Weak convergence and empirical processes*. New York: Springer.

Wichura, M. J. (1969). Inequalities with applications to the weak convergence of random processes with multi-dimensional time parameters. *Annals of Mathematical Statistics, 40*, 681–687.

Zhang, B. (1997). Estimating a distribution function in the presence of auxiliary information. *Metrika, 46*, 221–244.

On Quadratic Expansions of Log-Likelihoods and a General Asymptotic Linearity Result

Marc Hallin, Ramon van den Akker and Bas J.M. Werker

Abstract Irrespective of the statistical model under study, the derivation of limits, in the Le Cam sense, of sequences of local experiments (see, e.g., Jeganathan, Econometric Theory 11:818–887, 1995 and Strasser, *Mathematical Theory of Statistics: Statistical experiments and asymptotic decision theory*, Walter de Gruyter, Berlin, 1985) often follows along very similar lines, essentially involving differentiability in quadratic mean of square roots of (conditional) densities. This chapter establishes two abstracts but quite generally applicable results providing sufficient, and nearly necessary, conditions for (i) the existence of a quadratic expansion and (ii) the asymptotic linearity of local log-likelihood ratios. Asymptotic linearity is needed, for instance, when unspecified model parameters are to be replaced, in some statistic of interest, with some preliminary estimators. Such results have been established, for *locally asymptotically normal* (LAN) models involving independent and identically distributed observations, by, e.g., Bickel et al. (*Efficient and adaptive Estimation for semiparametric Models*, Johns Hopkins University Press, Baltimore, 1993), van der Vaart (*Statistical Estimation in Large Parameter Spaces*, CWI, Amsterdam, 1988; *Asymptotic Statistics*, Cambridge University Press, Cambridge, 2000). Similar results are provided here for models exhibiting serial dependencies which, so far, have been treated on a case-by-case basis (see Hallin and Paindaveine, *Journal of Statistical Planning and Inference* 136:1–32, 2005 and Hallin and Puri, *Journal of Multivariate Analysis* 50:175–237, 1994 for typical examples) and, in general, under stronger regularity assumptions. Unlike their i.i.d. counterparts, our results are established under LAQ conditions, hence extend beyond the context of LAN experiments, so

M. Hallin (✉)
ECARES, Université Libre de Bruxelles, Brussels, Belgium
e-mail: mhallin@ulb.ac.be

M. Hallin
ORFE, Princeton University, Princeton, USA

M. Hallin · R. van den Akker
Econometrics Group, Tilburg University, Tilburg, The Netherlands
e-mail: R.vdnAkker@TilburgUniversity.edu

B.J.M. Werker
Econometrics Group and Finance Group, Tilburg University, Tilburg, The Netherlands
e-mail: Werker@TilburgUniversity.edu

© Springer International Publishing Switzerland 2015
M. Hallin et al. (eds.), *Mathematical Statistics and Limit Theorems*,
DOI 10.1007/978-3-319-12442-1_9

that nonstationary unit-root time series and cointegration models, for instance, also can be handled (see Hallin et al., Optimal pseudo-Gaussian and rank-based tests of the cointegrating rank in semiparametric error-correction models, 2013).

1 Introduction

Asymptotic methods always have been a major tool in statistical inference, whenever exact optimality results are unavailable. The main justification for such fundamental daily practice procedures as maximum likelihood estimation or likelihood ratio testing is of intrinsically asymptotic nature. Despite this, a solid and mathematically rigorous treatment of asymptotics in statistics was not possible until the development of the *asymptotic theory of statistical experiments* attached, essentially, with the name of Lucien Le Cam. Even a short presentation of that theory is impossible in the limits of this contribution, and the interested reader is referred to Le Cam and Yang (1990), Ibragimov and Has'minskii (1991), Janssen et al. (1985), Le Cam (1960, 1986), Strasser (1985), Torgersen (1991), Shiryaev and Spokoiny (2000) or van der Vaart (2000) for background reading.

An essential ingredient in the asymptotic theory of statistical experiments is the characterization, in distribution, of the asymptotic behavior of the so-called *local log-likelihood processes*.

The most familiar case, from that point of view, is that of *Locally Asymptotically Normal* (LAN) experiments, the local likelihood processes of which are asymptotically the same, in distribution, as those of finite-dimensional *Gaussian shift* (*Gaussian location*) experiments. This includes, basically, all experiments/models where traditional maximal likelihood methods are valid and asymptotically efficient: smooth parametric models for independent, identically or nonidentically distributed observations, but also linear time-series models (Swensen 1985; Kreiss 1987, 1990; Hallin and Puri 1994; Drost et al. 1997; Taniguchi and Kakizawa 2000; Garel and Hallin 1995), possibly with long memory (Hallin and Serroukh 1998; Hallin et al. 1999), and some nonlinear ones (Linton 1993; Drost et al. 1997; Lee and Taniguchi 2005). Results for locally stationary processes have been obtained by Hirukawa and Taniguchi (2006). Continuous-time models such as diffusions also have been intensively studied from the LAN perspective, see Kutoyants (1984, 1994, 2004) and the references therein. Finally, LAN also appears in unit-root autoregressive processes with trend (Hallin et al. 2011) and, as explained in the sequel, in the context of cointegration models (Hallin et al. 2013). Optimality problems in the LAN context are well understood, and, thanks to the simplicity of Gaussian shift experiments, admit simple solutions: see, for instance, Hájek (1970, 1972), Jeganathan (1981, 1983), van der Vaart (1991), or Sect. 11.9 of Le Cam (1986).

While the LAN case yields the most familiar type of limiting experiments, more general cases, such as *Locally Asymptotically Mixed Normal* (LAMN) or *Locally Asymptotically Brownian Functional* (LABF) ones (see Jeganathan 1995) also are quite common, essentially for dependent observations, even in very classical settings. Examples of LAMN experiments are found in supercritical Galton–Watson processes (Davies 1985), explosive and unit-root autoregressive processes (Jeganathan 1995), null-recurrent diffusions (Kutoyants 2004), or cointegration models (Phillips 1991; Boswijk 2000; Hallin et al. 2013). Optimality properties are well studied in that context (see, for instance, Jeganathan 1982, 1983; Basawa and Brockwell 1984; Janssen 1991; Bhattacharya and Basu 2006).

The situation is less bright for LABF experiments, the optimality features of which remain largely unknown; see, however, Greenwood and Wefelmeyer (1993), Gushchin (1996), Jansson (2008), Lin and Lototsky (2013), as well as Kutoyants (2004). Again, LABF appears in the context of dependent data, either in continuous time (continuous-time Gaussian autoregressions: Lin and Lototsky 2013) or in discrete time (cointegration models: Hallin et al. 2013).

The need for establishing LAN, LAMN, or LABF in a variety of situations has stimulated the production of several sets of sufficient conditions, addressing more or less general situations. Here again, a complete review of those results is impossible, and we only quote some of them, focusing on the discrete-time time series context. In the i.i.d. context, the quadratic-mean differentiability of the root of the density plays the main role: see, e.g., pp. 101–104 of Ibragimov and Has'minskii (1991), Chapter II.2 of Le Cam and Yang (1990), or Pollard (1997). In the dependent data case, a pioneering role was played by Roussas (1965, 1979) for Markov processes. Building on these results, Akritas and Johnson (1982), then Kreiss (1987, 1990) derived conditions for AR(p) and AR(∞) processes. Adopting a different approach, Swensen (1985) obtained sufficient conditions for the case of AR processes with a linear trend; his approach was extended to the ARMA, then the multivariate VARMA case by Hallin and Puri (1994) and Garel and Hallin (1995), respectively. In a nonlinear context, Linton (1993) under rather stringent conditions discusses ARCH models, while Jeganathan (1995), and Drost et al. (1997) consider more general partially nonlinear time series models; Koul and Schick (1996, 1997) and Akharif and Hallin (2003) study the case of random coefficient autoregressive models Hallin and Paindaveine (2004) thta of elliptical VARMA models.

Now, all of those results either aim at establishing LAN, LAMN, or LABF. None of them can be used in models exhibiting those structures simultaneously. This is the case, for instance, of cointegration models, which at the same time are LAN, LAMN, and LABF, with, moreover, distinct contiguity rates, depending on the direction of local alternatives—see Hallin et al. (2013) for a complete picture. A common feature of LAN, LAMN, and LABF is the availability of a particular quadratic expansion of local log-likelihoods (see Sect. 3, and Eq. (6) in Proposition 1). Experiments for which such an expansion is valid are called *Locally Asymptotically Quadratic* (LAQ), a term coined by Jeganathan (1995). Establishing LAQ thus appears as an important and natural first step in the analysis of such complex experiments.

We do not know of any sufficient conditions for LAQ alone, and existing conditions for LAN, LAMN, or LABF unfortunately never clearly split into (i) sufficient conditions for LAQ, and (ii) whatever is needed on top of LAQ for log-likelihoods to exhibit the required asymptotic distributions associated either with LAN, LAMN, or LABF. Quite on the contrary, those two issues (i) and (ii), as a rule, are intimately intertwined, so that the case of complex experiments such as those appearing in cointegration models are not covered in the literature. The first objective of this contribution is to provide a sufficient set of conditions for LAQ alone, that can handle a multiplicity of contiguity rates, and therefore is tailor-made for those complex cases.

In Proposition 1, we give such a set of conditions for the validity of the quadratic expansion (characterizing LAQ). Specifying a particular experiment is not necessary at this point, and the proposition is stated in terms of the likelihood ratios associated with two general sequences of probability distributions; as a consequence, no parameter space, and no contiguity rates, are involved. LAQ will hold for a given experiment if those sufficient conditions are satisfied at all points of the parameter space, for some adequate collection of contiguity rates (defining the so-called *local alternatives*).

To the best of our knowledge, no such condition exists in the literature so far, which makes comparisons meaningless or somewhat unfair. Note, however, that our Assumptions (a)–(d) are in line with those made by Jansson (2008) when establishing LABF for unit-root processes (see his Lemma 2), as well as with Conditions A–E of Theorem 2.1 in Drost et al. (1997) that ensures LAN under very general settings. The sufficient condition in that latter paper being itself either less restrictive or more general than other conditions available in the literature, it can be considered that our Proposition 1, which we successfully applied (Hallin et al. 2013) to cointegration models, also compares favorably with the best possible (but so far not available) ones.

Establishing LAQ or, depending on the problem at hand, LAN, LAMN, or LABF, hardly can be seen as an end in itself. Rather, such structures constitute tools in the derivation of locally asymptotically optimal inference procedures—the most common of which are efficient estimation and testing. Efficient estimation (based on Le Cam's *one-step* procedure) and optimal testing (involving, except for the very special case of testing a fully specified parameter, the estimation of nuisances) both require the validity of another asymptotic expansion—namely, the *asymptotic linearity* of the *central sequence* involved in estimation, or the *asymptotic linearity* of the test statistic in which nuisances are to be estimated.

Most asymptotic linearity results in the literature are obtained on a case-by-case basis (see Hallin and Paindaveine 2005; Hallin and Puri 1994 for typical examples). They all are established under either LAN, LAMN, or LABF. And, while sophisticated regularity assumptions are generally invoked for quadratic expansions of log-likelihoods, the assumptions used for asymptotic linearity are usually, and quite unnecessarily so, more basic. An important exception is the elegant condition considered by van der Vaart (1988) in his Proposition A.10; that result, however, is limited to the LAN case with independent observations.

The second part of this contribution (Proposition 2) is devoted to a similar set of assumptions, to be used in parallel with the assumptions of Proposition 1, in a time-series context, and under much less -restrictive LAQ conditions (so that the result *a fortiori* applies under LAN, LAMN, or LABF, or any combination thereof).

2 Main Notation and Some Preliminary Results

For each $T \in \mathbb{N}$, let $(\Omega_T, \mathscr{F}_T)$ be a measurable space on which two probability measures, $\tilde{\mathbb{P}}_T$ and $\mathbb{P}_T$, are defined. Let $\mathscr{F}_{T0} \subset \cdots \subset \mathscr{F}_{TT} \subset \mathscr{F}_T$ be a sequence of increasing σ-fields. Still for $T \in \mathbb{N}$, define the restrictions $\tilde{\mathrm{P}}_T := \tilde{\mathbb{P}}_T|_{\mathscr{F}_{TT}}$ and $\mathrm{P}_T := \mathbb{P}_T|_{\mathscr{F}_{TT}}$ of $\tilde{\mathbb{P}}_T$ and $\mathbb{P}_T$, respectively, to $\mathscr{F}_{TT}$. Using obvious notation, similarly define, for $t = 0, \ldots, T$, the restrictions $\tilde{\mathrm{P}}_{Tt} := \tilde{\mathbb{P}}_T|_{\mathscr{F}_{Tt}}$ and $\mathrm{P}_{Tt} := \mathbb{P}_T|_{\mathscr{F}_{Tt}}$. The Lebesgue decomposition of $\tilde{\mathrm{P}}_{Tt}$ on P_{Tt} (with respect to $\mathscr{F}_{Tt}$) takes the form

$$\tilde{\mathrm{P}}_{Tt}(A) = \int_A L_{Tt} \mathrm{d}\mathrm{P}_{Tt} + \tilde{\mathrm{P}}_{Tt}(A \cap N_{Tt}) \qquad A \in \mathscr{F}_{Tt},$$

where $N_{Tt} \in \mathscr{F}_{Tt}$ is such that $\mathrm{P}_{Tt}(N_{Tt}) = 0$ and L_{Tt} is the Radon–Nikodym derivative of that part of $\tilde{\mathrm{P}}_{Tt}$ which is absolutely continuous with respect to P_{Tt}.

The likelihood ratio statistic LR_T for $\tilde{\mathrm{P}}_T$ with respect to P_T is, by definition, L_{TT}. Put $LR_{T0} := L_{T0}$, and define the conditional likelihood ratio contribution of observation t as

$$LR_{Tt} := L_{Tt}/L_{T,\,t-1}, \quad t = 1, \ldots, T,$$

with the convention $0/0 = 1$. Then, the likelihood ratio statistic LR_T factorizes into

$$LR_T = \prod_{t=0}^{T} LR_{Tt}, \quad \mathrm{P}_T\text{-a.s.}$$

This factorization follows from the fact that, under P_T, $\{L_{Tt} : 0 \leq t \leq T\}$ is a supermartingale with respect to the filtration $\{\mathscr{F}_{Tt} : 0 \leq t \leq T\}$ (which is easy to check) by repeated application of the following Lemma with $X = L_{Tt}$, $Y = L_{T,\,t-1}$, and $\mathscr{F} = \mathscr{F}_{T,\,t-1}, t = 1, \ldots, T$.

Lemma 1 *Let X be a nonnegative integrable random variable, and Y a $\mathscr{F}$-measurable random variable satisfying $Y \geq \mathrm{E}\,[X|\mathscr{F}]$. Then, $X\mathbb{1}_{\{Y=0\}} = 0$ a.s.*

Proof The claim readily follows from the fact that

$$0 \leq \mathrm{E}X\mathbb{1}_{\{Y=0\}} = \mathrm{E}\mathrm{E}\,[X|\mathscr{F}]\,\mathbb{1}_{\{Y=0\}} \leq \mathrm{E}Y\mathbb{1}_{\{Y=0\}} = 0. \qquad \square$$

We conclude this section with two lemmas that are needed in the sequel. The first one is a consequence of Theorem 2.23 and Corollary 3.1 in Hall and Heyde (1980). We refer to Lemma 2.2 in Drost et al. (1997) for additional details.

Lemma 2 *If, for all $T \in \mathbb{N}$, the square integrable process $\{X_{Tt} : 1 \leq t \leq T\}$ is adapted to the filtration $(\mathscr{F}_{Tt})_{0 \leq t \leq T}$ and satisfies $\sum_{t=1}^{T} \mathrm{E}\,[X_{Tt}^2 \mid \mathscr{F}_{T,\,t-1}] = o_\mathrm{P}(1)$ as $T \to \infty$, then,*

$$\sum_{t=1}^{T} X_{Tt}^2 = o_P(1) \quad \text{and} \quad \sum_{t=1}^{T} \left(X_{Tt} - \mathrm{E}\left[X_{Tt} \mid \mathscr{F}_{T,\,t-1} \right] \right) = o_P(1)$$

as $T \to \infty$.

This second lemma follows by an application of a result due to Dvoretzky (see the proof of Theorem 2.23 in Hall and Heyde 1980).

Lemma 3 *If, for all $T \in \mathbb{N}$, the process $\{X_{Tt} : 1 \leq t \leq T\}$ is adapted to the filtration $(\mathscr{F}_{Tt})_{0 \leq t \leq T}$ and satisfies, for all $\delta > 0$,*

$$\sum_{t=1}^{T} \mathrm{E}\left[X_{Tt}^2 \mathbb{1}_{\{|X_{Tt}|>\delta\}} \mid \mathscr{F}_{T,\,t-1} \right] = o_P(1)$$

as $T \to \infty$, then $\max_{t=1,\ldots,T} |X_{Tt}| = o_P(1)$ as $T \to \infty$.

3 Quadratic Expansions of Log-Likelihood Ratios

The following proposition provides a general sufficient condition for the existence of a quadratic expansion of local log-likelihood ratios. All limits, o_P, and O_P quantities are to be understood as $T \to \infty$ and, unless explicitly stated otherwise, under P_T.

Proposition 1 *Suppose that, for some $k \in \mathbb{N}$, there exist, for each $T \in \mathbb{N}$, $\mathscr{F}_{Tt}$-measurable mappings $S_{Tt} : \Omega_T \to \mathbb{R}^k$ and $R_{Tt} : \Omega_T \to \mathbb{R}$, $t = 1, \ldots, T$, such that the conditional likelihood ratio contribution LR_{Tt} can be written as*

$$LR_{Tt} = \left(1 + \frac{1}{2} \left(h_T' S_{Tt} + R_{Tt} \right) \right)^2, \tag{1}$$

where

(a) h_T is a bounded (deterministic) sequence in $\mathbb{R}^k$;
(b) for each $T \in \mathbb{N}$, $\{S_{Tt} : 1 \leq t \leq T\}$ is a P_T-square integrable martingale difference array with respect to the filtration $\{\mathscr{F}_{Tt} : 0 \leq t \leq T\}$, satisfying the conditional Lindeberg condition and with tight squared conditional moments, i.e., such that

$$\mathrm{E}_{\mathrm{P}_T}\left[S_{Tt} \mid \mathscr{F}_{T,\,t-1} \right] = 0, \quad t = 1, \ldots, T, \tag{2}$$

$$\sum_{t=1}^{T} \mathrm{E}_{\mathrm{P}_T}\left[\left(h_T' S_{Tt} \right)^2 \mathbb{1}_{\{|h_T' S_{Tt}|>\delta\}} \mid \mathscr{F}_{T,\,t-1} \right] = o_P(1)\, \text{for all}\ \delta > 0, \tag{3}$$

and

$$J_T := \sum_{t=1}^{T} \mathrm{E}_{\mathrm{P}_T} \left[S_{Tt} S_{Tt}' \mid \mathscr{F}_{T,\, t-1} \right] = O_{\mathrm{P}}(1);$$

(c) the remainder terms R_{Tt} and the null-sets N_{Tt} from the Lebesgue decomposition of $\tilde{\mathrm{P}}_T$ on P_T are sufficiently small, i.e.,

$$\sum_{t=1}^{T} \mathrm{E}_{\mathrm{P}_T} \left[R_{Tt}^2 \mid \mathscr{F}_{T,\, t-1} \right] = o_{\mathrm{P}}(1) \tag{4}$$

and

$$\sum_{t=1}^{T} \left(1 - \mathrm{E}_{\mathrm{P}_T} \left[LR_{Tt} \mid \mathscr{F}_{T,\, t-1} \right] \right) = o_{\mathrm{P}}(1); \tag{5}$$

(d) $\log LR_{T0} = o_{\mathrm{P}}(1)$.

Then the log-likelihood ratio admits the quadratic expansion

$$\log LR_T = h_T' \sum_{t=1}^{T} S_{Tt} - \frac{1}{2} h_T' J_T h_T + o_{\mathrm{P}}(1). \tag{6}$$

Proof Let $r : 2x \mapsto r(2x) := 2\left(\log(1+x) - x + x^2/2 \right)$, and rewrite the log-likelihood ratio statistic as

$$\log LR_T = \sum_{t=0}^{T} \log LR_{Tt} = o_{\mathrm{P}}(1) + \sum_{t=1}^{T} h_T' S_{Tt} - \frac{1}{2} h_T' J_T h_T$$

$$+ \frac{1}{4} \left(h_T' J_T h_T - \sum_{t=1}^{T} \left(h_T' S_{Tt} \right)^2 \right) + \sum_{t=1}^{T} \left(R_{Tt} - \mathrm{E}_{\mathrm{P}_T} \left[R_{Tt} \mid \mathscr{F}_{T,\, t-1} \right] \right)$$

$$- \frac{1}{4} \sum_{t=1}^{T} R_{Tt}^2 - \frac{1}{2} \sum_{t=1}^{T} h_T' S_{Tt} R_{Tt} + \left(\sum_{t=1}^{T} \mathrm{E}_{\mathrm{P}_T} \left[R_{Tt} \mid \mathscr{F}_{T,\, t-1} \right] \right.$$

$$\left. + \frac{1}{4} h_T' J_T h_T \right) + \sum_{t=1}^{T} r \left(h_T' S_{Tt} + R_{Tt} \right), \tag{7}$$

where we used Condition (d) to neglect the first term $\log LR_{T0}$. To establish (6), we show that the six remainder terms on the right-hand side of (7) all converge to zero in P_T-probability.

By Theorem 2.23 in Hall and Heyde (1980), Condition (a), and (1)–(2), we have

$$\sum_{t=1}^{T} \left(h_T' S_{Tt} \right)^2 - h_T' J_T h_T = o_{\mathrm{P}}(1), \tag{8}$$

which shows that the first remainder term is indeed $o_{\mathrm{P}}(1)$.

Since $(L_{Tt})_{0 \le t \le T}$ is a P_T-supermartingale, we have $\mathrm{E}_{\mathrm{P}_T} L R_{Tt} \le 1$. Since S_{Tt} is also P_T-square integrable, it follows from (1) that R_{Tt} is P_T-square integrable. From Lemma 2 and (4), we now immediately obtain

$$\sum_{t=1}^{T} \left(R_{Tt} - \mathrm{E}_{\mathrm{P}_T} \left[R_{Tt} \mid \mathscr{F}_{T,\, t-1} \right] \right) = o_{\mathrm{P}}(1) \text{ and } \sum_{t=1}^{T} R_{Tt}^2 = o_{\mathrm{P}}(1), \tag{9}$$

i.e., the second and third remainder terms also are negligible.

Next we show that the remainder term $(1/2) \sum_{t=1}^{n} h_T' S_{Tt} R_{Tt}$ vanishes asymptotically. First note that Condition (a), (1) and (8) jointly imply $\sum_{t=1}^{T} (h_T' S_{Tt})^2 = O_{\mathrm{P}}(1)$. Combined with (9), an application of the Cauchy–Schwarz inequality thus yields the convergence of the fourth remainder term.

To prove the negligibility of the fifth remainder term in (7), observe that (1), (2), and (4), combined with the Cauchy–Schwarz inequality again, entail

$$\sum_{t=1}^{T} \left(\mathrm{E}_{\mathrm{P}_T} \left[L R_{Tt} \mid \mathscr{F}_{T,\, t-1} \right] - 1 \right)$$

$$= \sum_{t=1}^{T} \mathrm{E}_{\mathrm{P}_T} \left[h_T' S_{Tt} \mid \mathscr{F}_{T,\, t-1} \right] + \sum_{t=1}^{T} \mathrm{E}_{\mathrm{P}_T} \left[R_{Tt} \mid \mathscr{F}_{T,\, t-1} \right]$$

$$+ \frac{1}{4} \sum_{t=1}^{T} \mathrm{E}_{\mathrm{P}_T} \left[\left(h_T' S_{Tt} \right)^2 \mid \mathscr{F}_{T,\, t-1} \right] + \frac{1}{4} \sum_{t=1}^{T} \mathrm{E}_{\mathrm{P}_T} \left[R_{Tt}^2 \mid \mathscr{F}_{T,\, t-1} \right]$$

$$+ \frac{1}{2} \sum_{t=1}^{T} \mathrm{E}_{\mathrm{P}_T} \left[\left(h_T' S_{Tt} \right) R_{Tt} \mid \mathscr{F}_{T,\, t-1} \right]$$

$$= \sum_{t=1}^{T} \mathrm{E}_{\mathrm{P}_T} \left[R_{Tt} \mid \mathscr{F}_{T,\, t-1} \right] + \frac{1}{4} h_T' J_T h_T + o_{\mathrm{P}}(1).$$

Now, the second part of (4) implies

$$\sum_{t=1}^{T} \mathrm{E}_{\mathrm{P}_T} \left[R_{Tt} \mid \mathscr{F}_{T,\, t-1} \right] + \frac{1}{4} h_T' J_T h_T = o_{\mathrm{P}}(1). \tag{10}$$

Thus, the fifth remainder term in (7) also is negligible.

Turning to the sixth and last remainder term, let us first show that

$$\max_{t=1,\dots,T} \left| h'_T S_{Tt} + R_{Tt} \right| = o_{\mathrm{P}}(1) \quad \text{and} \quad \sum_{t=1}^{T} \left| h'_T S_{Tt} + R_{Tt} \right|^3 = o_{\mathrm{P}}(1). \tag{11}$$

As (3) and (4) yield, for $\delta > 0$,

$$\sum_{t=1}^{T} \mathrm{E}_{\mathrm{P}_T} \left[(h'_T S_{Tt} + R_{Tt})^2 \mathbb{1}_{\{|h'_T S_{Tt}+R_{Tt}|>\delta\}} \mid \mathscr{F}_{T,\,t-1} \right]$$

$$\leq 4 \sum_{t=1}^{T} \mathrm{E}_{\mathrm{P}_T} \left[(h'_T S_{Tt})^2 \mathbb{1}_{\{|h'_T S_{Tt}|>\delta/2\}} \mid \mathscr{F}_{T,\,t-1} \right] + 4 \sum_{t=1}^{T} \mathrm{E}_{\mathrm{P}_T} \left[R^2_{Tt} \mid \mathscr{F}_{T,\,t-1} \right]$$

$$= o_{\mathrm{P}}(1),$$

the first part of (11) follows as an application of Lemma 3. The second part is obtained from the latter by taking out the maximum (which tends to zero) and by observing that the remaining quadratic term is bounded in probability. In view of the first part of (11), indeed, it is sufficient to study the behavior of the final remainder term on the event $\left\{ \left| h'_T S_{Tt} + R_{Tt} \right| \leq 1 \right\}$. On this set, this remainder term is bounded: using the fact that

$$\left| \log (1 + x) - x + \frac{1}{2} x^2 \right| \leq \frac{2}{3} x^3 \quad \text{for} \quad |x| \leq \frac{1}{2},$$

indeed, we obtain

$$\left| \sum_{t=1}^{T} r \left(h'_T S_{Tt} + R_{Tt} \right) \right| \leq \frac{4}{3} \sum_{t=1}^{T} \left(h'_T S_{Tt} + R_{Tt} \right)^3.$$

Convergence to zero is now obtained from the second part of (11). This completes the proof of the proposition. $\qquad\square$

4 Asymptotic Linearity: General Result

This section provides a sufficient condition for the asymptotic linearity of a fairly general class of statistics, extending and generalizing Proposition A.10 in van der Vaart (1988) to the case of serially dependent observations under possibly non-LAN limit experiments.

All limits are taken as $T \to \infty$ and, unless otherwise specified, under P_T.

Proposition 2 *Let, for each $T \in \mathbb{N}$, $\{\tilde{Z}_{Tt} : 1 \le t \le T\}$ and $\{Z_{Tt} : 1 \le t \le T\}$ be a $\tilde{P}_T$ and a P_T-square integrable martingale difference array, respectively. Suppose that Conditions (a)–(d) in Proposition 1 hold, as well as the following Conditions (e)–(h):*

(e) $(\sum_{t=1}^{T} S_{Tt}, J_T)$ converges in distribution to a limit (Δ, J) satisfying

$$\mathrm{E} \exp\left(a'\Delta - \frac{1}{2} a' J a \right) = 1 \quad \text{for all } a \in \mathbb{R}^k;$$

(f) $\displaystyle\sum_{t=1}^{T} \mathrm{E}_{\mathrm{P}_T} \left[\left(\tilde{Z}_{Tt} \sqrt{LR_{Tt}} - Z_{Tt} \right)^2 \mid \mathscr{F}_{T,\,t-1} \right] = o_\mathrm{P}(1);$

(g)
$$\sum_{t=1}^{T} \mathrm{E}_{\tilde{\mathrm{P}}_T} \left[\tilde{Z}_{Tt}^2 \mid \mathscr{F}_{T,\,t-1} \right] = O_\mathrm{P}(1) \text{ under } \tilde{\mathrm{P}}_T, \text{ and}$$
$$\sum_{t=1}^{T} \mathrm{E}_{\mathrm{P}_T} \left[Z_{Tt}^2 \mid \mathscr{F}_{T,\,t-1} \right] = O_\mathrm{P}(1) \text{ under } \mathrm{P}_T;$$

(h) the conditional Lindeberg condition holds for $\{\tilde{Z}_{Tt} : 1 \le t \le T\}$ under $\tilde{\mathrm{P}}_T$,

namely, for all $\delta > 0$, $\displaystyle\sum_{t=1}^{T} \mathrm{E}_{\tilde{\mathrm{P}}_T} \left[\tilde{Z}_{Tt}^2 \mathbb{1}_{\left\{ \left| \tilde{Z}_{Tt} \right| > \delta \right\}} \mid \mathscr{F}_{T,\,t-1} \right] = o_\mathrm{P}(1) \text{ under } \tilde{\mathrm{P}}_T.$

Then, letting
$$\tilde{I}_T := \sum_{t=1}^{T} \tilde{\iota}_{Tt} := \sum_{t=1}^{T} \mathrm{E}_{\mathrm{P}_T} \left[(h_T' S_{Tt}) Z_{Tt} \mid \mathscr{F}_{T,\,t-1} \right], \tag{12}$$

we have, under P_T,
$$\sum_{t=1}^{T} \tilde{Z}_{Tt} = \sum_{t=1}^{T} Z_{Tt} - \tilde{I}_T + o_\mathrm{P}(1). \tag{13}$$

Proof The proof decomposes into four parts. In Part 1, we show that (13) holds if
$$\sum_{t=1}^{T} \tilde{Z}_{Tt} \left(1 - \sqrt{LR_{Tt}} \right) + \frac{1}{2} \tilde{I}_T = o_\mathrm{P}(1). \tag{14}$$

In Part 2, we show that (14) holds provided that
$$\sum_{t=1}^{T} \tilde{Z}_{Tt} (h_T' S_{Tt}) - \tilde{I}_T = o_\mathrm{P}(1). \tag{15}$$

In Part 3, we introduce a new sequence of probability measures (P'_T) and show that it is contiguous to (P_T). In Part 4, we establish that (15) holds under the new sequence (P'_T). In view of contiguity, thus, it also holds under (P_T), which concludes the proof.

Note that Lemma 1, Condition (e), and Le Cam's first lemma imply that $(\tilde{P}_T)$ and (P_T) are contiguous. It follows that o_P's and O_P's under $(\tilde{P}_T)$ and (P_T) coincide; therefore, in the sequel, we safely can write o_P and O_P without worrying whether $(\tilde{P}_T)$ or (P_T) is the underlying sequence of probability measures.

Part 1. Recalling the definition (12) of $\tilde{I}_T$, we have

$$\sum_{t=1}^{T}\left\{\tilde{Z}_{Tt} - Z_{Tt} + \tilde{\iota}_{Tt}\right\} = \sum_{t=1}^{T}\tilde{Z}_{Tt}\left(1 - \sqrt{LR_{Tt}}\right) + \frac{1}{2}\tilde{I}_T$$

$$+ \sum_{t=1}^{T}\left\{\tilde{Z}_{Tt}\sqrt{LR_{Tt}} - Z_{Tt} - \mathrm{E}_{P_T}\left[\tilde{Z}_{Tt}\sqrt{LR_{Tt}} \mid \mathscr{F}_{T,\,t-1}\right]\right\}$$

$$+ \sum_{t=1}^{T}\left\{\mathrm{E}_{P_T}\left[\tilde{Z}_{Tt}\sqrt{LR_{Tt}} \mid \mathscr{F}_{T,\,t-1}\right] + \frac{1}{2}\tilde{\iota}_{Tt}\right\};$$

hence, (14) implies (13) in case

$$\sum_{t=1}^{T}\left\{\tilde{Z}_{Tt}\sqrt{LR_{Tt}} - Z_{Tt} - \mathrm{E}_{P_T}\left[\tilde{Z}_{Tt}\sqrt{LR_{Tt}} \mid \mathscr{F}_{T,\,t-1}\right]\right\} = o_P(1) \qquad (16)$$

and

$$\sum_{t=1}^{T}\left\{\mathrm{E}_{P_T}\left[\tilde{Z}_{Tt}\sqrt{LR_{Tt}} \mid \mathscr{F}_{T,\,t-1}\right] + \frac{1}{2}\tilde{\iota}_{Tt}\right\} = o_P(1). \qquad (17)$$

As (16) is implied by Condition (f) and Lemma 2 (recall $\mathrm{E}_{P_T}[Z_{Tt} \mid \mathscr{F}_{T,\,t-1}] = 0$), we only need to show that (17) holds in order to complete Part 1. We have

$$\sum_{t=1}^{T}\mathrm{E}_{P_T}\left[\tilde{Z}_{Tt}\sqrt{LR_{Tt}} \mid \mathscr{F}_{T,\,t-1}\right]$$

$$= \sum_{t=1}^{T}\mathrm{E}_{P_T}\left[Z_{Tt}(1 - \sqrt{LR_{Tt}}) \mid \mathscr{F}_{T,\,t-1}\right]$$

$$+ \sum_{t=1}^{T}\mathrm{E}_{P_T}\left[(\tilde{Z}_{Tt}\sqrt{LR_{Tt}} - Z_{Tt})(1 - \sqrt{LR_{Tt}}) \mid \mathscr{F}_{T,\,t-1}\right]$$

$$+ \sum_{t=1}^{T} \mathrm{E}_{\mathrm{P}_T} \left[\tilde{Z}_{Tt} LR_{Tt} \mid \mathscr{F}_{T,\, t-1} \right]$$

$$= -\frac{1}{2}\tilde{I}_T - \frac{1}{2} r_T^{(1)} + r_T^{(2)} + r_T^{(3)},$$

with

$$r_T^{(1)} = \sum_{t=1}^{T} \mathrm{E}_{\mathrm{P}_T} \left[Z_{Tt} R_{Tt} \mid \mathscr{F}_{T,\, t-1} \right],$$

$$r_T^{(2)} = \sum_{t=1}^{T} \mathrm{E}_{\mathrm{P}_T} \left[(\tilde{Z}_{Tt}\sqrt{LR_{Tt}} - Z_{Tt})(1 - \sqrt{LR_{Tt}}) \mid \mathscr{F}_{T,\, t-1} \right], \text{ and}$$

$$r_T^{(3)} = \sum_{t=1}^{T} \mathrm{E}_{\mathrm{P}_T} \left[\tilde{Z}_{Tt} LR_{Tt} \mid \mathscr{F}_{T,\, t-1} \right].$$

Starting with $r_T^{(1)}$, note that

$$|r_T^{(1)}|^2 \leq \left(\sum_{t=1}^{T} \sqrt{\mathrm{E}_{\mathrm{P}_T}\left[Z_{Tt}^2 \mid \mathscr{F}_{T,\, t-1}\right]}\sqrt{\mathrm{E}_{\mathrm{P}_T}\left[R_{Tt}^2 \mid \mathscr{F}_{T,\, t-1}\right]} \right)^2$$

$$\leq \sum_{t=1}^{T} \mathrm{E}_{\mathrm{P}_T}\left[Z_{Tt}^2 \mid \mathscr{F}_{T,\, t-1}\right] \sum_{t=1}^{T} \mathrm{E}_{\mathrm{P}_T}\left[R_{Tt}^2 \mid \mathscr{F}_{T,\, t-1}\right],$$

so that (4) and Condition (g) imply $r_T^{(1)} = o_{\mathrm{P}}(1)$. In the same way, (1), (4), and Condition (f) yield $r_T^{(2)} = o_{\mathrm{P}}(1)$. As for $r_T^{(3)}$, since $\mathrm{E}_{\tilde{\mathrm{P}}_T}\left[\tilde{Z}_{Tt} | \mathscr{F}_{T,\, t-1}\right] = 0$, we obtain, using (4) and Condition (g) again,

$$|r_T^{(3)}|^2 = \left| \sum_{t=1}^{T} \mathrm{E}_{\tilde{\mathrm{P}}_T}\left[\tilde{Z}_{Tt}\mathbb{1}_{N_{Tt}} | \mathscr{F}_{T,\, t-1}\right] \right|^2$$

$$\leq \sum_{t=1}^{T} \mathrm{E}_{\tilde{\mathrm{P}}_T}\left[\tilde{Z}_{Tt}^2 | \mathscr{F}_{T,\, t-1}\right] \sum_{t=1}^{T} (1 - \mathrm{E}_{\mathrm{P}_T}\left[LR_{Tt} \mid \mathscr{F}_{T,\, t-1}\right]) = o_{\mathrm{P}}(1).$$

Part 2. From Cauchy–Schwarz, we have

$$\left| \sum_{t=1}^{T} \tilde{Z}_{Tt}(1 - \sqrt{LR_{Tt}}) + \frac{1}{2} \sum_{t=1}^{T} \tilde{Z}_{Tt}(h_T' S_{Tt}) \right|$$

$$= \frac{1}{2} \left| \sum_{t=1}^{T} \tilde{Z}_{Tt} R_{Tt} \right| \leq \frac{1}{2} \sqrt{\sum_{t=1}^{T} \tilde{Z}_{Tt}^2} \sqrt{\sum_{t=1}^{T} R_{Tt}^2}.$$

Now, by (9), $\sum_{t=1}^{T} R_{Tt}^2 = o_P(1)$ and, by Conditions (g) and (h), and an application of Hall and Heyde (1980, Theorem 2.23), $\sum_{t=1}^{T} \tilde{Z}_{Tt}^2 = O_P(1)$. Hence, (14) follows from (15).

Part 3. For all $T \in \mathbb{N}$, define the new sequence $(P'_{Tt})_{t=1}^{T}$ of probability measures on $\mathscr{F}_{Tt}$, where P'_{Tt} is absolutely continuous with respect to P_{Tt}, with density

$$\frac{dP'_{Tt}}{dP_{Tt}} := \prod_{s=1}^{t} \sqrt{LR_{Ts}} \, c_{Ts}$$

with, for $s = 1, \ldots, T$, $c_{Ts}^{-1} := E_{P_T}\left[\sqrt{LR_{Ts}} \mid \mathscr{F}_{T,s-1} \right]$. Note that the probability that all c_{Ts}^{-1} are strictly positive tends to one, since (4) implies

$$\lim_{T \to \infty} P_T \left[\exists s \in \{1, \ldots, T\} : c_{Ts}^{-1} = 0 \right]$$

$$\leq \lim_{T \to \infty} P_T \left[\sum_{t=1}^{T} (1 - E_{P_T}[LR_{Tt} | \mathscr{F}_{T,\,t-1}]) \geq 1 \right] = 0.$$

In the sequel, we thus safely can ignore the event $\{\exists s \in \{1, \ldots, T\} : c_{Ts}^{-1} = 0\}$. Defining $P'_T := P'_{TT}$, note that P'_{Tt} is the restriction of P'_T to $\mathscr{F}_{T,t}$. Because of (2), we have $c_{Ts}^{-1} = 1 + \frac{1}{2} E_{P_T}\left[R_{Tt} \mid \mathscr{F}_{T,\,t-1} \right]$. This yields, using an expansion of $\log(1+x)$, (4), and (10),

$$\sum_{t=1}^{T} \log c_{Tt}^{-1} = -\frac{1}{8} h_T' J_T h_T + o_P(1).$$

Moreover, an application of Lemma 3 and (4) yields $\max_{t=1,\ldots,T} |c_{Tt}^{-1} - 1| = o_P(1)$, and thus also

$$\max_{t=1,\ldots,T} |c_{Tt} - 1| = o_P(1). \tag{18}$$

Inserting (6) and recalling that $\log LR_{T0} = o_{\mathrm{P}}(1)$, we obtain, under P_T,

$$\log \frac{\mathrm{dP}_T'}{\mathrm{dP}_T} = \frac{1}{2} \sum_{t=1}^{T} \log LR_{Tt} - \sum_{t=1}^{T} \log c_{Tt}^{-1} + o_{\mathrm{P}}(1)$$

$$= \frac{1}{2} \sum_{t=1}^{T} h_T' S_{Tt} - \frac{1}{8} h_T' \tilde{I}_T h_T + o_{\mathrm{P}}(1).$$

Condition (e) and Le Cam's first lemma entail that the sequences (P_T') and (P_T) are mutually contiguous. This completes Part 3 of the proof.

Part 4. Let us show that, under the measures (P_T'),

$$\sum_{t=1}^{T} \mathrm{E}_{\mathrm{P}_T'} \left[\tilde{Z}_{Tt}(h_T' S_{Tt}) \mid \mathscr{F}_{T, \, t-1} \right] = \tilde{I}_T + o_{\mathrm{P}}(1) \tag{19}$$

and

$$\sum_{t=1}^{T} \tilde{Z}_{Tt}(h_T' S_{Tt}) = \sum_{t=1}^{T} \mathrm{E}_{\mathrm{P}_T'} \left[\tilde{Z}_{Tt}(h_T' S_{Tt}) \mid \mathscr{F}_{T, \, t-1} \right] + o_{\mathrm{P}}(1). \tag{20}$$

Since $o_{\mathrm{P}}(1)$'s under (P_T') are $o_{\mathrm{P}}(1)$'s under the contiguous (P_T) too, a combination of these two results yields (15) and concludes the proof.

Starting with (19), we have

$$\sum_{t=1}^{T} \mathrm{E}_{\mathrm{P}_T'} \left[\tilde{Z}_{Tt}(h_T' S_{Tt}) \mid \mathscr{F}_{T, \, t-1} \right] = \sum_{t=1}^{T} c_{Tt} \mathrm{E}_{\mathrm{P}_T} \left[\tilde{Z}_{Tt}\sqrt{LR}_{Tt}(h_T' S_{Tt}) \mid \mathscr{F}_{T, \, t-1} \right]$$

$$= \tilde{I}_T + \sum_{t=1}^{T} (c_{Tt} - 1) \mathrm{E}_{\mathrm{P}_T} \left[Z_{Tt}(h_T' S_{Tt}) \mid \mathscr{F}_{T, \, t-1} \right]$$

$$+ \sum_{t=1}^{T} c_{Tt} \mathrm{E}_{\mathrm{P}_T} \left[(\tilde{Z}_{Tt}\sqrt{LR}_{Tt} - Z_{Tt})(h_T' S_{Tt}) \mid \mathscr{F}_{T, \, t-1} \right].$$

Condition (f) and (18) imply (19) since $\sum_{t=1}^{T} \mathrm{E}_{\mathrm{P}_T} \left[(h_T' S_{Tt})^2 \mid \mathscr{F}_{T, \, t-1} \right] = O_{\mathrm{P}}(1)$ (see (1)) and $\sum_{t=1}^{T} \mathrm{E}_{\mathrm{P}_T} \left[Z_{Tt}^2 \mid \mathscr{F}_{T, \, t-1} \right] = O_{\mathrm{P}}(1)$ (see Condition (g)).

Turning to (20), first note that $\sum_{t=1}^{T} (h_T' S_{Tt})^2 = O_{\mathrm{P}}(1)$ and $\sum_{t=1}^{T} \tilde{Z}_{Tt}^2 = O_{\mathrm{P}}(1)$ by an application of Hall and Heyde (1980, Theorem 2.23) and (3), (1), Condition (g) and Condition (h), respectively. Hence,

$$\sum_{t=1}^{T} |\tilde{Z}_{Tt}||h_T' S_{Tt}| = O_{\mathrm{P}}(1) \quad \text{and} \quad \sum_{t=1}^{T} \mathrm{E}_{\mathrm{P}_T'}[|\tilde{Z}_{Tt}||h_T' S_{Tt}| \mid \mathscr{F}_{T,\,t-1}] = O_{\mathrm{P}}(1).$$

Let $\varepsilon, \delta > 0$. In view of the previous remarks, we can find B and T_1 such that, for $T \geq T_1$,

$$\mathrm{P}_T'(\mathscr{A}_\delta^{(T)}) \leq \delta/6$$

with

$$\mathscr{A}_\delta^{(T)} := \left\{ \sum_{t=1}^{T} \left| (h_T' S_{Tt})\tilde{Z}_{Tt} - \mathrm{E}_{\mathrm{P}_T'}\left[(h_T' S_{Tt})\tilde{Z}_{Tt} \mid \mathscr{F}_{T,\,t-1}\right]\right| > B \right\}.$$

Setting $\eta := \min\{1, \sqrt{\delta}\varepsilon(108(B+2))^{-1/2}\}$ and

$$\mathscr{A}_{\eta,Tt} := \left\{ |Z_{Tt}| \leq \eta \right\} \bigcap \left\{ |h_T' S_{Tt}| \leq \eta \right\},$$

decompose

$$\sum_{t=1}^{T} \tilde{Z}_{Tt}(h_T' S_{Tt}) - \sum_{t=1}^{T} \mathrm{E}_{\mathrm{P}_T'}\left[\tilde{Z}_{Tt}(h_T' S_{Tt}) \mid \mathscr{F}_{T,\,t-1}\right] = p_T^{(1)} - p_T^{(2)} + p_T^{(3)},$$

with

$$p_T^{(1)} := \sum_{t=1}^{T} \tilde{Z}_{Tt}(h_T' S_{Tt})\mathbb{1}_{\mathscr{A}_{\eta,Tt}^c},$$

$$p_T^{(2)} := \sum_{t=1}^{T} \mathrm{E}_{\mathrm{P}_T'}\left[\tilde{Z}_{Tt}(h_T' S_{Tt})\mathbb{1}_{\mathscr{A}_{\eta,Tt}^c} \mid \mathscr{F}_{T,\,t-1}\right], \text{ and}$$

$$p_T^{(3)} := \sum_{t=1}^{T} \tilde{Z}_{Tt}(h_T' S_{Tt})\mathbb{1}_{\mathscr{A}_{\eta,Tt}} - \sum_{t=1}^{T} \mathrm{E}_{\mathrm{P}_T'}\left[\tilde{Z}_{Tt}(h_T' S_{Tt})\mathbb{1}_{\mathscr{A}_{\eta,Tt}} \mid \mathscr{F}_{T,\,t-1}\right].$$

Let us show that there exists $T^\star$ such that, for all $T \geq T^\star$, $\mathrm{P}_T'\left(|p_T^{(i)}| > \varepsilon/3 \right) \leq \delta/3$, which, as $\varepsilon > 0$ and $\delta > 0$ can be taken arbitrarily small, yields (20). Applying Theorem 2.23 in Hall and Heyde (1980), (1), (3), Condition (g), and Condition (h), we obtain

$$\sum_{t=1}^{T} \tilde{Z}_{Tt}^2 1\{|\tilde{Z}_{Tt}| > \eta\} + \sum_{t=1}^{T} (h_T' S_{Tt})^2 1\{|h_T' S_{Tt}| > \eta\} = o_{\mathrm{P}}(1).$$

This yields, using (1) and Condition (g) again,

$$
|p_T^{(1)}| \le \sqrt{\sum_{t=1}^{T} (h_T' S_{Tt})^2 \mathbb{1}_{\{|h_T' S_{Tt}|>\eta\}}} \sqrt{\sum_{t=1}^{T} \tilde{Z}_{Tt}^2}
$$
$$
+ \sqrt{\sum_{t=1}^{T} (h_T' S_{Tt})^2} \sqrt{\sum_{t=1}^{T} \tilde{Z}_{Tt}^2 \mathbb{1}_{\{|\tilde{Z}_{Tt}|>\eta\}}} = o_{\mathrm{P}}(1).
$$

From (3), (1), Condition (g) and Condition (h), we also obtain

$$
|p_T^{(2)}| \le \sqrt{\sum_{t=1}^{T} c_{Tt}^2 \mathrm{E}_{\tilde{\mathrm{P}}_T} \left[\tilde{Z}_{Tt}^2 \mathbb{1}_{\{|\tilde{Z}_{Tt}|>\eta\}} \mid \mathscr{F}_{T,\,t-1} \right]} \sqrt{\sum_{t=1}^{T} \mathrm{E}_{\mathrm{P}_T} \left[(h_T' S_{Tt})^2 \mid \mathscr{F}_{T,\,t-1} \right]}
$$
$$
+ \sqrt{\sum_{t=1}^{T} c_{Tt}^2 \mathrm{E}_{\tilde{\mathrm{P}}_T} \left[\tilde{Z}_{Tt}^2 \mid \mathscr{F}_{T,\,t-1} \right]} \sqrt{\sum_{t=1}^{T} \mathrm{E}_{\mathrm{P}_T} \left[(h_T' S_{Tt})^2 \mathbb{1}_{\{|h_T' S_{Tt}|>\eta\}} \mid \mathscr{F}_{T,\,t-1} \right]}
$$
$$
= o_{\mathrm{P}}(1).
$$

Hence, there exists T_2 such that, for all $T \ge T_2$, $\mathrm{P}_T' \left(|p_T^{(j)}| > \varepsilon/3 \right) \le \delta/3$ for $j = 1, 2$. Next, define the P_T'-martingales

$$
\left\{ A_{Tt} := \sum_{s=1}^{t} \left\{ \tilde{Z}_{Tt}(h_T' S_{Tt}) \mathbb{1}_{\mathscr{A}_{\eta,Tt}} - \mathrm{E}_{\mathrm{P}_T'}[\tilde{Z}_{Tt}(h_T' S_{Tt}) \mathbb{1}_{\mathscr{A}_{\eta,Tt}} \mid \mathscr{F}_{T,s-1}] \right\} : 1 \le t \le T \right\},
$$

the stopping times

$$
\mathscr{S}^{(T)} := \inf \left\{ t \in \mathbb{N} \mid \sum_{s=1}^{t} |\Delta A_{Ts}| > B \right\},
$$

and the processes

$$
\left\{ M_{Tt} := A_{T,t \wedge \mathscr{S}^{(T)}} : 1 \le t \le T \right\},
$$

namely, the stopped versions of the martingales $\{A_{Tt} : 1 \le t \le T\}$—which thus also are martingales. Note that $|\Delta A_{Tt}| \le 2\eta^2$. We obtain

$$
\mathrm{E}_{\mathrm{P}_T'} M_{TT}^2 = \sum_{t=1}^{T} \mathrm{E}_{\mathrm{P}_T'} (M_{Tt} - M_{T,\,t-1})^2 \le \mathrm{E}_{\mathrm{P}_T'} \left[\sum_{t=1}^{\mathscr{S}^{(T)}} (\Delta A_{Tt})^2 \right]
$$
$$
\le 2\eta^2 \mathrm{E}_{\mathrm{P}_T'} \left[\sum_{t=1}^{\mathscr{S}^{(T)}} |\Delta A_{Tt}| \right] \le 2\eta^2 (B + 2\eta^2).
$$

So, for $T \geq T_1$, we have

$$P'_T \left(|p_T^{(3)}| > \varepsilon/3 \right) = P'_T \left(|A_{TT}| > \varepsilon/3 \right) \leq P'_T \left(M_{TT} \neq A_{TT} \right) + P'_T \left(|M_{TT}| > \varepsilon/3 \right)$$

$$\leq P'_T (\mathscr{S}^{(T)} \leq T) + P'_T \left(|M_{TT}| > \varepsilon/3 \right)$$

$$\leq P'_T (\mathscr{A}_\delta^{(T)}) + P'_T \left(|M_{TT}| > \varepsilon/3 \right) \leq \frac{\delta}{6} + \frac{18\eta^2 (B+2)}{\varepsilon^2} \leq \frac{\delta}{3}.$$

Letting $T^\star := \max\{T_1, T_2\}$ completes the proof. $\qquad\square$

Acknowledgments The research of Marc Hallin is supported by the Belgian Science Policy Office (2012–2017) Interuniversity Attraction Poles and a crédit aux chercheurs of the FNRS.

References

Akharif, A., & Hallin, M. (2003). Efficient detection of random coefficients in autoregressive models. *Annals of Statistics, 31*, 675–704.

Akritas, M. G., & Johnson, R. A. (1982). Efficiencies of tests and estimators in autoregressions under nonnormal error distribution. *Annals of the Institute of Statistical Mathematics, 34*, 579–589.

Basawa, I. V., & Brockwell, P. J. (1984). Asymptotic conditional inference for regular nonergodic models with an application to autoregressive processes. *Annals of Statistics, 12*, 161–171.

Bhattacharya, D., & Basu, A. K. (2006). Local asymptotic minimax risk bounds in a locally asymptotically mixture of normal experiments under asymmetric loss. *In Optimality: The Second Erich L. Lehmann Symposium. IMS Lecture Notes-Monograph Series* (Vol. 49, pp. 312–321).

Bickel, P. J., Klaassen, C. A. J., Ritov, Y., & Wellner, J. A. (1993). *Efficient and adaptive estimation for semiparametric models*. Baltimore: Johns Hopkins University Press.

Boswijk, H. P. (2000). Mixed normality and ancillarity in $I(2)$ systems. *Econometric Theory, 16*, 878–904.

Davies, R. B. (1985). Asymptotic inference when the amount of information is random. In L. M. Le Cam, R. A. Olshen (Eds.), *Proceedings of the Berkeley Conference in Honor of Jerzy Neyman and Jack Kiefer* (Vol. 2). Belmont: Wadsworth

Drost, F. C., Klaassen, C. A. J., & Werker, B. J. M. (1997). Adaptive estimation in time series models. *Annals of Statistics, 25*, 786–818.

Garel, B., & Hallin, M. (1995). Local asymptotic normality of multivariate ARMA processes with a linear trend. *Annals of the Institute of Statistical Mathematics, 47*, 551–579.

Greenwood, P. E., & Wefelmeyer, W. (1993). Asymptotic minimax results for stochastic process families with critical points. *Stochastic Processes and Their Applications, 44*, 107–116.

Gushchin, A. A. (1996). Asymptotic optimality of parameter estimators under the LAQ condition. *Theory of Probability and Its Applications, 40*, 261–272.

Hájek, J. (1970). A characterization of limiting distributions of regular estimates. *Zeitschrift für Wahrscheinlichkeitstheorie und Verwandte Gebiete, 14*, 323–330.

Hájek, J. (1972). Local asymptotic minimax and admissibility in estimation. In *Proceedings of the Sixth Berkeley Symposium on Mathematical Statistics and Probability, Volume 1: Theory of Statistics* (pp. 175–194). Berkeley: University of California Press.

Hall, P., & Heyde, C. C. (1980). *Martingale limit theory and its application*. New York: Academic Press.

Hallin, M., & Paindaveine, D. (2004). Rank-based optimal tests of the adequacy of an elliptic VARMA model. *Annals of Statistics, 32*, 2642–2678.

Hallin, M., & Paindaveine, D. (2005). Asymptotic linearity of serial and nonserial multivariate signed rank statistics. *Journal of Statistical Planning and Inference, 136*, 1–32.

Hallin, M., & Puri, M. L. (1994). Aligned rank tests for linear models with autocorrelated errors. *Journal of Multivariate Analysis, 50*, 175–237.

Hallin, M., & Serroukh, A. (1998). Adaptive estimation of the lag of a long-memory process. *Statistical Inference for Stochastic Processes, 1*, 111–129.

Hallin, M., Taniguchi, M., Serroukh, A., & Choy, K. (1999). Local asymptotic normality for regression models with long-memory disturbance. *Annals of Statistics, 26*, 2054–2080.

Hallin, M., van den Akker, R., & Werker, B. J. M. (2011). A class of simple distribution-free rank-based unit root tests. *Journal of Econometrics, 163*, 200–214.

Hallin, M., van den Akker, R., & Werker, B. J. M. (2013). Optimal pseudo-Gaussian and rank-based tests of the cointegrating rank in semiparametric error-correction models, available at SSRN: http://papers.ssrn.com/sol3/papers.cfm?abstract_id=2320810

Hirukawa, J., & Taniguchi, M. (2006). LAN theorem for non-Gaussian locally stationary processes and its applications. *Journal of Statistical Planning and Inference, 136*, 640–688.

Ibragimov, I. A., & Has'minskii, R. S. (1981). *Statistical estimation: Asymptotic theory. Application of Mathematics* (Vol. 16). New York: Springer.

Janssen, A. (1991). Asymptotically linear and mixed normal sequences of statistical experiments, *Sankhyā: Series A, 53*, 1–26.

Janssen, A., Milbrodt, H., & Strasser, H. (1985). *Infinitely divisible statistical experiments. Lecture Notes in Statistics*. New York: Springer.

Jansson, M. (2008). Semiparametric power envelopes for tests of the unit root hypothesis. *Econometrica, 76*, 1103–1142.

Jeganathan, P. (1981). On a decomposition of the limit distribution of a sequence of estimators. *Sankhyā: Series A, 43*, 26–36.

Jeganathan, P. (1982). On the asymptotic theory of estimation when the limit of the log-likelihood ratios is mixed normal. *Sankhyā: Series A, 44*, 173–212.

Jeganathan, P. (1983). Some asymptotic properties of risk functions when the limit of the experiment is mixed normal. *Sankhyā: Series A, 45*, 66–87.

Jeganathan, P. (1995). Some aspects of asymptotic theory with applications to time series models. *Econometric Theory, 11*, 818–887.

Koul, H. L., & Schick, A. (1996). Adaptive estimation in a random coefficient autoregressive model. *Annals of Statistics, 24*, 1025–1052.

Koul, H. K., & Schick, A. (1997). Efficient estimation in nonlinear autoregressive time-series models. *Bernoulli, 3*, 247–277.

Kreiss, J.-P. (1987). On adaptive estimation in stationary ARMA processes. *Annals of Statistics, 15*, 112–133.

Kreiss, J.-P. (1990). Testing linear hypotheses in autoregressions. *Annals of Statistics, 18*, 1470–1482.

Kutoyants, Y. (1984). Parameter estimation for diffusion type processes of observations. *Mathematische Operationsforschung und Statistik. Series Statistics, 15*, 541–551.

Kutoyants, Y. (1994). *Identification of dynamical systems with small noise*. Dordrecht: Kluwer.

Kutoyants, Yu. A. (2004). *Statistical inference for Ergodic diffusion processes*. London: Springer Series in Statistics.

Lee, S., & Taniguchi, M. (2005). Asymptotic theory for ARCH-SM models: LAN and residual empirical processes. *Statistica Sinica, 15*, 215–234.

Le Cam, L. (1960). Locally asymptotically normal families of distributions. *University of California Publications in Statistics, 3*, 27–98.

Le Cam, L. (1986). *Asymptotic Methods in Statistical Decision Theory*. New York: Springer.

Le Cam, L., & Yang, G. L. (1990). *Asymptotics in Statistics—Some basic concepts*. New York: Springer.

Linton, O. (1993). Adaptive estimation in ARCH models. *Econometric Theory, 9*, 539–569.

Lin, N., & Lototsky, S. V. (2013). Second-order continuous-time non-stationary Gaussian autoregression, forthcoming. *Statistical Inference for Stochastic Processes*.

Phillips, P. C. B. (1991). Optimal inference in cointegrated systems. *Econometrica, 59*, 283–306.

Pollard, D. (1997). Another look at differentiability in quadratic mean. In D. Pollard, E. Torgersen, & G. L. Yang (Eds.), *Festschrift for Lucien Le Cam: Research papers in probability and statistics* (pp. 305–314). New York: Springer.

Roussas, G. G. (1965). Asymptotic inference in Markov processes. *Annals of Mathematical Statistics, 36*, 978–992.

Roussas, G. G. (1979). Asymptotic distribution of the log-likelihood function for stochastic processes. *Zeitschrift für Wahrscheinlichkeitstheorie und Verwandte Gebiete, 47*, 31–46.

Shiryaev, A. N., & Spokoiny, V. G. (2000). *Statistical experiments and decisions: Asymptotic theory. Advanced Series on Statistical Science & Applied Probability* (Vol. 8). River Edge: World Scientific.

Strasser, H. (1985). *Mathematical theory of statistics: Statistical experiments and asymptotic decision theory*. Berlin: Walter de Gruyter.

Swensen, A. R. (1985). The asymptotic distribution of the likelihood ratio for autoregressive time series with a regression trend. *Journal of Multivariate Analysis, 16*, 54–70.

Taniguchi, M., & Kakizawa, Y. (2000). *Asymptotic Theory of Statistical inference for time series. Springer Series in Statistics*. New York: Springer.

Torgersen, E. (1991). *Comparison of statistical experiments. Encyclopedia of Mathematics and Its Applications* (Vol. 36). Cambridge: Cambridge University Press.

van der Vaart, A. W. (1988). *Statistical Estimation in Large Parameter Spaces. CWI Tract 44*. Amsterdam: CWI.

van der Vaart, A. W. (1991). An asymptotic representation theorem. *International Statistical Review, 59*, 97–121.

van der Vaart, A. W. (2000). *Asymptotic Statistics*. Cambridge: Cambridge University Press.

Asymptotic Normality of Binned Kernel Density Estimators for Non-stationary Dependent Random Variables

Michel Harel, Jean-François Lenain and Joseph Ngatchou-Wandji

Abstract We establish the asymptotic normality of binned kernel density estimators for a sequence of dependent and nonstationary random variables *converging* to a sequence of stationary random variables. We compute the asymptotic variance of a suitably normalized binned kernel density estimator and study its absolute third-order moment. Then, we show that its characteristic function tends to that of a zero-mean Gaussian random variable (rv). We illustrate our results with a simulation experiment.

1 Introduction

Let $\{X_1, \ldots, X_n\}$ be a sequence of rvs from a univariate distribution with density f. The Rosenblatt (1971) kernel density estimator (KDE) $\widehat{f}$ of f is defined by

$$\widehat{f}(x) = \frac{1}{nh} \sum_{i=1}^{n} K\left(\frac{x - X_i}{h}\right), \tag{1}$$

where $h = h(n)$ is the smoothing parameter and K is the kernel, usually a density function. The most popular K are, the triangular, the Gaussian, and the Epanechnikov kernels (see, e.g., Lenain et al. (2011)), while optimal values of h have the general form $Cn^{-1/5}$, for some generic constant C.

M. Harel
ÉSPÉ de Limoges, Limoges, 87036 Cedex, France
e-mail: michel.harel@unilim.fr

M. Harel
Université Paul Sabatier, Toulouse, 31062 Cedex, France

J.-F. Lenain
Faculté des Sciences et Techniques, Limoges, 87060 Cedex, France
e-mail: jflenain@gmail.com

J. Ngatchou-Wandji (✉)
EHESP de Rennes, Institut Élie Cartan de Lorraine, Université de Lorraine,
Vandoeuvre-lès-Nancy, 54506 Cedex, France
e-mail: joseph.ngatchou-wandji@univ-lorraine.fr

© Springer International Publishing Switzerland 2015

M. Hallin et al. (eds.), *Mathematical Statistics and Limit Theorems*,
DOI 10.1007/978-3-319-12442-1_10

For reducing the computational time of KDE, binned kernel estimators are sometimes used. They are also used in the situation where the data suffer some kind of binning or rounding on a grid. These estimators, which can be regarded as approximations of KDE, or as direct estimators of f have the general form

$$\widehat{f_B}(x) = \frac{1}{nh} \sum_{j \in \mathbb{Z}} K\left(\frac{x - a_j}{h}\right) \sum_{i=1}^{n} T\left(\frac{X_i - a_j}{\delta}\right), \tag{2}$$

where $\{a_j\}_{j \in \mathbb{Z}} = \{a_0 + j\delta\}_{j \in \mathbb{Z}}$ are given grid points with an arbitrary origin $a_0 \in \mathbb{R}$, T is a kernel with window width δ, and h and K are as above.

Denote by $\lfloor x \rceil$ the integer closest to x (if x is a half integer, we assume for definiteness that $\lfloor x \rceil = x - 1/2$), and by $[x]$ the largest integer less than or equal to x. For $\delta > 0$, define the real-valued function a by $a(y) = \delta \lfloor (y - a_0)/\delta \rceil + a_0$, for rounding $y \in \mathbb{R}$ to its nearest value, or $a(y) = \delta [(y - a_0)/\delta] + a_0$, for rounding y down.

In many practical situations, only the rounded $a(X_i)$ values of the X_i's are available. Consequently, some information is lost and the differences $X_i - a(X_i)$'s can be considered as missing variables. Following a previous result of Hall (1983), Hall and Wand (1996) show the convergence of $\{Z_i = (X_i - a(X_i))/\delta\}_{i \in \mathbb{N}}$ to a rv uniformly distributed over $[0, 1]$. This result is extended by Lenain et al. (2011) to a class of nonstationary rvs. In such situations where only the $a(X_i)$'s are at hand, estimating the stationary distribution of $\{X_i\}_{i \in \mathbb{N}}$ by a KDE is tantamount to using a binned kernel estimator for some T. The estimator $\widehat{f_B}$ associated with some particular kernels T has been investigated in the literature. As example, one can cite the ones associated with the rounding kernels $T(y) = I(y \in [0, 1])$ or $T(y) = I(y \in [-1/2, 1/2))$ given by

$$\widehat{f_B}(x) = \frac{1}{nh} \sum_{i=1}^{n} K\left(\frac{x - X_i}{h} + \frac{\delta}{h} Z_i\right), \tag{3}$$

where for all $i \in \mathbb{N}$, $Z_i = (X_i - a(X_i))/\delta \in [0, 1)$ or $Z_i \in [-1/2, 1/2)$. These estimators, sometimes called kernel density estimator with binned data, have been investigated by Scott and Sheather (1985) for independent and identically distributed (iid) rvs and for stationary rvs. Scott and Sheather (1985) obtained the leading terms in the asymptotic expansion of the bias and also that of the mean square error (MSE) as a function of both the window widths h and δ.

The estimators (3) are more easier to compute than the following one obtained with the triangular kernel $T(y) = (1 - |y|)I(y \in [-1, 1])$ advised by Jones and Lotwick (1984):

$$\widehat{f_B}(x) = \frac{1}{nh} \sum_{i=1}^{n} \left[(1 - Z_i) K\left(\frac{x - X_i}{h} + \frac{\delta}{h} Z_i\right) + Z_i K\left(\frac{x - X_i}{h} + \frac{\delta}{h}(Z_i - 1)\right)\right], \tag{4}$$

where for all $i \in \mathbb{N}$, $Z_i \in [0, 1)$. The estimator (4), also called discretized kernel density estimator, has been investigated by Jones (1989), for both iid and stationary rvs. Similarly, Jones (1989) obtained the leading terms in the asymptotic expansion of the integrated mean square error (IMSE) as a function of both the window widths h and δ. Under some assumptions on the derivatives of f and K, his work improves the results of Scott and Sheather (1985) in the asymptotic expansion, as the informations obtained are more precise on the orders of magnitude of the errors. However, neither Scott and Sheather (1985) nor Jones (1989) established the asymptotic normality of their estimators.

Let $\{X_i\}_{i \in \mathbb{N}}$ be an alpha-mixing process *locally nonstationary* in the sense that there exist two indices i_0, $i_1 \in \mathbb{N}$ such that the sequence $\{X_t\}_{i_0 \le i \le i_1}$ is possibly nonstationary and the sequences $\{X_i\}_{i < i_0}$ and $\{X_i\}_{i > i_1}$ behave like stationary series with the same stationary density function f^*. Simple examples of such a series are those that can be decomposed into a stationary series plus a trend, say, $\vartheta(i)$, that vanishes as i grows. For such nonstationary series, several consistency results of $\widehat{f_B}$ to f^* are established in Lenain et al. (2011). In this paper, we aim to establish a CLT for $\widehat{f_B}$. It is well known that it can allow for the construction of confidence bands for f^*. In Sect. 2, we discuss some motivating examples. We give the notation and list the sequence of assumptions considered in Sect. 3, where we state and prove our central limit theorem for the two simple classes of $\widehat{f_B}$ given by (3) and (4). In Sect. 4, we state the CLT for the more general $\widehat{f_B}$ defined in (2). In Sect. 5, we present the results of a simulation experiment. The last section is devoted to the proofs of our theoretical results.

2 Examples, Notation and Assumptions

In this section, we discuss some motivating examples to our work and explain the reason of our study of binned kernel estimators. We also precise the notation and we list the assumptions needed for the proofs of our result.

2.1 Motivating Examples

Assume we are interested in a stationary and weakly dependent latent time series $\{X_i^*\}_{i \in \mathbb{N}}$ (which is not observed). At the place, one observes a time series $\{X_i\}_{i \in \mathbb{N}}$, nonstationary in the sense given in the introduction. The link between the two series can be on the form:

$$X_i = \vartheta(i) + X_i^*, \tag{5}$$

where the trend $\vartheta(i)$ vanishes as i grows.

In this example, the series $\{X_i^*\}_{i \in \mathbb{Z}}$ can be a usual stationary time series as ARMA, bilinear, TAR, ARCH, GARCH or any other stationary time series. The weak dependence and the stationarity properties of most of them are widely studied in the literature. Another example can be taken from the so-called state-space models:

$$\begin{cases} X_i = \xi X_i^* + \varepsilon_i \\ X_i^* = \zeta X_{i-1}^* + \eta_i, \end{cases} \tag{6}$$

where ξ and ζ are constants, $\{\eta_i\}_{i \in \mathbb{N}}$ is a white noise with finite variance σ_η^2 and $\{\varepsilon_i\}_{i \in \mathbb{N}}$ is a sequence of rvs with mean $\vartheta(i)$ and/or variance $(\zeta(i) + 1)\sigma_\varepsilon^2$ depending on i, and both decreasing to 0 and $\sigma_\varepsilon^2 (> 0)$, respectively, leading to the usual state-space models encountered in the literature. Some other examples can be found in the time-varying coefficients models also largely studied in the literature. Figure 1 displays the graphs of four nonstationary time series $X_1, X_2, \ldots, X_{1000}$. The first three graphs are those of series sampled from (5) for trends 0.99^i, $(-0.99)^i$ and -0.99^i, respectively. The fourth graph is that of a series generated from (6) for zero-mean Gaussian errors $\varepsilon_1, \varepsilon_2, \ldots, \varepsilon_{1000}$ with standard deviation 0.08, and zero-mean Gaussian $\eta_1, \eta_2, \ldots, \eta_{1000}$ with standard deviations $0.08 + 0.98^i$. For all of them, the sequence $X_1^*, X_2^*, \ldots, X_{1000}^*$ is simulated from an AR model with autoregressive coefficient 0.5 and a zero-mean Gaussian white noise with standard deviation 0.08.

Some domains where the above models can be used are, among others, epidemiology, finance, meteorology, agriculture, and social science. A concrete example in epidemiology is what happens during an epidemic : the number of sufferers is high at the beginning and generally decreases toward the end of the epidemic until it becomes stationary. A concrete example in finance is the situation where in a financial market,

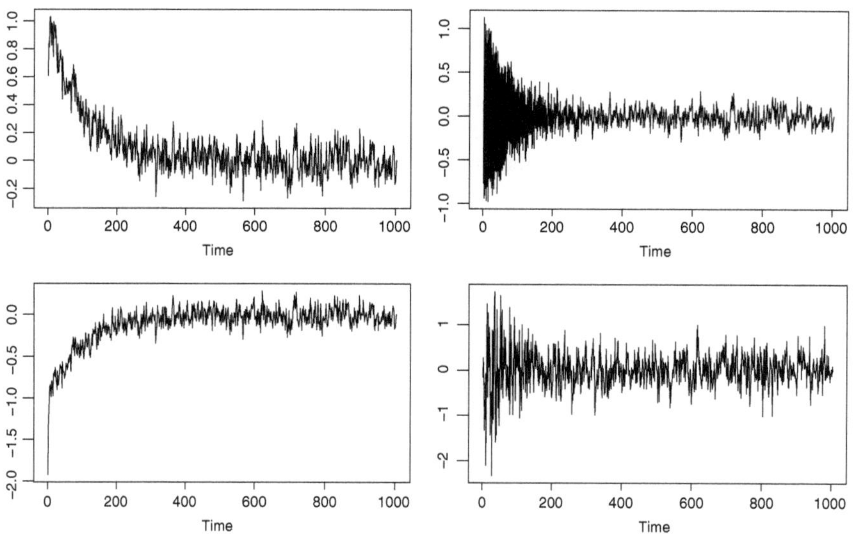

Fig. 1 Time series plots of $X_1, X_2, \ldots, X_{1000}$ sampled from (5) (the first three graphs) or (6) (the fourth graph)

the price of a stock option is low at the opening and starts increasing at a certain time until its becomes stationary. Such phenomena, which can be also encountered in many other domains, can be modeled by (5) and/or (6). It is clear that any statistical inference on the stationary distribution of the latent series $\left\{X_i^*\right\}_{i \in \mathbb{N}}$ will be based on the observed series $\{X_i\}_{i \in \mathbb{N}}$ which has the same asymptotic behavior.

2.2 The Binned Kernel Estimator Based on Non-stationary Data

The Rosenblatt and the binned kernel estimators are widely studied in the literature for iid data, and for stationary data. Simonoff (1996) gives a digest of this huge work. Tran (1990) obtains a central limit theorem (CLT) for the Rosenblatt KDE $\widehat{f}$ for strong mixing random fields. A recent concrete example where KDE is used can be found in Geange et al. (2011) where the Niche Overlap is estimated for iid data. Nevertheless, it is well known that many sequences of rvs encountered in practice are neither independent nor stationary. This is why we wish to study these estimators in a dependent and nonstationary data context. To the best of our knowledge, the only studies in this direction are Harel and Puri (see 1996; 1999), where a CLT is established for $\widehat{f}$ under absolutely regular assumption of the data, and Lenain et al. (2011) where the bias and the MSE of $\widehat{f_B}$ and its particular forms given by (3) and (4) are investigated under strong mixing conditions of the nonstationary data considered. To complete these studies, we focus here on the CLT for $\widehat{f_B}$ based on dependent and nonstationary data. Such a result can lead to the construction of a confidence band for f^*.

2.3 Assumptions and Notation

After the motivating examples given in the previous subsection, we give the notation we use and list and discuss our assumptions.

(H1):

- Let $\{X_i\}_{i \in \mathbb{N}}$, be a sequence of nonstationary rvs, such that each X_i has a cumulative distribution function F_i with density function f_i. Denote by $F_{i,j}$ the cumulative distribution function of (X_i, X_j), and by $f_{i,j}$ its density function. Denote the cumulative distribution function of (X_i, Z_i) by G_i and g_i its density function. Denote the cumulative distribution function of (X_i, X_j, Z_i, Z_j) by $G_{i,j}$ and $g_{i,j}$ its density function.
- Let $\left\{X_i^*\right\}_{i \in \mathbb{N}}$ be a sequence of strictly stationary rvs with stationary cumulative distribution function F^* and density f^*. Denote by F_{i-j}^* the cumulative distribution of (X_i^*, X_j^*) and by f_{i-j}^* its density function.

- Assume that all the above cumulative distribution and density functions are continuous, with f^* and f_{i-j} absolutely continuous and differentiable. Assume that these functions satisfy

$$\sup_{i\in\mathbb{N}}\sup_{x\in\mathbb{R}} f_i(x) < \infty, \qquad \sup_{i\in\mathbb{N}}\sup_{x,y\in\mathbb{R}} f_{i,j}(x,y) < \infty, \qquad \sup_{x\in\mathbb{R}} f^{*\prime}(x) < \infty,$$

$$\int \left| f^{*\prime}(x) \right| dx < \infty, \qquad \sup_{\substack{i,j\in\mathbb{N}\, x,y\in\mathbb{R} \\ z=x,y}}\sup \left\{ \left| \frac{\partial f^*_{j-i}(x,y)}{\partial z} \right|, \int \left| \frac{\partial f^*_{j-i}(x,y)}{\partial z} \right| dxdy \right\} < \infty,$$

$$\forall j > i, \ \left\| F_{i,j} - F^*_{j-i} \right\|_V = O\left(\eta\left(i\right)\right) \to 0 \ \text{ and } \ \left\| F_i - F^* \right\|_V = O\left(\eta\left(i\right)\right) \to 0 \ \text{ as } i \to \infty,$$

where g' denotes the derivative of a function g and $\|G\|_V$ stands for the total variation norm of a function G.

Remark 1 Consider $\{X_i\}_{i\in\mathbb{N}}$ and $\left\{X_i^*\right\}_{i\in\mathbb{N}}$ linked by (5) for a decreasing function ϑ, and assume that $\left\{X_i^*\right\}_{i\in\mathbb{N}}$ is a zero-mean Gaussian process with variance σ^2. Then f^* is a univariate Gaussian density with parameters 0 and σ^2, and f^*_{j-i} is a bivariate Gaussian density with parameters $(0,0)'$ and covariance matrix $\Omega_{i,j} = \begin{pmatrix} \sigma^2 & \rho_{j-i} \\ \rho_{j-i} & \sigma^2 \end{pmatrix}$, where for $j > i$, $\rho_{j-i} = Cov(X_i^*, X_j^*)$. It is an easy matter that $f_{i,j}$ is a Gaussian density with mean $(\vartheta(i), \vartheta(j))'$ and covariance matrix $\Omega_{i,j}$. For this example, all the points of assumption **(H1)** hold. Using the version of the total variation norm with densities and using first-order Taylor expansions of f^* and f^*_{j-i} with respect to appropriate variables (the parameters), it can be checked easily that $O\left(\eta\left(i\right)\right) = O\left(\vartheta\left(i\right)\right)$.

(H2):

- The function K is nonnegative, bounded, symmetric, absolutely continuous and piecewise differentiable such that:

$$\int K(x)dx = 1, \ \int x^2 K(x)dx < \infty, \ \int xK'(x)dx < \infty, \ \sup_{x\in\mathbb{R}} \left|K'(x)\right| < \infty, \ \int \left|K'(x)\right| dx < \infty.$$

- The bandwidths $h = h(n)$ and $\delta = \delta(n)$ are positive real numbers such that $h \to 0$, $\delta \to 0$ and $\delta/h \to 0$, $nh \to \infty$ as $n \to \infty$.
- The nonstationarity rate η satisfies $\sum_{i=1}^{\infty} \eta\left(i\right) < \infty$.

Remark 2 Recalling that for a function $\varpi \in L^p$, $\|\varpi\|_p = \left(\int \varpi^p(x)dx\right)^{1/p}$, in view of **(H2)** and $\int \left|K'(t)\right|^p dt \le \sup_{x\in\mathbb{R}} \left(\left|K'(x)\right|^{p-1}\right) \int \left|K'(t)\right| dt$, both $\|K\|_p$ and $\left\|K'\right\|_p$ are finite for any $p \ge 1$.

Remark 3 The condition $\sum_{i=1}^{\infty} \eta(i) < \infty$ is satisfied in particular for $\eta(i) = O(\tau^i)$, $\tau \in (0, 1)$ and for $\eta(i) = O(i^{-\tau})$, $\tau > 1$.

(H3):

- The sequence $\{X_i\}_{i \in \mathbb{N}}$ is strongly mixing with mixing coefficient $\alpha(n)$ such that $\alpha(n) \to 0$ as $n \to \infty$.
- There exist $v > 0$, $c \in 2\mathbb{N}$ such that for $c > 3$, $\sum_{i=1}^{n} (i+1)^{c-2} [\alpha(i)]^{1/\lambda} < \infty$ with $\lambda = (v+c)/v$ and for $h = h(n)$, there exists a nondecreasing sequence $m = m(n)$ such that $hm \to 0$ and $h(n)^{-1/\lambda} \sum_{i=m}^{n} \alpha(i)^{1/\lambda} \to 0$ as $n \to \infty$.

Remark 4 Assumption **(H3)** is satisfied in the following situations :

- $\alpha(i) = O(\rho^i)$ with $\rho \in (0, 1)$ and $h \log(n) \to 0$ as $n \to \infty$.
- $\alpha(i) = O(i^{-\rho})$ with $\rho > 1 + \gamma/\varepsilon > 2, 0 < \varepsilon < \gamma < 1$ and $hn^{\varepsilon} \to 0, n^{\gamma}h \to \infty$, $n^{(\rho-\gamma/\varepsilon)/3} h \to \infty$ as $n \to \infty$.

3 The Central Limit Theorem

In this section, we establish a CLT for $\widehat{f}_B(x)$. We first recall some recent results on the bias and the MSE of the binned kernel estimators we are studying, and we study the asymptotic behavior of their variance and their absolute third-order moment.

3.1 Some Existing Results

Denote by $\widehat{f}^*$ the Rosenblatt estimator of f^*. For the estimator $\widehat{f}_B(x)$ given by (3) and written for $Z_i \in \Delta$, where $\Delta = [0, 1)$ for rounding down, or $\Delta = [-1/2, 1/2)$ for rounding to the nearest values, Lenain et al. (2011) prove that under **(H1)** and **(H2)**, the bias of $\widehat{f}_B$ is asymptotically equal to that of the Rosenblatt estimator up to an $O(1/nh) + O(\delta^2/h^2 + \delta \int_{\Delta} u du + \delta^2/h)$ term. For the functions f^* and K having third-order derivatives with that of K in L^1, they give a more explicit expression of the bias. They do the same for $\widehat{f}_B(x)$ defined in (4). Next, under additional assumptions on the mixing coefficient, Lenain et al. (2011) show that $\widehat{f}_B(x)$ on its forms (3) and (4) converges in the MSE to $f^*(x)$. Finally, by doing similar work for the more general form of $\widehat{f}_B(x)$ given by (2), they mainly show that the bias is asymptotically equal to that of the Rosenblatt estimator up to a term of order $O(\delta) + O(\delta/h) + O(1/nh)$.

Many authors have established central limit theorems for the Rosenblatt estimator $\widehat{f}(x)$ defined in (1) and have investigated the IMSE $\int [\widehat{f}(x) - f(x)]^2 dx$ for iid rvs or for stationary-dependent rvs. Harel and Puri (1996) generalized these results for nonstationary and dependent rvs. For this, they established the convergence of the characteristic function of $\sqrt{nh}[\widehat{f}(x) - f(x)]$ to that of a zero-mean Gaussian rv, by splitting the sequence of the involved rvs into two sequences such that one is negligible and the other is asymptotically independent. They proved also the convergence of the variance of $\sqrt{nh}\,\widehat{f}(x)$ to some real value, and study the absolute third-order moment. These techniques are generalized in this paper to get the asymptotic normality of the binned kernel density.

Harel and Puri (1996) gave applications to Markov processes (aperiodic, geometrically ergodic and Harris recurrent or aperiodic and Doeblin recurrent) and ARMA processes for which the initial measure is not necessarily the invariant measure. Later, Harel and Puri (1999) proved the weak invariance of the conditional nearest neighbor regression function estimators (Rosenblatt version) called the conditional empirical process for nonstationary absolutely regular rvs.

3.2 Preliminaries to the Central Limit Theorem

For establishing our central limit theorem, we first show that the variance of $\sqrt{nh}\,\widehat{f}_B(x)$ tends to some real value $\sigma^2(x)$. Next, we study its absolute third-order moment. These results are then used for establishing the convergence of the characteristic function of $\sqrt{nh}[\widehat{f}_B(x) - E\,\widehat{f}_B(x)]$ to that of a zero-mean Gaussian rv with variance $\sigma^2(x)$, by making use of a suitable blocking of the sequence of the rvs involved. This blocking could have been avoided by using the techniques of Bardet et al. (2008). But it seems not easy to apply them in this nonstationary context.

Lemma 1 *Assume that* **(H1)–(H3)** *hold. Then, as n tends to ∞,* $Var\left(\sqrt{nh}\,\widehat{f}_B(x)\right)$ *tends to* $\sigma^2(x) = f^*(x)\int K^2(t)dt$.

Proof See Sect. 6.

For some function ε defined on Δ with values in $(0, 1)$ and for all $i \in \mathbb{N}$, define the rvs $A_i = K\left[(x - X_i)/h\right] + (\delta/h)Z_i K'\left[(x - X_i)/h + (\delta/h)Z_i\varepsilon\,(Z_i)\right]$ and $H_i = A_i - EA_i$. Let ℓ be a positive integer. Consider any subsequence $H_{\ell_i}, i = 1\ldots,\ell$ of $(H_i)_{i\in\mathbb{N}}$. We have the following lemma:

Lemma 2 *Assume that* **(H1)–(H3)** *hold. Then,*

$$E\left(\left|\sum_{i=1}^{\ell} H_{\ell_i}\right|\right)^3 \leq C\,\ell^{3/2}h.$$

Proof See Sect. 6.

For all $i \in \mathbb{N}$, letting $G_i = K\left[(x - X_i)/h + \delta Z_i/h\right]$ and $W_i = G_i - EG_i$, decompose $(nh)^{-1/2}\sum_{i=1}^{n} W_i$ into two sequences U_j and V_j, $j = 1,\ldots,q$ such that $q = n/(\ell + m)$, with $\ell = \lfloor n^{1-\beta}\rfloor$, $\beta \in (0, 1)$ and $m = o(\ell)$. Using the sequence of increasing numbers $\{a_1, b_1, \ldots, a_q, b_q, a_{q+1}\}$, one can write

$$S_n = \sum_{j=1}^{q} \left(U_j + V_j \right) = \frac{1}{\sqrt{nh}} \sum_{j=1}^{q} \left(\sum_{i=a_j}^{b_j - 1} W_i + \sum_{i=b_j}^{a_{j+1}-1} W_i \right)$$

with $a_{j+1} - b_j = m$ and $b_k - a_k = \ell$ for all $k, j \in \{1, 2, \ldots, q\}$.

The following lemma states that (V_j) is a sequence of negligible rvs and that (U_j) is a sequence of asymptotically independent rvs.

Lemma 3 *Assume that* **(H1)**–**(H3)** *hold. Then for larger values of n, one has*

$$\left| E \left\{ \exp \left[it\sqrt{nh} \left(\widehat{f}_B(x) - E\widehat{f}_B(x) \right) \right] \right\} - \prod_{j=1}^{q} E \left[\exp \left(it\, U_j \right) \right] \right| \leq C\, q\, \alpha\,(m), \quad (7)$$

where C is a generic constant and i stands for the complex number such that $i^2 = -1$.

Proof See Sect. 6.

3.3 Convergence in Distribution

In this subsection, we state our main result.

Theorem 1 *Assume that* **(H1)**–**(H3)** *hold, and that there exist $\beta \in (0, 1)$ and $m = o\left(n^{1-\beta}\right)$ such that $\alpha\,(m) = o\left(n^{-\beta}\right)$. Then, for all $\beta_0 \in (0, \beta)$ and $h = Cn^{-\beta_0}$, $\sqrt{nh} \left[\widehat{f}_B(x) - E\widehat{f}_B(x) \right]$ converges in distribution to a zero-mean Gaussian rv with variance $\sigma^2(x) = f^*(x) \int K^2(t)dt$.*

Proof As can be seen in Sect. 6, the proof uses the techniques of Takahata and Yoshihara (1987), and Lemmas 1 and 2.

4 The General Binned Kernel Estimator

In this section, **(H1)**–**(H2)** still hold. Since the assumptions on cumulative distribution and density functions of (X_i, Z_i) and (X_i, X_j, Z_i, Z_j) are no more in force, **(H3)** is replaced by

(H3)′:

- The sequence $\{X_i\}_{i \in \mathbb{N}}$ is strongly mixing with mixing coefficients $\alpha(n) \to 0$ as $n \to \infty$.
- There exist $v > 0$, $c \in 2\mathbb{N}$ such that for $c > 3$, $\sum_{i=1}^{n}(i+1)^{c-2}[\alpha(i)]^{1/\lambda} < \infty$ with $\lambda = (v+c)/v$, and for $h = h(n)$, there exists a nondecreasing sequence $m = m(n)$ such that $hm \to 0$ and $h\delta^{-1/\lambda-1}\sum_{i=m}^{n}\alpha(i)^{1/\lambda} \to 0$ as $n \to \infty$.

The proof of a central limit theorem for the more general binned kernel estimator given by (2) can be handled in the same way as in the previous case. The remaining thing to do is the computation of the asymptotic variance of $\sqrt{nh}\,\widehat{f}_B(x)$. This is given by

Proposition 1 *Assume that* **(H1)**–**(H2)** *and* **(H3)′** *hold. Then* $Var\left(\sqrt{nh}\,\widehat{f}_B(x)\right)$ *tends to*

$$\sigma^2(x) = \left\{\|K\|_2^2\left[\|T\|_2^2 + 2\int_\Delta T(t)\,T(t-1)\,dt\right]\right\}f^*(x) \qquad (8)$$

as $n \to \infty$, where T is the kernel function defined in (2).

Proof See Sect. 6.

Theorem 2 *Assume that* **(H1)**−**(H2)** *and* **(H3)′** *hold, and that there exist $\beta \in (0,1)$ and $m = o\left(n^{1-\beta}\right)$ such that $\alpha(m) = o\left(n^{-\beta}\right)$. Then for all $\beta_0 \in (0,\beta)$ and $h = Cn^{-\beta_0}$, $\sqrt{nh}\left[\widehat{f}_B(x) - E\,\widehat{f}_B(x)\right]$ converges in distribution to a zero-mean Gaussian rv with variance $\sigma^2(x)$ whose expression is given by (8).*

Proof The proof can be handled along the same lines as those of Theorem 1.

5 Simulations

In this section, we use the software R to do a simulation experiment for illustrating our results. We limit ourselves to (1) and (3), based on models (5) and (6) with $a_0 = 0$ in the expression of $a(y)$. For both estimators, we consider a Gaussian kernel with either $h = n^{-1/2}$ or $h = n^{-1/3}$ and $\delta = 0.5h$, where n is the sample size.

Figure 2 displays the graphs of f^*, $\widehat{f}$, and $\widehat{f}_B$ as well as those of two functions LB and UB delimiting a 5% confidence band of f^*. The computation of all these functions is based on nonstationary time series $X_1, X_2, \ldots, X_n$ of lengths $n = 300, 500, 800,$ and 1000, sampled from (5). The trend used was $\vartheta(i) = 0.8^i$ and the corresponding stationary time series $X_1^*, X_2^*, \ldots, X_n^*$ was from an AR model with autoregressive coefficient 0.5 and a zero-mean Gaussian white noise with standard

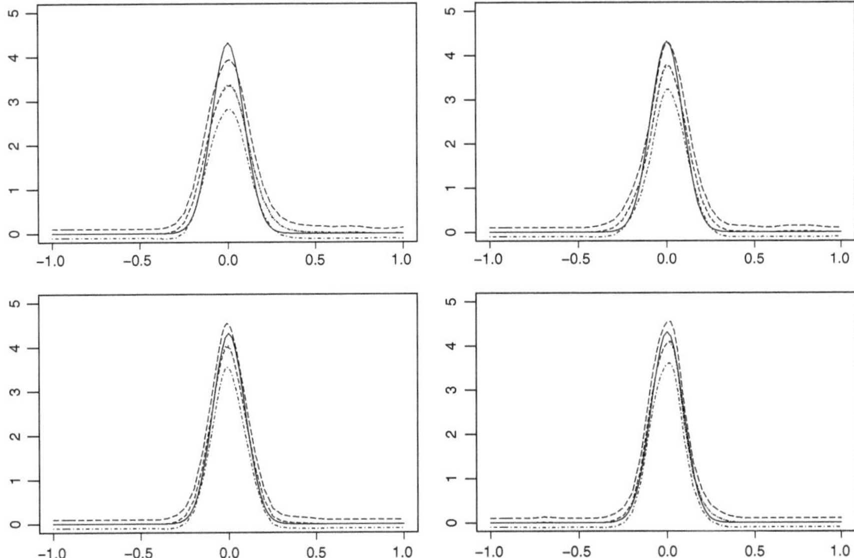

Fig. 2 Stationary density f^* (*solid line*), KDE $\widehat{f}$ (*short-dashed line*), binned kernel estimator $\widehat{f}_B$ (*dotted line*), confidence band delimited by the functions LB (*long-dashed line*) and UB (*dashed-dotted line*), based on (5)

deviation 0.08. For this model, the density of the stationary distribution f^* is a zero-mean Gaussian density with standard deviation $0.08 \times 0.75^{-1/2} \approx 0.0924$.

It can be seen from the graphs that KDE and binned estimators have the same behavior, and both are generally consistent to the stationary density which, generally lies within its confidence band. One can also see that this consistency is more accurate as n grows.

Figure 3 also displays the same graphs based on nonstationary time series $X_1, X_2, \ldots, X_n$ generated from (6) for zero-mean Gaussian errors $\varepsilon_1, \varepsilon_2, \ldots, \varepsilon_n$ with standard deviation 0.08, and zero-mean Gaussian errors η_i with standard deviations $0.08 + 0.98^i$, $i = 1, \ldots, n$. The corresponding $X_1^*, X_2^*, \ldots, X_n^*$ was generated from an AR model with autoregressive coefficient 0.3 and a zero-mean Gaussian white noise with standard deviation 0.8. For this model, the density of the stationary distribution f^* is a zero-mean Gaussian density with standard deviation $[(0.24^2)/0.75 + 0.08^2]^{1/2} \approx 0.29$.

As in the preceding case, one can see from the graphs that KDE and binned estimators are consistent to f^* and are more accurate as n grows. Here also, f^* generally lies in its confidence band.

The trials with an Epanechnikov kernel gave similar results. We tried many other models, in particular (5) for trends $\vartheta(i) = 0.9^i, 0.95^i$ and 0.99^i. From the results that we do not present, we noted that both estimators still had the same behavior, but

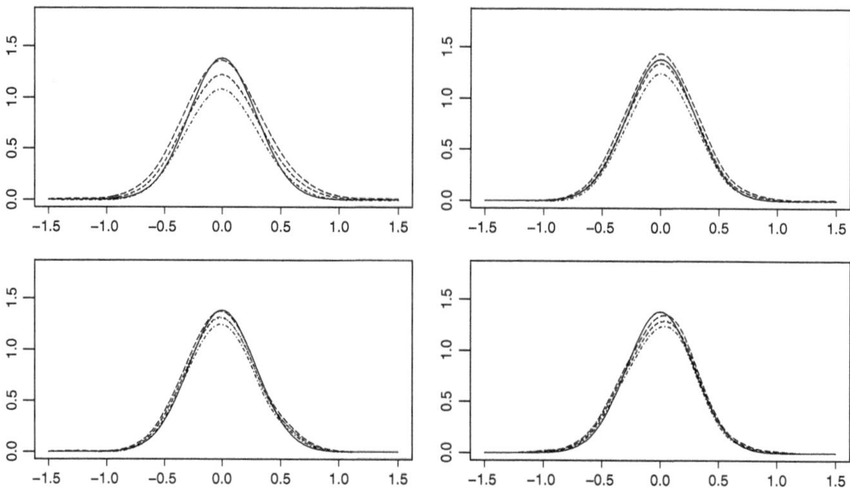

Fig. 3 Stationary density f^* (*solid line*), KDE $\widehat{f}$ (*short-dashed line*), binned kernel estimator $\widehat{f_B}$ (*dotted line*), confidence band delimited by the functions LB (*long-dashed line*) and UB (*dashed-dotted line*), based on (6)

they were less consistent to the stationary density for $\vartheta(i) = 0.99^i$ and $\dot{n} < 800$. This is likely due to the fact that the trend converges more slowly to 0, and by this, induces a large proportion of nonstationary observations in the data.

6 Proof of the Results

Proof of Lemma 1: One shows that for $m = o(n)$,

$$Var\left(\sqrt{nh}\,\widehat{f_B}(x)\right) = \|K\|_2^2\, f^*(x) - hf^*(x)^2 + O\left(h^2\right)$$

$$+O\left(\frac{\delta^2}{h^2} + \delta \int_\Delta u\,du + \frac{\delta^2}{h}\right) + O\left(\frac{1}{nh}\right) + O\left(h\,m + h^{-1/\lambda}\sum_{i=m}^{n}\alpha\,(i)^{1/\lambda}\right).$$

For $\widehat{f_B}(x)$ defined in (3) with $Z_i \in \Delta(= [0, 1)$ or $[-1/2, 1/2))$, by a Taylor expansion, there exists a function ε defined on Δ, with values in $(0, 1)$ such that

$$\widehat{f_B}(x) = \frac{1}{nh}\sum_{i=1}^{n}\left\{K\left(\frac{x - X_i}{h}\right) + \frac{\delta}{h}Z_i K'\left[\frac{x - X_i}{h} + \frac{\delta}{h}Z_i\varepsilon\,(Z_i)\right]\right\},\ Z_i \in \Delta$$

$$= \frac{1}{nh}\sum_{i=1}^{n}A_i.$$

Also, for all $i \in \mathrm{IN}$, denoting by $A_i^* = K\left[(x - X_i^*)/h\right]$ and by H_i and H_i^* the centered rvs $A_i - EA_i$, and $A_i^* - E(A_i^*)$, respectively, it is easy to see that by a classical decomposition, one has

$$Var\left(\sqrt{nh}\,\widehat{f}_B(x)\right) = \frac{1}{nh}\sum_{i=1}^{n}\left(E\left(H_i^2\right) - E\left(H_i^{*2}\right)\right) + \frac{1}{nh}\sum_{i=1}^{n}E\left(H_i^{*2}\right)$$
$$+ \frac{2}{nh}\sum_{i=1}^{n-1}\sum_{j=i+1}^{n}E\left(H_iH_j\right).$$

Writing the right-hand side of the above equality as $(VAR - VAR^*) + VAR^* + COV$, we first evaluate $VAR - VAR^*$. One can write

$$VAR - VAR^* = \frac{1}{nh}\sum_{i=1}^{n}\left(\left[E\left(A_i^2\right) - E\left(A_i^{*2}\right)\right] - (EA_i)^2 + \left(EA_i^*\right)^2\right).$$

Now it is easy to see that

$$EA_i^* \le h\sup_x f^*(x)\int K(t)\,dt = O(h), \quad EA_i - EA_i^* = O\left(\eta(i) + \frac{\delta^2}{h}\right) \quad \text{and}$$

$$(EA_i)^2 - \left(EA_i^*\right)^2 = \left(EA_i - EA_i^*\right)^2 + 2EA_i^*\left(EA_i - EA_i^*\right).$$

This entails the following equalities:

$$\frac{1}{nh}\sum_{i=1}^{n}\left[(EA_i)^2 - \left(EA_i^*\right)^2\right] = O\left(\frac{1}{nh} + \frac{\delta^4}{h^3} + \delta^2\right)$$

$$\frac{1}{nh}\sum_{i=1}^{n}\left[E\left(A_i^2\right) - E\left(A_i^{*2}\right)\right]$$

$$= \frac{1}{nh}\sum_{i=1}^{n}\left[\int_{\Delta}\int K\left(\frac{x - y_i}{h}\right)^2 g_i(y_i, u_i)dy_i du_i - \int K\left(\frac{x - y_i}{h}\right)^2 f^*(y_i)dy_i\right]$$

$$+ \frac{\delta}{nh^2}\sum_{i=1}^{n}\int_{\Delta}\int\left\{u_i K\left(\frac{x - y_i}{h}\right)K'\left[\frac{x - y_i}{h} + \frac{\delta}{h}u_i\varepsilon(u_i)\right]\right.$$

$$\left. + \frac{\delta}{h}u_i^2 K'\left[\frac{x - y_i}{h} + \frac{\delta}{h}u_i\varepsilon(u_i)\right]^2\right\}g_i(y_i, u_i)dy_i du_i.$$

Writing the right-hand side of the above equality as $E_1 + E_2$, by the convergence with respect to the total variation norm, one can write:

$$|E_1| \le \frac{1}{nh} \sup_y \left| K\left(\frac{x-y}{h}\right) \right|^2 \sum_{i=1}^{n} \int \left| f_i(y_i) - f^*(y_i) \right| dy_i = \frac{1}{nh} C \sum_{i=1}^{n} O(\eta(i)).$$

Letting E^* be the stationary counterpart of E, and observing that $E_2 = (E_2 - E_2^*) + E_2^*$, by a slight extension of Lemma 3 of Hall (1983) to nonstationary rvs, one has

$$\begin{aligned}
\left| E_2 - E_2^* \right| &\le \frac{\delta}{nh^2} \sup_{y,u} \left| K\left(\frac{x-y}{h}\right) \, u K'\left[\frac{x-y}{h} + \frac{\delta}{h} u\varepsilon(u)\right] + O\left(\frac{\delta}{h}\right) \right| \\
&\quad \times \sum_{i=1}^{n} \int_\Delta \int \left| g_i(y_i, u_i) - f^*(y_i) \right| dy_i du_i \\
&= \frac{\delta}{nh^2} C \sum_{i=1}^{n} \left[O(\eta(i)) + O(\delta) \right].
\end{aligned}$$

It results from above that

$$|E_1| = O\left(1/(nh)\right) \quad \text{and} \quad \left| E_2 - E_2^* \right| = O\left(\delta/(nh^2)\right) + O\left(\delta^2/h^2\right).$$

By a Taylor expansion, there exists a function $\tilde{\varepsilon}(t) \in (0, 1)$, $t \in \mathbb{R}$ such that

$$\begin{aligned}
E_2^* &= -\frac{\delta}{h} f^*(x) \int_\Delta u \left\{ \int K(t) K'\left[t - \frac{\delta}{h} u\varepsilon(u)\right] dt \right\} du \\
&\quad + \frac{\delta^2}{h^2} f^*(x) \int_\Delta u^2 \left\{ \int K'\left[t - \frac{\delta}{h} u\varepsilon(u)\right]^2 dt \right\} du \\
&\quad - \delta \int_\Delta u \int tK(t) K'\left[t - \frac{\delta}{h} u\varepsilon(u)\right] f^{*\prime}(x + ht\,\tilde{\varepsilon}(t)) \, dtdu \\
&\quad - \frac{\delta^2}{h} f^*(x) \int_\Delta u^2 \int K'\left[t - \frac{\delta}{h} u\varepsilon(u)\right]^2 f^{*\prime}(x + ht\tilde{\varepsilon}(t)) \, dtdu.
\end{aligned}$$

With our assumptions, it is easy to see that

$$E_2^* = O\left(\frac{\delta^2}{h^2} + \delta + \frac{\delta}{h} + \frac{\delta^2}{h}\right).$$

Finally,

$$VAR - VAR^* = O\left(\frac{1}{nh}\right) + O\left(\frac{\delta^2}{h^2} + \delta + \frac{\delta}{h} + \frac{\delta^2}{h}\right),$$

which shows that $VAR - VAR^*$ is asymptotically negligible.

For the study of VAR^*, simple computations give

$$VAR^* = \|K\|_2^2 \, f^*(x) - hf^*(x)^2 + O(h^2).$$

We now turn to the study of COV. This term can be splitted into two terms :

$$COV = \frac{2}{nh}\left(\sum\sum_{1<j-i\leq m} + \sum\sum_{j-i>m}\right) E\left(H_i H_j\right) = CV_1 + CV_2.$$

It remains to show that CV_1 and $CV_2 \to 0$ as $n \to \infty$. Since $EA_i = O\left(h\right)$, one can write

$$CV_1 = \frac{2}{nh}\sum\sum_{1<j-i\leq m} E\left(A_i A_j\right) + O\left(hm\right).$$

Moreover, using the mean value theorem (MVT) one has

$$E\left(A_i A_j\right) = h^2 \int_\Delta \int_\Delta \iint K(t_i - \frac{\delta}{h}u_i)K(t_j - \frac{\delta}{h}u_j)\, g_{i,j}(x + ht_i, x + ht_j, u_i, u_j)\, dt_i dt_j du_i du_j$$

$$\leq h^2 \max_{i,j} \sup_{s,t} f_{i,j}(x + hs, x + ht) \max_{i,j} \iint K\left(t_i - \frac{\delta}{h}c_i\right) K\left(t_j - \frac{\delta}{h}c_j\right)\, dt_i dt_j.$$

It is clear from this that $E\left(A_i A_j\right) = O\left(h^2\right)$ and $CV_1 = O\left(hm\right).$

For evaluating CV_2, in view of **(H3)**, one can use the covariance inequality of Doukhan and Portal (1983) to obtain, for p such that $1/\lambda = 1 - 2/p,$

$$|CV_2| \leq \frac{C}{nh}\sum\sum_{j-i>m} \alpha\left(j-i\right)^{1/\lambda}\left|\int_\Delta \int \left|K\left(\frac{x-y_i}{h} + \frac{\delta}{h}u_i\right)\right|^p g_i(y_i, u_i)dy_i du_i\right|^{1/p}$$

$$\times \left|\int_\Delta \int \left|K\left(\frac{x-y_j}{h} + \frac{\delta}{h}u_j\right)\right|^p g_j(y_j, u_j)dy_j du_j\right|^{1/p}$$

$$\leq \frac{C}{nh}\sum_{i=m}^{n}\left(n-i\right)\alpha\left(i\right)^{1/\lambda}\, h^{2/p}$$

$$\times \left(\max_k \left|\int_\Delta \int \left|K(t_k - \frac{\delta}{h}u_k)\right|^p g_k(x + ht_k, u_k)\, dt_k du_k\right|^{1/p}\right)^2.$$

Again, by the MVT, one has

$$|CV_2| \leq C\, n^{-1} h^{-1/\lambda} \sum_{i=m}^{n} (n-i)\, \alpha\,(i)^{1/\lambda}$$

$$\times \left\{ \max_k \left| \int \left| K(t_k - \frac{\delta}{h} c_k) \right|^p \left[\int_\Delta g_k(x + ht_k, u_k)\, du_k \right] dt_k \right|^{1/p} \right\}^2$$

$$\leq C\, h^{-1/\lambda} \sum_{i=m}^{n} \alpha\,(i)^{1/\lambda} \left[\|K\|_p^p + O\left(\frac{\delta}{h}\right) \right]^{2/p} \max_k \sup_y f_k(y)^{2/p}$$

$$= O\left(h^{-1/\lambda} \sum_{i=m}^{n} \alpha\,(i)^{1/\lambda} \right).$$

Thus, CV_2 and CV_1 tend to zero, as $n \to \infty$. Consequently, the variance of $\sqrt{nh}\,\left(\widehat{f}_B(x)\right)$ tends to $\sigma^2(x) = \|K\|_2^2\, f^*(x)$ as $n \to \infty$. $\qquad\square$

Proof of Lemma 2: For $\theta > 0$, $\mu > 0$ and $0 < \ell \leq n$, $\ell \in \mathrm{IN}$, define

$$L(\mu, \theta, \ell) = \sum_{i=1}^{\ell} \left(E\,|H_{\ell_i}|^{\mu+\theta} \right)^{\frac{\mu}{\mu+\theta}} \quad \text{and} \quad D(\mu, \theta, \ell) = \max\{ L\,(\mu, \theta, \ell), [L(2, \theta, \ell)]^{\frac{\mu}{2}} \}.$$

From Theorem 2 of Doukhan (1983), p. 26, one can write

$$E\left(\left| \sum_{i=1}^{\ell} H_{\ell_i} \right|^3 \right) \leq C D(3, 1, \ell),$$

where C is some positive generic constant. We have now to evaluate $D(3, 1, \ell)$. Using the identity $|a + b|^k \leq 2^k(|a|^k + |b|^k)$, one has

$$\sum_{i=1}^{\ell} \left(E\,|H_{\ell_i}|^4 \right)^{3/4} \leq C \sum_{i=1}^{\ell} \left[\int K^4 \left(\frac{x - y_i}{h} \right) f_i(y_i) dy_i \right.$$

$$\left. + (\delta/h)^4 \int_\Delta \int u_i^4 K'^4 [(x - y_i)/h + (\delta/h) u_i \varepsilon(u_i)]\, g_i(y_i, u_i) dy_i du_i + O(h^4) \right]^{3/4}.$$

Making use of the change of variable $t_i = (x - y_i)/h$, by assumption **(H2)**, one has

$$\sum_{i=1}^{\ell} \left(E\,|H_{\ell_i}|^4 \right)^{3/4} \leq C\ell \left[hC + h(\delta/h)^4 + O(h^4) \right]^{3/4},$$

which implies that

$$L(3, 1, \ell) \le C\ell h^{3/4}.$$

Similarly, one can show that

$$\sum_{i=1}^{\ell} \left(E\, |H_i|^3 \right)^{2/3} \le C\ell h^{2/3} \text{ and } [L(2, 1, \ell)]^{3/2} \le C\ell^{3/2} h.$$

Finally, we get

$$E\left(\left| \sum_{i=1}^{\ell} H_{\ell_i} \right|^3 \right) \le C \max\{\ell h^{\frac{3}{4}}, \ell^{3/2} h\} \le C\ell^{3/2} h.$$

$\square$

Proof of Lemma 3: Let i be the complex number such that $i^2 = -1$. It is easy to show that

$$\left| E\left[\exp(it\, S_n)\right] - E\left[\exp\left(it \sum_{j=1}^{q} U_j \right) \right] \right| \le |t| \sqrt{ E\left(\sum_{j=1}^{q} V_j \right)^2 }.$$

Now, for evaluating the term in the right-hand side of the above inequality, write

$$E\left(\sum_{j=1}^{q} V_j \right)^2 = VAR + \widetilde{CV}_1 + \widetilde{CV}_2.$$

Since $q = n/(\ell + m)$ is asymptotically equal to $n\ell^{-1}$, one has asymptotically

$$VAR = n^{-1}\, q\, m\, \left[\|K\|_2^2\, f^*(x) + O\left(\frac{\delta}{h} \right) + O\,(h) \right] = O\left(\ell^{-1}\, m \right),$$

which tends to zero as m is smaller than ℓ, while the terms $\widetilde{CV}_1$ and $\widetilde{CV}_2$ can be shown to be asymptotically nil, as CV_1 and CV_2 in the proof of Lemma 1. Thus, for larger values of n,

$$\left| E\left[\exp(it\, S_n)\right] - E\left[\exp\left(it \sum_{j=1}^{q} U_j \right) \right] \right| = O\left(\sqrt{\ell^{-1}\, m} \right),$$

which tends to zero as n tends to infinity. Now, using the version of the moment inequality of Nahapetian (1987) given in Tran (1990), one has

$$\left| E\left[\exp\left(it\sum_{j=1}^{q}U_j\right)\right] - \prod_{j=1}^{q}E\left[\exp\left(it\,U_j\right)\right] \right| \leq C\,q\,\alpha\,(m),$$

which yields (7). $\qquad\qquad\qquad\qquad\qquad\qquad\qquad\qquad\qquad\qquad\qquad\qquad\qquad\qquad\quad\square$

Proof of Theorem 1: Let (U_j) and (V_j) be the sequences constructed in the previous subsection. Denote by $\varphi(t) = E\left\{\exp\left[it\sum_{j=1}^{q}(U_j + V_j)\right]\right\}$ the characteristic function of $\sqrt{nh}\left[\widehat{f}_B(x) - E\widehat{f}_B(x)\right]$. One wishes to show that

$$\lim_{n\to\infty}\varphi(t) = \exp\left[-\frac{1}{2}t^2\sigma^2(x)\right].$$

To this end, we use the same techniques as Takahata and Yoshihara (1987) also used by Harel and Puri (1996). Denote by $H_{j,k} = H_{aj+k-1}$. By a second-order Taylor expansion, one has, for all $\ell \leq n$,

$$E\left[\exp\left(it\,U_j\right)\right] = E\left[\exp\left(it\,(nh)^{-1/2}\sum_{k=1}^{\ell}H_{j,k}\right)\right]$$

$$= E\left[1 - \frac{1}{2}t^2\,(nh)^{-1}\left(\sum_{k=1}^{\ell}H_{j,k}\right)^2 + O\left(|t^3|\,(nh)^{-3/2}\left|\sum_{k=1}^{\ell}H_{j,k}\right|^3\right)\right]$$

$$= 1 - \frac{1}{2}t^2\,E\left((nh)^{-1/2}\sum_{k=1}^{\ell}H_{j,k}\right)^2 + O\left(|t^3|\,E\left|(nh)^{-1/2}\sum_{k=1}^{\ell}H_{j,k}\right|^3\right).$$

Using the result obtained for the variance in Lemma 1, one has

$$E\left\{\left[(nh)^{-1/2}\sum_{k=1}^{\ell}H_{j,k}\right]^2\right\} \underset{n\to\infty}{\sim} \frac{\ell}{n}\sigma^2(x).$$

It follows from Lemma 2 that

$$E\left(\left|(nh)^{-1/2}\sum_{k=1}^{\ell}H_{j,k}\right|^3\right) \leq C(nh)^{-3/2}\ell^{3/2}h = C\frac{\ell}{n}\left(\frac{\ell}{nh}\right)^{1/2}.$$

Recalling that $\ell/nh = Cn^{\beta_0-\beta}$ with $0 < \beta_0 < \beta < 1$, one has

$$E\left(\left|(nh)^{-1/2}\sum_{k=1}^{\ell}H_{j,k}\right|^3\right) = o\left(\frac{\ell}{n}\right).$$

From the above results, one then obtains

$$\prod_{j=1}^{q} E\left[\exp\left(it\,U_j\right)\right] = \prod_{j=1}^{q} E\left\{\exp\left[it\,(nh)^{-1/2}\sum_{k=1}^{\ell} H_{j,k}\right]\right\}$$

$$\underset{n\to\infty}{\sim}\left[1 - \frac{1}{2}t^2\,\frac{\ell}{n}\,\sigma^2(x) + o\left(\frac{\ell}{n}\right)\right]^q.$$

Now, since m is negligible compared to ℓ, one has $q = n/(\ell + m) \sim n^{\beta}$ and $\ell/n \sim n^{-\beta}$, as n tends to infinity, which gives

$$\prod_{j=1}^{q} E\left[\exp\left(it\,U_j\right)\right] \underset{n\to\infty}{\sim} \left[1 - n^{-\beta}\frac{1}{2}t^2\sigma^2(x) + o\left(n^{-\beta}\right)\right]^{n^{\beta}}.$$

From this, (7) yields

$$\varphi(t) \underset{n\to\infty}{\sim} \exp\left[-\frac{1}{2}t^2\sigma^2(x) + o\,(1)\right] + O\left(n^{\beta}\,\alpha\,(m)\right).$$

Consequently, $\varphi(t)$ tends to the characteristic function of a zero-mean Gaussian distribution with variance $\sigma^2(x)$, as n tends to infinity. This concludes the proof of Theorem 1. $\qquad\square$

Proof of Proposition 1: Let T be the kernel defined in (2). One shows that for $m = o\,(n)$:

$$Var\left(\sqrt{nh}\,\widehat{f_B}(x)\right) = \left[\|K\|_2^2\left(\|T\|_2^2 + 2\int_{\Delta} T\,(t)\,T\,(t-1)\,dt\right)\right]f^*\,(x) + O\,(h)$$

$$+O\left(\delta + \frac{\delta^2}{h^2}\right) + O\left(\frac{1}{nh}\right) + O\left((h+\delta)\,m + \left(h\delta^{-1-1/\lambda} + \delta^{-1/\lambda}\right)\sum_{i=m}^{n}\alpha\,(i)^{1/\lambda}\right),$$

where λ is the real number given in **(H3)**′.

For all $i \in$ IN, let $A_i = \sum_{j\in\mathbb{Z}} K\left[(x - a_j)/h\right] T\left[(X_i - a_j)/\delta\right]$. Then, the function $\widehat{f_B}(x)$ can be written as $\widehat{f_B}(x) = (1/nh)\sum_{i=1}^{n} A_i$, and the variance of $\sqrt{nh}\,\widehat{f_B}(x)$ can be decomposed into

$$Var\left(\sqrt{nh}\,\widehat{f_B}(x)\right) = \left(VAR - VAR^*\right) + VAR^* + COV.$$

Here VAR^* stands for the stationary counterpart of VAR. After some algebra, using the TVN, one has

$$VAR - VAR^* = O\,(1/(nh))$$

and, by Lemma 3.3 of Carbon et al. (1997),

$$VAR^* = \left\{ \|K\|_2^2 \left[\|T\|_2^2 + 2 \int_\Delta T(t) \, T(t-1) \, dt \right] \right\} f^*(x) + O(\delta) + O(h) + O\left(\frac{\delta^2}{h^2}\right).$$

Let $m = m(n) = o(n)$. In view of $(\mathbf{H3})'$, by the covariance inequality of Doukhan and Portal (1983) and Lemma 3.3 of Carbon et al. (1997), one has, after some computations,

$$|COV| \leq \frac{C}{nh} \, nm \left[\iint h^2 K(u) \, K(v) \, du \, dv + O\left(h\delta \, \|K'\|_1\right) + O\left(\delta^2 \, \|K'\|_1^2\right) \right]$$
$$+ 16 \frac{\delta^{2/p}}{nh} \frac{1}{\delta^2} \left[\iint h^2 K(u) \, K(v) \, du \, dv + O\left(h\delta \, \|K'\|_1\right) + O\left(\delta^2 \, \|K'\|_1^2\right) \right]$$
$$\times \max_i \sup_x f_i(x) \left[\|T\|_p^p + O(\delta) \right]^{2/p} \, O\left(\sum_{i=m}^n (n-i) \, \alpha(i)^{1/\lambda} \right),$$

which leads to :

$$COV = O\left((h+\delta)\, m\right) + O\left(\left(h\delta^{-1/\lambda-1} + \delta^{-1/\lambda}\right) \sum_{i=m}^n \alpha(i)^{1/\lambda} \right).$$

$\square$

Remark 5 The final expression of the variance depends essentially on T. In particular, for rounding kernels T, as $\|T\|_2^2 = 1$ and the other terms are nil, one has

$$VAR^* = \|K\|_2^2 \, f^*(x) + O(\delta) + O(h).$$

For the triangular kernel T and $\Delta = [0, 1[$, one has $\|T\|_2^2 = 2/3$, and $\int_\Delta T(t) \, T(t-1) \, dt = 1/6$. Moreover, if K is at least twice differentiable with absolutely continuous derivatives such that $K' \in L^2$, and if it admits a bounded third-order derivative, one can show that

$$VAR^* = \left(\|K\|_2^2 - \frac{1}{3} \frac{\delta^2}{h^2} \|K'\|_2^2 \right) f^*(x) + O(h) + O(\delta) + O\left(\frac{\delta^3}{h^3}\right).$$

Acknowledgments We would like to thank the Editor and two anonymous referees whose comments led to improve the paper.

References

Bardet, J.-M., Doukhan, P., Lang, G., & Ragache, N. (2008). Dependent Lindeberg central limit theorem and some applications. *ESAIM:PS, 12*, 154–172.

Carbon, M., Garel, B., & Tran, L. T. (1997). Frequency polygons for weakly dependent processes. *Statistics and Probability Letters, 33*, 1–13.

Doukhan, P. (1983). *Mixing; properties and examples (142 p.)*. New York: Springer.

Doukhan, P., & Portal, F. (1983). Moments de variables aléatoires mélangeantes. *C.R.A.S., 297*, 129–132.

Geange, S. W., Pledger, S., Burns, K. C., & Shima, J. S. (2011). A unified analysis of niche overlap incorporating data of different types. *Methods in Ecology and Evolution, 2*, 175–184.

Hall, P., & Wand, M. P. (1996). On the accuracy of binned kernel density estimators. *Journal of Multivariate Analysis, 56*, 165–184.

Hall, P. (1983). Edgeworth expansions of the distribution of Stein's statistic. *Mathematical Proceedings of the Cambridge Philosophical Society, 93*, 163–175.

Harel, M., & Puri, M. L. (1996). Nonparametric density estimators based on nonstationary absolutely regular random sequences. *Journal of Applied Mathematics and Stochastic Analysis, 9*(3), 233–254.

Harel, M., & Puri, M. L. (1999). Conditional empirical processes defined by nonstationary absolutely regular sequences. *Journal of Multivariate Analysis, 70*, 250–285.

Jones, M. C. (1989). Discretized and interpolated kernel density estimates. *Journal of the American Statistical Association, 84*, 733–741.

Jones, M. C., & Lotwick, H. W. (1984). Remark ASR50. A remark on algorithm AS176: Kernel density estimation using the fast Fourier transform. *Applied Statistics, 33*, 120–122.

Lenain, J. F., Harel, M., & Puri, M. L. (2011). Asymptotic behavior of the kernel density estimator for dependent and nonstationary random variables with binned data. *Festschrift in Honor of Professor P.K. Bhattacharya* (pp. 396–420). World Scientific

Nahapetian, B. S. (1987). An approach to proving limit theorems for dependent random variables. *Theory of Probability and Its Applications, 32*, 535–539.

Rosenblatt, M. (1971). Curve estimates. *Annals of Mathematical Statistics, 42*(6), 1815–1842.

Scott, D. W., & Sheather, S. J. (1985). Kernel density estimation with binned data. *Communications in Statistics Theory Methods, 14*, 1353–1359.

Simonoff, J. S. (1996). *Smoothing methods in statistics*. New York: Springer.

Takahata, H., & Yoshihara, K. I. (1987). Central limit theorems for integrated square error of nonparametric density estimators based on absolutely regular sequences. *Yokohama Mathematical Journal, 35*, 95–111.

Tran, L. T. (1990). Kernel density estimation on random fields. *Journal of Multivariate Analysis, 34*(1), 37–53.

Approximation of a Random Process with Variable Smoothness

Enkelejd Hashorva, Mikhail Lifshits and Oleg Seleznjev

Abstract We consider the rate of piecewise constant approximation to a locally stationary process $X(t)$, $t \in [0, 1]$, having a variable smoothness index $\alpha(t)$. Assuming that $\alpha(\cdot)$ attains its unique minimum at zero and satisfies

$$\alpha(t) = \alpha_0 + bt^\gamma + \mathrm{o}(t^\gamma) \quad \text{as } t \to 0,$$

we propose a method for construction of observation points (composite dilated design) such that the integrated mean square error

$$\int_0^1 \mathbb{E}\{(X(t) - X_n(t))^2\}dt \sim \frac{K}{n^{\alpha_0}(\log n)^{(\alpha_0+1)/\gamma}} \quad \text{as } n \to \infty,$$

where a piecewise constant approximation X_n is based on $N(n) \sim n$ observations of X. Further, we prove that the suggested approximation rate is optimal, and then show how to find an optimal constant K.

E. Hashorva
Actuarial Department, HEC Lausanne, University of Lausanne,
1015 Lausanne, Switzerland
e-mail: enkelejd.hashorva@unil.ch

M. Lifshits (✉)
Department of Mathematics and Mechanics, St. Petersburg State University,
198504 St. Petersburg, Russia
e-mail: mikhail@lifshits.org

M. Lifshits
MAI, Linköping university, Linköping, Sweden

O. Seleznjev
Department of Mathematics and Mathematical Statistics, Umeå University,
SE-901 87 Umeå, Sweden
e-mail: oleg.seleznjev@matstat.umu.se

© Springer International Publishing Switzerland 2015

M. Hallin et al. (eds.), *Mathematical Statistics and Limit Theorems*,
DOI 10.1007/978-3-319-12442-1_11

1 Introduction

Probabilistic models based on the locally stationary processes with variable smoothness became recently an object of interest for applications in various areas (e.g., Internet traffic, financial records, natural landscapes) due to their flexibility for matching local regularity properties (see, e.g., Dahlhaus 2012; Echelard et al. 2010 and references therein). The most known representative random process of this class is a multifractional Brownian motion (mBm) independently introduced in Benassi et al. (1997) and Peltier and Lévy Véhel (1995). We refer to Ayache (2001) for a survey and to Ayache and Bertrand (2010), Ayache et al. (2000), Falconer and Lévy Véhel (2009), Surgailis (2006) for studies of particular aspects of mBm.

A more general class of $\alpha(\cdot)$-locally stationary Gaussian processes with a variable smoothness index $\alpha(t)$, $t \in [0, 1]$, was elaborated in Debicki and Kisowski (2008). This class generalizes also the class of locally stationary Gaussian processes with index α (introduced by Berman 1974). It is worthwhile, however, to notice that another approach to "local stationarity" is possible whenever the processes with time varying parameters are considered. This direction leads to interesting models and applications in Statistics, long memory theory, etc., see Dahlhaus (2012) for more information. The two approaches are technically different but describe, in our opinion, the same phenomena. In this paper we stick to $\alpha(\cdot)$-local stationarity, as defined in (1) below. Whenever we need to model such processes with a given accuracy, the approximation (time discretization) accuracy has to be evaluated.

More specifically, consider a random process $X(t)$, $t \in [0, 1]$, with finite second moment and variable quadratic mean smoothness (see precise definition (1) below). The process X is observed at $N = N(n)$ points and a piecewise constant approximation X_n is built upon these observations. The approximation performance on the entire interval is measured by the integrated mean square error (IMSE) $\int_0^1 \mathbb{E}\{(X(t) - X_n(t))^2\}dt$. We construct a sequence of sampling designs (i.e., sets of observation points) taking into account the varying smoothness of X such that on a class of processes, the IMSE decreases faster when compared to conventional regular sampling designs (see, e.g., Seleznjev 2000) or to quasi-regular designs, Abramowicz and Seleznjev (2011), used for approximation of locally stationary random processes and random processes with an isolated singularity point, respectively.

The approximation results obtained in this paper can be used in various problems in signal processing, e.g., in optimization of compressing digitized signals, (see, e.g., Cohen et al. 2002), in numerical analysis of random functions (see, e.g., Benhenni and Cambanis 1992; Creutzig et al. 2007; Creutzig and Lifshits 2006), in simulation studies with controlled accuracy for functionals on realizations of random processes (see, e.g., Abramowicz and Seleznjev 2008; Eplett 1986). It is known that a piecewise constant approximation gives an optimal rate for certain class of continuous random processes satisfying a Hölder condition (see, e.g., Buslaev and Seleznjev 1999; Seleznjev 2000). In this paper we develop a technique improving this rate for a certain class of locally stationary processes with variable smoothness. The developed technique can be generalized for more advanced approximation methods (e.g., Hermite splines) and various classes of random processes and fields. Some related approximation results for continuous and smooth random functions can be found in Hüsler et al. (2003), Kon and Plaskota (2005), Seleznjev (1996). The book Ritter (2000) contains a very detailed survey of various random function approximation problems.

The paper is organized as follows. In Sect. 2 we specify the problem setting. We recall a notion of a locally stationary process with variable smoothness, introduce a class of piecewise constant approximation processes, and define integrated mean square error (IMSE) as a measure of approximation accuracy. Furthermore, we introduce a special method of *composite dilated sampling designs* that suggests how to distribute the observation points sufficiently densely located near the point of the lowest smoothness. The implementation of this design depends on some functional and numerical parameters, and we set up a certain number of mild assumptions about these parameters. In Sect. 3, our main results are stated. Namely, for a locally stationary process with known smoothness, we consider the piecewise constant interpolation related to dilated sampling designs (adjusted to smoothness parameters) and find the asymptotic behavior of its approximation error. In the second part of that section, the approximation for conventional regular and some quasi-regular sampling designs are studied. In Sect. 4, the results and conjectures related to the optimality of our bounds are discussed. Section 5 contains the proofs of the statements from Sect. 3.

2 Variable Smoothness Random Processes and Approximation Methods: Basic Notation

2.1 Approximation Problem Setting

Let $X = X(t), t \in [0, 1]$, be an $\alpha(\cdot)$-locally stationary random process, i.e., $\mathbb{E}\{X(t)^2\} < \infty$ and

$$\lim_{s \to 0} \frac{||X(t+s) - X(t)||^2}{|s|^{\alpha(t)}} = c(t) \quad \text{uniformly in } t \in [0, 1], \tag{1}$$

where $||Y|| := (\mathbb{E}Y^2)^{1/2}$, $\alpha(\cdot), c(\cdot) \in C([0, 1])$, and $2 \geq \alpha(t) > 0$, $c(t) > 0$.

We assume that the following conditions hold for the function $\alpha(\cdot)$ describing the smoothness of X:

(C1) $\alpha(\cdot)$ attains its global minimum $\alpha_0 := \alpha(0)$ at the unique point $t_0 = 0$.

(C2) There exist $b, \gamma > 0$ such that

$$\alpha(t) = \alpha_0 + bt^\gamma + o(t^\gamma) \quad \text{as } t \to 0.$$

The choice $t_0 = 0$ in $(C1)$ is made only for notational convenience. The results are essentially the same for any location of the unique minimum of $\alpha(\cdot)$.

Let X be sampled at the distinct design points $T_n = (t_0(n), \ldots, t_N(n))$ (also referred to as knots), where $0 = t_0(n) < t_1(n) < \cdots < t_N(n) = 1$, $N = N(n)$. We suppress the auxiliary integer argument n for design points $t_j = t_j(n)$ and for the number of points $N = N(n)$ when doing so causes no confusion. The corresponding piecewise constant approximation is defined by

$$X_n(t) := X(t_{j-1}), \qquad t_{j-1} \le t < t_j, \ j = 1, \ldots, N.$$

In this article, we consider the accuracy of the approximation to X by X_n with respect to the integrated mean square error (IMSE)

$$e_n^2 = ||X - X_n||_2^2 := \int_0^1 ||X(t) - X_n(t)||^2 dt.$$

We shall describe below a construction of sampling designs $\{T_n\}$ providing the fastest decay of e_n^2.

2.2 Sampling Design Construction

The construction idea is as follows. In order to achieve a rate-optimal approximation of X by X_n, we introduce a sequence of *dilated* sampling designs $\{T_n\}$.

Recall first that any probability density $f(t), t \in [0, 1]$, generates a sequence of associated conventional sampling designs, (cf., e.g., Sacks and Ylvisaker 1966; Benhenni and Cambanis 1992; Seleznjev 2000) defined by

$$\int_0^{t_j} f(t)dt = \frac{j}{n}, \qquad j = 0, \ldots, n, \tag{2}$$

i.e., the corresponding sampling points are (j/n)-percentiles of the distribution having density $f(\cdot)$. We call $f(\cdot)$ a *sampling density*.

Let $p(\cdot)$ be a probability density on $\mathbb{R}_+ := [0, \infty)$; we shall refer to it as the *design density*. In our problem, it turns out to be useful to *dilate* the design density $p(\cdot)$ by replacing it with a *dilated sampling density*

$$p_n(t) := d_n\, p(d_n t), \qquad t \in [0, 1], \tag{3}$$

where $d_n \nearrow \infty$ is a *dilation coefficient*. Note, that formally $p_n(\cdot)$ is not a probability density, but

$$\int_0^1 p_n(t)dt = \int_0^{d_n} p(u)du \to 1 \text{ as } n \to \infty.$$

The idea of dilation is obvious: we wish to put more knots near the point of the worst smoothness. The dilation coefficient should be chosen according to the smoothness behavior at this critical point. In our case, $(C2)$ requires the choice

$$d_n := (\log n)^{1/\gamma}$$

that will be maintained in the sequel. As in (2), we define the knots by

$$\int_0^{t_j} p_n(t)dt = \frac{j}{n}. \tag{4}$$

Further optimization of the approximation accuracy bound requires one more adjustment: it turns out to be useful to choose the knots t_j as in (4) using different densities in a neighborhood of the critical point and outside of it. We call *composite* such sampling design constructions operating differently on two disjoint domains.

Now we pass to the rigorous description of our sampling designs. Let $p(u)$ and $\widetilde{p}(u)$, $u \in [0, \infty)$, be two probability densities. Let the dilated sampling densities $p_n(\cdot)$ be defined as in (3). Similarly, $\widetilde{p}_n(t) := d_n \, \widetilde{p}(d_n t)$.

For $0 < \rho < 1$, we define the *composite dilated $(p, \rho, \widetilde{p})$-designs* T_n by choosing t_j according to (4) for

$$0 \le j \le J(p, \rho, n) := n \int_0^{\rho} p_n(t)dt = n \int_0^{\rho d_n} p(u)du \le n.$$

Notice that for these knots, we have $0 \le t_j \le \rho$. Furthermore, we fill the interval $[\rho, 1]$ with analogous knots t_i using the probability density $\widetilde{p}(\cdot)$,

$$\int_0^{t_i} \widetilde{p}_n(t)dt = \frac{j}{n}, \tag{5}$$

where

$$J(\widetilde{p}, \rho, n) < j \le J(\widetilde{p}, 1, n),$$

$$i = j + J(p, \rho, n) - J(\widetilde{p}, \rho, n).$$

For these knots we clearly have $\rho < t_i \le 1$. Note, it follows by definition that

$$J(p, \rho, n) = n \int_0^{\rho d_n} p(u)du \sim n \qquad \text{as } n \to \infty,$$

and similarly, in the interval $[\rho, 1]$, the number of points does not exceed

$$n - J(\widetilde{p}, \rho, n) = n \int_{\rho d_n}^{\infty} \widetilde{p}(u)du = o(n) \qquad \text{as } n \to \infty,$$

that is the total number of sampling points satisfies

$$N(n) \sim J(p, \rho, n) \sim n \quad \text{as } n \to \infty. \tag{6}$$

In the sequel, we will use $(p, \rho, \widetilde{p})$-designs satisfying the following additional assumptions on $p(\cdot)$, ρ, and $\widetilde{p}(\cdot)$:

(A1) The design density $p(\cdot)$ is bounded, nonincreasing, and

$$p(u) \geq q_1 \exp\{-q_2 u^\gamma\}, \quad u \geq 0, \; q_1 > 0, \; \tfrac{b}{\alpha_0} > q_2 > 0. \tag{7}$$

(A2) We assume that $\widetilde{p}$ is *regularly varying* at infinity with some index $r \leq -1$. This means that for all $\lambda > 0$,

$$\frac{\widetilde{p}(\lambda u)}{\widetilde{p}(u)} \to \lambda^r \quad \text{as } u \to \infty. \tag{8}$$

(A3) Finally, we assume that the parameter ρ is small enough. Namely, applying $q_2 < b/\alpha_0$ and using $(C2)$ we may choose ρ satisfying

$$q_2 \sup_{0 \leq t \leq \rho} \alpha(t) < \inf_{0 \leq t \leq \rho} \frac{\alpha(t) - \alpha(0)}{t^\gamma} \tag{9}$$

and

$$q_2 \rho^\gamma < 1. \tag{10}$$

For example, let $\alpha(t) = 1 + t^\gamma$. Then $(C1)$, $(C2)$ hold, and (A3) corresponds to $\rho < (1/q_2 - 1)^{1/\gamma}$, where $0 < q_2 < 1$.

Regularly varying probability densities satisfy (7) for large u, thus we could simplify the design construction by letting $p = \widetilde{p}$. For example, the choice

$$p(u) = \widetilde{p}(u) := (1 + u)^{-2}$$

agrees with (A1) and (A2).

Moreover, in this case the knots may be easily calculated explicitly, as $t_j = \frac{j}{d_n(n-j)}$. However, this kind of the simplified choice does not provide an optimal constant K in the main approximation error asymptotics (11) below.

3 Main Results

3.1 Dilated Approximation Designs

In the following theorem, we give the principal result of the paper and consider IMSE e_n^2 of approximation to X by X_n for the proposed sequence of composite dilated sampling designs T_n, $n \geq 1$. It follows from (A1) that the following constant is finite,

$$K = K(c, \alpha, (p, \rho, \widetilde{p})) := \frac{c_0}{\alpha_0 + 1} \int_0^\infty p(u)^{-\alpha_0} e^{-bu^\gamma} du < \infty,$$

where $c_0 := c(0)$.

Theorem 1 *Let $X(t)$, $t \in [0, 1]$, be an $\alpha(\cdot)$-locally stationary random process such that assumptions $(C1)$, $(C2)$ hold. Let X_n be the piecewise constant approximations corresponding to composite dilated $(p, \rho, \widetilde{p})$-designs $\{T_n\}$ satisfying $(A1)$–$(A3)$. Then $N(n) \sim n$ and*

$$\|X - X_n\|_2^2 \sim \frac{K}{n^{\alpha_0} (\log n)^{(\alpha_0+1)/\gamma}} \sim \frac{K}{N^{\alpha_0} (\log N)^{(\alpha_0+1)/\gamma}} \qquad \text{as } n \to \infty. \quad (11)$$

Remark 1 Among the assumptions of Theorem 1, the monotonicity of $p(\cdot)$ is worth of a discussion. Of course, it agrees with the heuristics to put more knots at places where the smoothness of the process is worse. However, this assumption may be easily replaced by some mild regularity assumptions on $p(\cdot)$.

Remark 2 The following probability density $p^*(\cdot)$

$$p^*(u) = Ce^{-bu^\gamma/(\alpha_0+1)}, \quad C = \frac{b^{1/\gamma}}{(\alpha_0 + 1)^{1/\gamma} \Gamma(1/\gamma + 1)}$$

minimizes the constant K in Theorem 1 and generates the asymptotically optimal sequence of designs T_n^*. For the optimal T_n^*,

$$K^* := \frac{c_0}{\alpha_0 + 1} \left(\int_0^\infty e^{-bu^\gamma/(\alpha_0+1)} du \right)^{\alpha_0+1} = \frac{c_0}{\alpha_0 + 1} \left(\frac{(\alpha_0 + 1)^{1/\gamma} \Gamma(1/\gamma + 1)}{b^{1/\gamma}} \right)^{\alpha_0+1},$$

see, e.g., Seleznjev (2000). We emphasize that $p^*(\cdot)$ satisfies assumption (7) but it is not regularly varying. In other words, a simple design based on $\widetilde{p} = p = p^*$ does not fit in theorem's assumptions.

Remark 3 The idea of considering composite designs might seem to be overcomplicated at first glance. However, in some sense it cannot be avoided. The previous remark shows that if we want to get the optimal constant K, we must handle the exponentially decreasing densities. Assume that

$$p(u) \leq q_1 \exp\{-q_2 u^\gamma\}. \tag{12}$$

If we would simplify the design by defining $t_j(n)$ as in (4) for the entire interval, i.e., with $\rho = 1$, then we would have

$$\int_0^{t_j} p_n(t)dt = \frac{j}{n},$$

hence,

$$\frac{1}{n} = \int_{t_j}^{t_{j+1}} p_n(t)dt = \int_{t_j}^{t_{j+1}} d_n p(d_n t)dt \leq d_n q_1 \int_{t_j}^{t_{j+1}} \exp\{-q_2(d_n t)^\gamma\}dt$$
$$\leq d_n q_1(t_{j+1} - t_j) \exp\{-q_2(d_n t_j)^\gamma\}.$$

Let $a \in (0, 1)$ and $t_j \in [1 - a, 1]$. Then for the length of the corresponding intervals, we have

$$t_{j+1} - t_j \geq \frac{\exp\{q_2(d_n t_j)^\gamma\}}{nd_n q_1} \geq \frac{\exp\{q_2 \log n(1 - a)^\gamma\}}{nd_n q_1}.$$

If $q_2 > 1$ and a is so small that $q_2(1 - a)^\gamma > 1$, we readily obtain $t_{j+1} - t_j > a$ for large n which is impossible. Therefore, for $q_2 > 1$ there are no sampling points t_j in $[1 - a, 1]$, i.e., clearly $e_n^2 \geq C > 0$ for any n, i.e., IMSE does not tend to zero at all.

The confusion described above may really appear in practice because $q_2 > 1$ is compatible with the assumption $q_2 < b/\alpha_0$ from (7) whenever $b > \alpha_0$.

Theorem 1 shows that for the design densities with regularly varying tails, we may define all knots by (4) without leaving empty intervals as above. However, we cannot achieve the optimal constant K on this simpler way.

Remark 4 Actually, the choice of knots outside of $[0, \rho]$ is not relevant for the approximation rate. One can replace the knots from (5) with a uniform grid of knots $t_i = in^{-\mu}$ with appropriate $\mu < 1$.

3.2 Regular Sampling Designs

The approximation algorithm investigated in Theorem 1 is based upon the assumption that we know the point where $\alpha(\cdot)$ attains its minimum, as well as the index γ in (C2). If for the same process neither the critical point nor the index γ are known, a conventional *regular design* can be used.

Let $X(t), t \in [0, 1]$, be an $\alpha(\cdot)$-locally stationary random process, i.e., (1) holds. Consider now sampling designs $T_n = \{t_j(n), j = 0, 1, \ldots, n\}$ generated by a regular positive continuous density $p(t), t \in [0, 1]$, (see, e.g., Sacks and Ylvisaker 1966; Seleznjev 2000) through (13), i.e.,

$$\int_0^{t_j} p(t)dt = \frac{j}{n} \, , \qquad 0 \le j \le n. \tag{13}$$

Let the constant

$$K_1 := \frac{c_0}{\alpha_0 + 1} \, \frac{\Gamma(1/\gamma + 1)}{p_0^{\alpha_0} b^{1/\gamma}} \, , \quad p_0 := p(0).$$

Theorem 2 *Let $X(t), t \in [0, 1]$, be an $\alpha(\cdot)$-locally stationary random process such that $(C1)$ and $(C2)$ hold. Let X_n be the piecewise constant approximations corresponding to the (regular) sampling designs $\{T_n\}$ generated by $p(\cdot)$. Then*

$$\|X - X_n\|_2^2 \sim \frac{K_1}{n^{\alpha_0}(\log n)^{1/\gamma}} \qquad as\ n \to \infty.$$

Remark 5 If the point where $\alpha(\cdot)$ attains its minimum is known but γ is unknown, we may build the designs without dilating the design density. Instead, one could use quasi-regular sampling designs generated by a possibly unbounded design density $p(t), t \in (0, 1]$, at the singularity point $t_0 = 0$ (cf., Abramowicz and Seleznjev 2011). For example, if $p(\cdot)$ is a probability density on $(0, 1]$ such that

$$p(t) \sim A t^{-\kappa} \qquad as\ t \searrow 0, \qquad 0 < \kappa < 1,$$

and $t_j(n)$ are chosen through (13), then for an $\alpha(\cdot)$-locally stationary random process X satisfying $(C1)$ and $(C2)$, it is possible to show a slightly weaker asymptotics than that of Theorem 1, namely,

$$e_n^2 \sim \frac{K_2}{n^{\alpha_0}(\log n)^{(1+\kappa\alpha_0)/\gamma}} \qquad as\ n \to \infty,$$

with $K_2 := c_0 A^{-\alpha_0} \Gamma(1/\gamma + 1)/((\alpha_0 + 1)b^{1/\gamma})$.

Of course, all above-mentioned asymptotics differ only by a degree of logarithm while the polynomial rate is determined by the minimal regularity index α_0. But for some large-scale approximation problems and certain regularity properties $(C1)$, $(C2)$ of $\alpha(t)$ this gain could be significant.

Remark 6 As an anonymous referee pointed out to us, it would be interesting to extend the results to the case of more general behavior of the function $\alpha(\cdot)$ at the critical point by replacing $(C2)$ with

$$\alpha(t) = \alpha_0 + g(t) + o(g(t)) \qquad as\ t \to 0,$$

with a given function $g(\cdot)$ from an appropriate class. For example, it is clear that the results will be pretty much the same if we allow g to be γ-regularly varying at zero. Yet we preferred to consider here only the simplest (and arguably the most important) polynomial case and leave the general case for further research.

4 Optimality

4.1 Optimality of the Rate for Piecewise Constant Approximations

We explain here that the approximation rate $l_n^{-1}, l_n := n^{\alpha_0} d_n^{(\alpha_0+1)}$, achieved in Theorem 1 is optimal in the class of piecewise constant approximations for every $\alpha(\cdot)$-locally stationary random process satisfying $(C1)$ and $(C2)$. For a sampling design T_n, let the mesh size $|T_n| := \max\{(t_j - t_{j-1}), j = 1, \ldots, n\}$.

Proposition 1 *Let X_n be piecewise constant approximations to an $\alpha(\cdot)$-locally stationary random process X satisfying $(C1)$ and $(C2)$ constructed according to designs $\{T_n\}$ such that $N_n \sim n$ and $|T_n| \to 0$ as $n \to \infty$. Then*

$$\liminf_{n \to \infty} \, l_n \, e_n^2 > 0. \tag{14}$$

Proof Let $r_n := d_n^{-1} = (\log n)^{-1/\gamma}$ and $J_n := \inf\{j : t_j = t_j(n) \geq r_n\}$. Then (19) entails

$$e_n^2 \geq \sum_{j=1}^{J_n} e_{n,j}^2 = \sum_{j=1}^{J_n} B_{j-1} w_j^{\alpha(t_{j-1})+1}(1 + o(1)) = B \sum_{j=1}^{J_n} w_j^{a_n+1}(1 + o(1)),$$

where $a_n := \sup_{0 \leq t \leq r_n} \alpha(t)$ and $w_j = t_j - t_{j-1}$. By using the convexity of the power function $w \to w^{a_n+1}$, we obtain

$$\frac{1}{J_n} \sum_{j=1}^{J_n} w_j^{a_n+1} \geq \left(\frac{1}{J_n} \sum_{j=1}^{J_n} w_j\right)^{a_n+1} \geq \left(\frac{r_n}{J_n}\right)^{a_n+1},$$

hence,

$$\sum_{j=1}^{J_n} w_j^{a_n+1} \geq \frac{r_n^{a_n+1}}{J_n^{a_n}} \geq \frac{r_n^{a_n+1}}{N_n^{a_n}},$$

whereas

$$\begin{aligned}
e_n^2 &\geq B \frac{r_n^{a_n+1}}{N_n^{a_n}} (1 + o(1)) = B \frac{1}{d_n^{a_n+1}} \frac{1}{N_n^{a_n}} (1 + o(1)) \\
&= B \frac{1}{d_n^{\alpha_0+1} n^{\alpha_0}} \left(\frac{1}{d_n n}\right)^{a_n-\alpha_0} \left(\frac{n}{N_n}\right)^{a_n} (1 + o(1)) \\
&= B \, l_n^{-1} \left(\frac{1}{d_n n}\right)^{a_n-\alpha_0} (1 + o(1)).
\end{aligned}$$

Recall that by $(C2)$, $a_n - \alpha_0 = O(r_n^\gamma) = O((\log n)^{-1})$ and thus (14) follows. $\qquad\square$

4.2 Optimality of the Rate in a Class of Linear Methods

We explain here that the approximation rate l_n^{-1} achieved in Theorem 1 is optimal not only in the class of piecewise constant approximations but in a much wider class of linear methods,—at least for some $\alpha(\cdot)$-locally stationary random processes satisfying $(C1)$ and $(C2)$. The corresponding setting is based on the notion of Gaussian approximation numbers, or ℓ-numbers, that we recall here.

Gaussian approximation numbers of a Gaussian random vector X taking values in a normed space $\mathscr{X}$ are defined by

$$\ell_n(X;\,\mathscr{X})^2 = \inf_{\substack{x_1,\ldots,x_{n-1} \\ \xi_1,\ldots,\xi_{n-1}}} \mathbb{E}\left\{\left\|X - \sum_{j=1}^{n-1}\xi_j x_j\right\|_{\mathscr{X}}^2\right\}, \tag{15}$$

where infimum is taken over all $x_j \in \mathscr{X}$ and all Gaussian vectors $\xi = (\xi_1,\ldots,\xi_{n-1})$ $\in \mathbb{R}^{n-1}, n \geq 2$, see Kühn and Linde (2002), Lifshits (2012). If $\mathscr{X}$ is a Hilbert space, then

$$\ell_n(X;\,\mathscr{X})^2 = \sum_{j=n}^{\infty}\lambda_j,$$

where λ_j is a decreasing sequence of eigenvalues of the covariance operator of X.

Recall that a multifractional Brownian motion (mBm) with a variable smoothness index (or fractality function) $\alpha(\cdot) \in (0, 2)$ introduced in Benassi et al. (1997), Peltier and Lévy Véhel (1995) and studied in Ayache (2001), Ayache and Bertrand (2010), Ayache et al. (2000) is a Gaussian process defined through its white noise representation

$$X(t) = \int_{-\infty}^{\infty}\frac{e^{itu} - 1}{|u|^{(\alpha(t)+1)/2}}\,dW(u),$$

where $W(t), t \in \mathbb{R}$, is a standard Brownian motion. Notice that mBm is a typical example of a locally stationary process whenever $\alpha(\cdot)$ is a continuous function.

In the particular case of the constant fractality $\alpha(t) \equiv \alpha$, we obtain an ordinary fractional Brownian motion $B^\alpha, \alpha \in (0, 2)$. For $X = B^\alpha$ considered as an element of $\mathscr{X} = L_2[0, 1]$, the behavior of its eigenvalues λ_j is well known, cf. Bronski (2003). Namely,

$$\lambda_j \sim c_\alpha j^{-\alpha-1} \qquad \text{as } j \to \infty,$$

with some $c_\alpha > 0$ continuously depending on $\alpha \in (0, 2)$. It follows that

$$\ell_n(B^\alpha;\,L_2[0, 1])^2 \sim \alpha^{-1}c_\alpha\,n^{-\alpha} \qquad \text{as } n \to \infty.$$

Hence, for all $n \geq 1$,

$$\ell_n(B^\alpha; L_2[0, 1])^2 \geq C_\alpha n^{-\alpha}, \quad C_\alpha > 0.$$

Furthermore, since B^α is a self-similar process, we can scale this estimate from $\mathcal{X} = L_2[0, 1]$ to $\mathcal{X} = L_2[0, r]$ with arbitrary $r > 0$. An easy computation shows that

$$\ell_n(B^\alpha; L_2[0, r])^2 = r^{\alpha+1}\ell_n(B^\alpha; L_2[0, 1])^2 \geq C_\alpha r^{\alpha+1} n^{-\alpha}.$$

Let us now consider a multifractional Brownian motion X parameterized by a fractality function $\alpha(\cdot)$ satisfying $(C2)$. For example, let

$$\alpha(t) := \alpha_0 + b\, t^\gamma, \qquad 0 \leq t \leq 1, \tag{16}$$

with $\alpha_0, b > 0$ chosen so small that $\alpha_0 + b < 2$. This choice secures the necessary condition $0 < \alpha(t) < 2, \quad 0 \leq t \leq 1$. Then, letting $r = r_n := d_n^{-1}$, we have

$$\begin{aligned}
\ell_n(X; L_2[0, 1])^2 &\geq \ell_n(X; L_2[0, r_n])^2 \geq M\ell_n(B^{\alpha(r_n)}; L_2[0, r_n])^2 \\
&\geq MC_{\alpha(r_n)} r_n^{\alpha(r_n)+1} n^{-\alpha(r_n)} = MC_{\alpha(r_n)} d_n^{-\alpha(r_n)-1} n^{-\alpha(r_n)} \\
&\geq Cl_n^{-1}(d_n n)^{\alpha_0-\alpha(r_n)} = Cl_n^{-1}(d_n n)^{-br_n^\gamma} = Cl_n^{-1}(d_n n)^{-b(\log n)^{-1}} \geq \tilde{C}\, l_n^{-1},
\end{aligned}$$

for some positive M, $C_{\alpha(r_n)}$, C, $\tilde{C}$. All bounds here are obvious except for the second inequality comparing approximation rate of multifractional Brownian motion with that of a standard fractional Brownian motion. We state this fact as a separate result.

Proposition 2 *Let $X(t), a \leq t \leq b$, be a multifractional Brownian motion corresponding to a continuous fractality function $\alpha : [a, b] \to (0, 2)$. Let B^β be a fractional Brownian motion such that $\inf_{a \leq t \leq b} \alpha(t) \leq \beta < 2$. Then there exists $M = M(\alpha(\cdot), \beta) > 0$ such that*

$$\ell_n(X; L_2[a, b]) \geq M\ell_n(B^\beta, L_2[a, b]), \quad n \geq 1.$$

The proof of this proposition requires different methods from those used in this article. We relegate it to another publication.

Our conclusion is that a multifractional Brownian motion with fractality function (16) provides an example of an $\alpha(\cdot)$-locally stationary random process satisfying assumptions $(C1)$ and $(C2)$ such that no linear approximation method provides a better approximation rate than l_n^{-1}.

5 Proofs

Proof of Theorem 1 We represent the IMSE $e_n^2 = ||X(t) - X_n(t)||_2^2$ as the following sum

$$e_n^2 = \sum_{j=1}^{N} \int_{t_{j-1}}^{t_j} ||X(t) - X_n(t)||^2 dt = \sum_{j=1}^{N} \int_{t_{j-1}}^{t_j} ||X(t) - X(t_{j-1})||^2 dt =: \sum_{j=1}^{N} e_{n,j}^2 . \tag{17}$$

Next, for a large $U > 0$, let

$$e_n^2 = \sum_{j=1}^{N} e_{n,j}^2 = S_1 + S_2 + S_3,$$

where the sums S_1, S_2, S_3 include the terms $e_{n,j}^2$ such that $[t_{j-1}, t_j]$ belongs to $[0, U/d_n], [U/d_n, \rho]$, and $[\rho, 1]$, respectively. Let J_1 and J_2 denote the corresponding boundaries for the index j. Recall that $l_n = n^{\alpha_0} d_n^{\alpha_0+1} = n^{\alpha_0} (\log n)^{(\alpha_0+1)/\gamma}$ and l_n^{-1} is the approximation rate announced in the theorem. We show that only S_1 is relevant to the asymptotics of e_n^2, namely, that $l_n S_3 = o(1)$ as $n \to \infty$, while

$$\limsup_{n\to\infty} l_n S_2 = o(1) \qquad \text{as } U \to \infty. \tag{18}$$

Let $w_j := t_j - t_{j-1}$, $u_j := d_n t_j$ be the normalized knots and denote by $v_j := u_j - u_{j-1} = d_n w_j$ the corresponding dilated interval lengths. It follows by the definition of $\alpha(\cdot)$-local stationarity (1) that for large n,

$$e_{n,j}^2 = c(t_{j-1}) \int_{t_{j-1}}^{t_j} (t - t_{j-1})^{\alpha(t_{j-1})} dt \, (1 + r_{n,j})$$

$$= B_{j-1} \, (t_j - t_{j-1})^{\alpha(t_{j-1})+1} \, (1 + r_{n,j})$$

$$= B_{j-1} \, (v_j/d_n)^{\alpha(t_{j-1})+1} \, (1 + r_{n,j}), \tag{19}$$

where $|T_n| = \max_j w_j = o(1)$ and $\max_j r_{n,j} = o(1)$ as $n \to \infty$ and

$$B_j := \frac{c(t_j)}{\alpha(t_j) + 1}, \qquad j = 1, \dots, N.$$

First, we evaluate S_3. Recall that for $j > J_2$ we have $\rho d_n \leq u_{j-1} < u_j \leq d_n$. We use now the following property of regularly varying functions (see, e.g., Bingham et al. (1987)): convergence in (8) is uniform for all intervals $0 < a \leq \lambda \leq b < \infty$. Using this uniformity we obtain, for some $C_1 > 0$,

$$\inf_{u_{j-1} \leq u \leq u_j} \widetilde{p}(u) \geq \inf_{\rho d_n \leq u \leq d_n} \widetilde{p}(u) \geq C_1 \, \widetilde{p}(d_n).$$

It follows by (5) that

$$\int_{u_{j-1}}^{u_j} \widetilde{p}(u)\,du = \int_{t_{j-1}}^{t_j} \widetilde{p}_n(t)\,dt = \frac{1}{n}.$$

Hence, for some $C_2 > 0$,

$$v_j \le \left(\inf_{u_{j-1} \le u \le u_j} \widetilde{p}(u) \right)^{-1} \int_{u_{j-1}}^{u_j} \widetilde{p}(u)\,du \le \frac{1}{n} \left(\inf_{u_{j-1} \le u \le u_j} \widetilde{p}(u) \right)^{-1}$$

$$\le \frac{1}{C_1\, n\, \widetilde{p}(d_n)} \le C_2 \frac{d_n^{|r|+1}}{n}, \quad j = J_2 + 1, \ldots, N, \tag{20}$$

and $\max_{j > J_2} w_j = d_n^{|r|}/n$. Recall that by assumption $(C1)$,

$$\alpha_1 := \inf_{t \in [\rho,1]} \alpha(t) > \alpha_0.$$

Therefore, for large n, we get by (19) and (20), $C_3, C_4 > 0$,

$$S_3 \le n \max_{j > J_2} e_{n,j}^2 \le nC_3(v_j/d_n)^{\alpha_1+1} \le C_4 \frac{d_n^{|r|(\alpha_1+1)}}{n^{\alpha_1}} = o(l_n^{-1}) \text{ as } n \to \infty. \tag{21}$$

Now consider the first two zones corresponding to S_1, S_2. We have by definition

$$\int_0^{u_j} p(u)\,du = \frac{j}{n}, \qquad 0 \le j < J.$$

Since the function $p_n(t), t \in [0,1]$, is nonincreasing, the sequence $\{v_j\}$ is nondecreasing. In fact,

$$\frac{1}{n} = \int_{u_{j-1}}^{u_j} p(u)\,du \in [p(u_j)v_j, \, p(u_{j-1})v_j],$$

and therefore,

$$\frac{1}{np(u_{j-1})} \le v_j \le \frac{1}{np(u_j)} \le v_{j+1} \tag{22}$$

and it follows by (A1) that $\max_{j \le J_2} w_j = o(1)$ as $n \to \infty$. For $j \le J_2$, the bounds (19) and (22) yield for n large,

$$\begin{aligned}
e_{n,j}^2 &= B_{j-1}(v_j/d_n)^{\alpha(t_{j-1})}\frac{v_j}{d_n}(1+o(1)) \\
&\le B_{j-1}(nd_n p(u_j))^{-\alpha(t_{j-1})}\frac{v_j}{d_n}(1+o(1)) \\
&\le B_{j-1}(np(u_j))^{-\alpha(t_{j-1})}d_n^{-\alpha_0-1}v_j(1+o(1)) \\
&= B_{j-1}l_n^{-1}n^{-(\alpha(t_{j-1})-\alpha_0)}p(u_j)^{-\alpha(t_{j-1})}v_j(1+o(1)).
\end{aligned} \tag{23}$$

From now on, we proceed differently in the first and in the second zone.

For the second zone, $J_1 \le j \le J_2$, we do not care about the constant by using

$$B_j \le B_* := \max_{0\le t\le 1}\frac{c(t)}{\alpha(t)+1}. \tag{24}$$

Next, (7) and (9) give

$$p(u_j)^{-\alpha(t_{j-1})} \le C\exp\{q_2\alpha(t_{j-1})u_j^\gamma\} \le C\exp\{\beta_1 u_j^\gamma\},\quad C>0, \tag{25}$$

where $\beta_1 := q_2\sup_{0\le t\le\rho}\alpha(t)$. On the other hand, we infer from (9) that

$$n^{-(\alpha(t_{j-1})-\alpha_0)} = n^{-\frac{\alpha(t_{j-1})-\alpha_0}{t_{j-1}^\gamma}t_{j-1}^\gamma} \le n^{-\beta_2\frac{u_{j-1}^\gamma}{\log n}} = \exp\{-\beta_2 u_{j-1}^\gamma\}, \tag{26}$$

where $\beta_2 := \inf_{0\le t\le\rho}(\alpha(t)-\alpha_0)/t^\gamma > \beta_1$ by (9).

Recall that by (10), we have $1 - q_2\rho^\gamma > 0$. Moreover, for $U \le u_j \le \rho d_n$, we derive from (7) and (22)

$$v_j \le n^{-1}p(\rho d_n)^{-1} \le Cn^{-1}\exp\{q_2(\rho d_n)^\gamma\} = Cn^{-(1-q_2\rho^\gamma)},\quad C>0,$$

and it follows

$$u_{j+1}^\gamma - u_{j-1}^\gamma = u_{j-1}^\gamma\left(\left(\frac{u_{j+1}}{u_{j-1}}\right)^\gamma - 1\right) = O\left(d_n^\gamma n^{-(1-q_2\rho^\gamma)}\right) = o(1) \text{ as } n\to\infty \tag{27}$$

uniformly in $J_1 \le j \le J_2$.

Since $\{v_j\}$ is nondecreasing, (27) implies an integral bound

$$\begin{aligned}
&\exp\{-\beta_2 u_{j-1}^\gamma\}\exp\{\beta_1 u_j^\gamma\}v_j \\
&= \exp\{\beta_2[u_{j+1}^\gamma - u_{j-1}^\gamma]\}\exp\{\beta_1 u_j^\gamma - \beta_2 u_{j+1}^\gamma\}v_j \\
&\le C\inf_{u_j\le u\le u_{j+1}}\exp\{\beta_1 u^\gamma - \beta_2 u^\gamma\}v_{j+1} \\
&\le C\int_{u_j}^{u_{j+1}}e^{-(\beta_2-\beta_1)u^\gamma}du,\quad C>0.
\end{aligned} \tag{28}$$

By plugging (24), (26), and (28) into (23), and summing up the resulting bounds over $J_1 < j \le J_2$, we obtain

$$S_2 \le B_* l_n^{-1} \int_U^\infty e^{-(\beta_2 - \beta_1)u^\gamma} \, du \, (1 + o(1)) \qquad \text{as } n \to \infty. \tag{29}$$

Therefore, (18) is valid.

In the first zone, $j \le J_1$, $t_j \le U/d_n$, the knots are uniformly small. Hence, B_{j-1} are uniformly close to B due to the continuity of the functions $\alpha(\cdot)$ and $c(\cdot)$. Moreover, by $(C2)$ for any $\varepsilon > 0$, we have for all n large enough

$$\alpha_0 + (b - \varepsilon)t_{j-1}^\gamma \le \alpha(t_{j-1}) \le \alpha_0 + (b + \varepsilon)t_{j-1}^\gamma, \qquad j \le J_1. \tag{30}$$

Hence (23) yields

$$\begin{aligned}
e_{n,j}^2 &\le (B + \varepsilon)l_n^{-1} n^{-(b-\varepsilon)t_{j-1}^\gamma} \, p(u_j)^{-\alpha_0} p(u_j)^{-(\alpha(t_{j-1})-\alpha_0)} \, v_j \\
&= (B + \varepsilon)l_n^{-1} \, n^{-(b-\varepsilon)(u_{j-1}/d_n)^\gamma} \, p(u_j)^{-\alpha_0} p(u_j)^{-(\alpha(t_{j-1})-\alpha_0)} \, v_j.
\end{aligned} \tag{31}$$

Recall that by the definition of d_n, we have

$$n^{-(b-\varepsilon)(u_{j-1}/d_n)^\gamma} = n^{-(b-\varepsilon)u_{j-1}^\gamma / \log n} = \exp\{-(b - \varepsilon)u_{j-1}^\gamma\}.$$

Since $p(\cdot)$ is nonincreasing and $\{v_j\}$ is nondecreasing, we also have an integral bound

$$\begin{aligned}
&\exp\{-(b - \varepsilon)u_{j-1}^\gamma\} \, p(u_j)^{-\alpha_0} v_j \\
&= \exp\{(b - \varepsilon)[u_{j+1}^\gamma - u_{j-1}^\gamma]\} \, p(u_j)^{-\alpha_0} \exp\{-(b - \varepsilon)u_{j+1}^\gamma\} v_j \\
&\le \exp\{(b - \varepsilon)[u_{j+1}^\gamma - u_{j-1}^\gamma]\} \inf_{u_j \le u \le u_{j+1}} (p(u)^{-\alpha_0} e^{-(b-\varepsilon)u^\gamma}) \, v_j \\
&\le \exp\{(b - \varepsilon)[u_{j+1}^\gamma - u_{j-1}^\gamma]\} \int_{u_j}^{u_{j+1}} p(u)^{-\alpha_0} e^{-(b-\varepsilon)u^\gamma} \, du.
\end{aligned} \tag{32}$$

Moreover, for $u_j \le U$, we derive from (A1) and (22)

$$v_j \le n^{-1} p(U)^{-1}.$$

By using the convexity and the concavity of the power function for $\gamma \ge 1$ and $\gamma \le 1$, respectively, we get

$$\begin{aligned}
u_{j+1}^\gamma - u_{j-1}^\gamma &\le \gamma U^{\gamma-1}(u_{j+1} - u_j) = \gamma U^{\gamma-1}(v_j + v_{j+1}) \\
&\le 2\gamma U^{\gamma-1}v_{j+1} = o(1) \qquad \text{as } n \to \infty \qquad (\gamma \ge 1); \\
u_{j+1}^\gamma - u_{j-1}^\gamma &\le (u_{j+1} - u_{j-1})^\gamma \\
&= (v_j + v_{j+1})^\gamma = o(1) \qquad \text{as } n \to \infty \qquad (\gamma \le 1).
\end{aligned}$$

Therefore, the exponential factor in (32) turns out to be negligible.

Finally, for $u_j \leq U$, the property $d_n \to \infty$ yields

$$p(u_j)^{-(\alpha(t_{j-1})-\alpha_0)} \leq \max\{1, \, p(U)^{-\max_{0 \leq t \leq U/d_n}(\alpha(t)-\alpha_0)}\} = 1 + o(1). \qquad (33)$$

By plugging (32) and (33) into (31), and summing up the resulting bounds over $j \leq J_1$, we obtain

$$S_1 \leq (B + 2\varepsilon)l_n^{-1} \int_0^\infty p(u)^{-\alpha_0} e^{-(b-\varepsilon)u^\gamma} du \qquad \text{as } n \to \infty.$$

Since ε can be chosen arbitrarily small, we arrive at

$$\limsup_{n \to \infty} l_n S_1 \leq B \int_0^\infty p(u)^{-\alpha_0} e^{-bu^\gamma} du = K. \qquad (34)$$

Combining (21), (29), and (34) gives the desired upper bound.

The lower bound is obtained along the same lines: since S_2 and S_3 are asymptotically negligible, we shall evaluate only S_1 starting again from (19). As in (23), we have

$$
\begin{aligned}
e_{n,j}^2 &= B_{j-1}(v_j/d_n)^{\alpha(t_{j-1})} \frac{v_j}{d_n}(1 + o(1)) \\[2mm]
&\geq B_{j-1}(nd_n p(u_{j-1}))^{-\alpha(t_{j-1})} \frac{v_j}{d_n}(1 + o(1)) \\[2mm]
&= B_{j-1} n^{-\alpha_0} n^{-(\alpha(t_{j-1})-\alpha_0)} p(u_{j-1})^{-\alpha(t_{j-1})} d_n^{-\alpha_0-1} d_n^{\alpha_0-\alpha(t_{j-1})} v_j(1 + o(1)) \\[2mm]
&= B_{j-1} l_n^{-1} n^{-(\alpha(t_{j-1})-\alpha_0)} p(u_{j-1})^{-\alpha(t_{j-1})} d_n^{\alpha_0-\alpha(t_{j-1})} v_j(1 + o(1)). \qquad (35)
\end{aligned}
$$

Recall that for $j \leq J_1$ and the constants B_{j-1} are uniformly close to B. Moreover, by using (30), we have for large n,

$$d_n^{\alpha_0-\alpha(t_{j-1})} \geq d_n^{-(b+\varepsilon)t_{j-1}^\gamma} \geq d_n^{-(b+\varepsilon)(U/d_n)^\gamma} = 1 + o(1).$$

Hence, (35) yields

$$
\begin{aligned}
e_{n,j}^2 &\geq (B - \varepsilon)l_n^{-1} n^{-(b+\varepsilon)t_{j-1}^\gamma} p(u_{j-1})^{-\alpha_0} p(u_{j-1})^{-(\alpha(t_{j-1})-\alpha_0)} v_j \\[2mm]
&= (B - \varepsilon)l_n^{-1} \, n^{-(b+\varepsilon)(u_{j-1}/d_n)^\gamma} \, p(u_{j-1})^{-\alpha_0} p(u_{j-1})^{-(\alpha(t_{j-1})-\alpha_0)} v_j, \qquad (36)
\end{aligned}
$$

where as before

$$n^{-(b+\varepsilon)(u_{j-1}/d_n)^\gamma} = n^{-(b+\varepsilon)u_{j-1}^\gamma/\log n} = \exp\{-(b+\varepsilon)u_{j-1}^\gamma\}.$$

Since $p(\cdot)$ is nonincreasing and $\{v_j\}$ is nondecreasing, we also have an integral bound

$$\exp\{-(b+\varepsilon)u^{\gamma}_{j-1}\}p(u_{j-1})^{-\alpha_0}v_j$$

$$= \exp\{(b+\varepsilon)[u^{\gamma}_{j-2} - u^{\gamma}_{j-1}]\}p(u_{j-1})^{-\alpha_0}\exp\{-(b+\varepsilon)u^{\gamma}_{j-2}\}v_j$$

$$\geq \exp\{(b+\varepsilon)[u^{\gamma}_{j-2} - u^{\gamma}_{j-1}]\}\inf_{u_{j-2}\leq u\leq u_{j-1}}(p(u)^{-\alpha_0}e^{-(b+\varepsilon)u^{\gamma}})\,v_{j-1}$$

$$\geq \exp\{(b+\varepsilon)[u^{\gamma}_{j+1} - u^{\gamma}_{j-1}]\}\int_{u_{j-2}}^{u_{j-1}}p(u)^{-\alpha_0}e^{-(b+\varepsilon)u^{\gamma}}\,du. \tag{37}$$

We have already seen that the exponential factor in (37) is negligible.

Finally, for $u_j \leq U$, the fact that $d_n \to \infty$ implies (cf. (33))

$$p(u_{j-1})^{-(\alpha(t_{j-1})-\alpha_0)} \geq \min\{1, p(0)^{-\max_{0\leq t\leq U/d_n}(\alpha(t)-\alpha_0)}\} = 1 + o(1). \tag{38}$$

By plugging (37) and (38) into (36), and summing up the resulting bounds over $j \leq J_1$, we obtain

$$S_1 \geq (B - 2\varepsilon)l_n^{-1}\int_0^U p(u)^{-\alpha_0}e^{-(b+\varepsilon)u^{\gamma}}\,du \qquad \text{as } n \to \infty.$$

Since $\varepsilon > 0$ can be chosen arbitrarily small, we arrive at

$$\liminf_{n\to\infty} l_n S_1 \geq B\int_0^U p(u)^{-\alpha_0}e^{-bu^{\gamma}}\,du.$$

Finally,

$$\liminf_{n\to\infty} l_n e_n^2 \geq \sup_{U>0}\liminf_{n\to\infty} l_n S_1 \geq B\int_0^{\infty}p(u)^{-\alpha_0}e^{-bu^{\gamma}}\,du = K. \tag{39}$$

This is the desired lower bound. $\qquad\square$

Proof of Theorem 2 Applying the notation of Theorem 1, we have for an interval approximation error

$$e_{n,j}^2 = B_{j-1}\,w_j^{\alpha(t_{j-1})+1}\,(1+r_{n,j}), \quad w_j = t_j - t_{j-1}, j = 1, \ldots, n,$$

where $\max_j r_{n,j} = o(1)$ as $n \to \infty$. Now for a small enough $\rho > 0$, similarly to Theorem 1, we get

$$\int_{\rho}^1 e_n(t)^2 dt \leq C/n^{\alpha_1}, \ C > 0, \qquad \alpha_1 := \inf_{t\in[\rho,1]}\alpha(t) > \alpha_0,$$

that is only $e_{n,j}$ such that $[t_{j-1}, t_j] \subset [0, \rho]$ are relevant for the asymptotics, say, $e_{n,j}, j = 1, \ldots, J = J(\rho, n)$. Let us denote the approximation rate $L_n := n^{\alpha_0}(\log n)^{1/\gamma}$. Next, for $S_1 := \sum_{j=1}^J e_{n,j}$ and for a small enough ρ, we have by continuity of the design density $p(\cdot)$ and by the mean value theorem

$$e_{n,j}^2 = B_{j-1}\,(np(\eta_j))^{-\alpha(t_{j-1})}w_j\,(1+o(1))$$

$$\le \frac{B}{p(0)^{\alpha_0}}(1+\varepsilon)\,n^{-\alpha_0}\int_{t_{j-2}}^{t_{j-1}} e^{-(b-\varepsilon)t^\gamma \log n}dt\,(1+o(1))$$

$$= L_n^{-1}\,\frac{B}{p_0^{\alpha_0}}(1+\varepsilon)\int_{u_{j-2}}^{u_{j-1}} e^{-(b-\varepsilon)u^\gamma}du\,(1+o(1)),$$

where $p_0 := p(0)$. By summing up, we obtain

$$\limsup_{n\to\infty} L_n S_1 \le \frac{B}{p_0^{\alpha_0}}(1+\varepsilon)\int_0^\infty e^{-(b-\varepsilon)u^\gamma}du = \frac{B}{p_0^{\alpha_0}}(1+\varepsilon)\frac{\Gamma(1/\gamma+1)}{(b-\varepsilon)^{1/\gamma}}.$$

Hence,

$$\limsup_{n\to\infty} L_n e_n^2 = \limsup_{n\to\infty} L_n S_1 \le \frac{B}{p_0^{\alpha_0}}\,(1+\varepsilon)\,\frac{\Gamma(1/\gamma+1)}{(b-\varepsilon)^{1/\gamma}}.$$

Since ε can be chosen arbitrary small, we get

$$\limsup_{n\to\infty} L_n e_n^2 \le \frac{B}{p_0^{\alpha_0}}\,\frac{\Gamma(1/\gamma+1)}{b^{1/\gamma}} = K_1.$$

The lower bound follows from similar arguments. This completes the proof. $\square$

Acknowledgments We are grateful to both anonymous referees for useful advice concerning this work and eventual future development of the subject.

Research of E. Hashorva was partially supported by the Swiss National Science Foundation grant 200021-140633/1 and RARE-318984 (an FP7 Marie Curie IRSES Fellowship). Research of M. Lifshits was supported by RFBR grant 13-01-00172 and by SPbSU grant 6.38.672.2013. Research of O. Seleznjev was supported by the Swedish Research Council grant 2009-4489.

References

Abramowicz, K., & Seleznjev, O. (2008). On the error of the Monte Carlo pricing method. *Journal of Numerical Applied Mathematics, 96,* 1–10.

Abramowicz, K., & Seleznjev, O. (2011). Spline approximation of a random process with singularity. *Journal of Statistical Planning and Inference, 141,* 1333–1342.

Ayache, A. (2001). Du mouvement Brownien fractionnaire au mouvement Brownien multifractionnaire. *Technique et Science Informatiques, 20,* 1133–1152.

Ayache, A., & Bertrand, P. R. (2010). A process very similar to multifractional Brownian motion. *Recent developments in fractals and related fields* (pp. 311–326). Boston: Birkhäuser.

Ayache, A., Cohen, S., & Lévy Véhel, J. (2000). The covariance structure of multifractional Brownian motion. In *Proceedings of the IEEE International Conference on Acoustics, Speech, and Signal Processing,* (vol. 6, pp. 3810–3813).

Benassi, A., Jaffard, S., & Roux, D. (1997). Gaussian processes and pseudodifferential elliptic operators. *Revista Matemática Iberoamericana, 13*(1), 19–81.

Benhenni, K., & Cambanis, S. (1992). Sampling designs for estimating integrals of stochastic process. *Annals of Statistics, 20*, 161–194.

Berman, S. M. (1974). Sojourns and extremes of Gaussian process. *Annals of Probability, 2*, 999–1026; corrections: *8*, 999 (1980); *12*, 281 (1984).

Bronski, J. C. (2003). Small ball constants and their tight eigenvalue asymptotics for fractional Brownian motions. *Journal of Theoretical Probability, 16*, 87–100.

Bingham, N. H., Goldie, C. M., & Teugels, J. L. (1987). *Regular variation.* Cambridge: Cambridge University Press.

Buslaev, A. P., & Seleznjev, O. (1999). On certain extremal problems in theory of approximation of random processes. *East Journal on Approximations, 5*, 467–481.

Cohen, A., Daubechies, I., Guleryuz, O. G., & Orchard, M. T. (2002). On the importance of combining wavelet-based nonlinear approximation with coding strategies. *IEEE Transactions on Information Theory, 48*, 1895–1921.

Creutzig, J., & Lifshits, M. (2006). Free-knot spline approximation of fractional Brownian motion. In A. Keller, S. Heinrich, & H. Neiderriter (Eds.), *Monte Carlo and Quasi Monte Carlo methods* (pp. 195–204). Berlin: Springer.

Creutzig, J., Müller-Gronbach, T., & Ritter, K. (2007). Free-knot spline approximation of stochastic processes. *Journal of Complexity, 23*, 867–889.

Dahlhaus, R. (2012). Locally stationary processes. *Handbook of statistics: Time series analysis: Methods and applications* (Vol. 30, pp. 351-413). Elsevier.

Debicki, K., & Kisowski, P. (2008). Asymptotics of supremum distribution of $\alpha(t)$-locally stationary Gaussian processes. *Stochastic Processes and Their Applications, 118*, 2022–2037.

Echelard, A., Lévy Véhel, J., & Barriére, O. (2010). Terrain modeling with multifractional Brownian motion and self-regulating processes. *Computer vision and graphics. Lecture Notes in Computer Science* (Vol. 6374, pp. 342–351). Berlin: Springer.

Eplett, W. T. (1986). Approximation theory for simulation of continuous Gaussian processes. *Probability Theory and Related Fields, 73*, 159–181.

Falconer, K. J., & Lévy Véhel, J. (2009). Multifractional, multistable and other processes with prescribed local form. *Journal of Theoretical Probability, 22*, 375–401.

Hüsler, J., Piterbarg, V., & Seleznjev, O. (2003). On convergence of the uniform norms for Gaussian processes and linear approximation problems. *Annals of Applied Probability, 13*, 1615–1653.

Kon, M., & Plaskota, L. (2005). Information-based nonlinear approximation: an average case setting. *Journal of Complexity, 21*, 211–229.

Kühn, T., & Linde, W. (2002). Optimal series representation of fractional Brownian sheets. *Bernoulli, 8*, 669–696.

Lifshits, M. A. (2012). *Lectures on Gaussian Processes.* Heidelberg: Springer.

Peltier, R. F., & Lévy Véhel, J. (1995). *Multifractional Brownian motion: definition and preliminary results.* Rapport de recherche de l'INRIA, 2645.

Ritter, K. (2000). *Average-case analysis of numerical problems. Lecture Notes in Mathematics* (Vol. 1733). Berlin: Springer.

Sacks, J., & Ylvisaker, D. (1966). Design for regression problems with correlated errors. *Annals of Mathematical Statistics, 37*, 66–89.

Seleznjev, O. (1996). Large deviations in the piecewise linear approximation of Gaussian processes with stationary increments. *Advances in Applied Probability, 28*, 481–499.

Seleznjev, O. (2000). Spline approximation of stochastic processes and design problems. *Journal of Statistical Planning and Inference, 84*, 249–262.

Surgailis, D. (2006). Non-homogeneous fractional integration and multifractional processes. *Stochastic Processes and Their Applications, 116*, 200–221.

A Cramér–von Mises Test for Gaussian Processes

Gennady Martynov

Abstract We propose a statistical method for testing the null hypothesis that an observed random process on the interval [0, 1] is a mean zero Gaussian process with specified covariance function. Our method is based on a finite number of observations of the process. To test this null hypothesis, we develop a Cramér–von Mises test based on an infinite-dimensional analogue of the empirical process. We also provide a method for computing the critical values of our test statistic. The same theory also applies to the problem of testing multivariate uniformity over a high-dimensional hypercube. This investigation is based upon previous joint work by Paul Deheuvels and the author.

Keywords Goodness-of-fit test · Cramér–von Mises test · Gaussianity test · Hilbert space

1 Introduction

This work proposes a method for testing the *null hypothesis* that a random process $x(t)$, $t \in [0, 1]$, is Gaussian with zero mean and specified covariance function $K_x(t, \tau)$ satisfying the condition

$$\int_D K_x(t, t)dt < \infty.$$

We shall represent this random process x as a point in a Hilbert space with axes defined by the eigenfunctions of the covariance operator K_x. Under the null hypothesis, the coordinates of the random process $x(t)$, $t \in [0, 1]$, are independent mean zero normal variables with variances equal to the eigenvalues of the covariance operator.

G. Martynov (✉)
Kharkevich Institute for Information Transmission Problems, Moscow, Russia
e-mail: magevl@gmail.com

G. Martynov
Higher School of Economics, Moscow, Russia

© Springer International Publishing Switzerland 2015
M. Hallin et al. (eds.), *Mathematical Statistics and Limit Theorems*,
DOI 10.1007/978-3-319-12442-1_12

We also consider the general Gaussian distribution in a Hilbert space H, where the random element is represented accordingly. In both cases, each coordinate is transformed to a uniform random variable on $[0, 1]$. We shall construct an empirical process on $[0, 1]^\infty$ of a special form. This process converges weakly in $L_2([0, 1]^\infty)$ to a Gaussian process. Our test statistic will be represented as an infinite-dimensional integral on $[0, 1]^\infty$ of this empirical process with respect to some suitable measure. The main task of this work is to determine the eigenvalues of the covariance function of this empirical process.

Our test was first proposed in Martynov (1979). The present work uses essentially results for one-dimensional weighted Cramér–von Mises statistics, obtained in Deheuvels and Martynov (2003). The main results of this work are briefly presented in Sect. 2. In Sect. 3 we describe the problem of testing the null hypothesis that a random process x is Gaussian and propose a statistic to test this null hypothesis. In Sect. 4, the theory is developed for determining the eigenvalues of the covariance function of the empirical process on which the Cramér–von Mises statistic (introduced in Sect. 3) is based. Methods for calculating the asymptotic critical values of our test statistic are described briefly in Sect. 4.4. The exact quantiles of the distribution for the proposed statistic are given in Sect. 4.5, quantiles obtained by Monte Carlo simulation are presented in Sect. 4.6.

2 Weighted Cramér–von Mises Statistics for Finite-Dimensional Samples

2.1 One-Dimensional Samples

2.1.1 Cramér–von Mises Statistics with General Weight Function

In dimension one, the Cramér–von Mises statistic with a general weight function is given by

$$\bar{\omega}_n^2 = n \int_0^1 \psi^2(t)(F_n(t) - t)^2 dt, \tag{1}$$

where $F_n(t)$ is the empirical distribution function based on an i.i.d. sample $X_1, X_2, \ldots, X_n$ with common cumulative distribution function [cdf] F, which is assumed to be continuous, and $\psi(t)$ is a weight function. The statistic (1) is designed to test the null hypothesis

$$H_0 : \quad F \text{ is the uniform cdf on } [0, 1]$$

against the alternative

$$H_1 : \quad H_0 \text{ is not true.}$$

If the condition

$$\int_0^1 \psi^2(t)\, t\, (1-t)\, dt < \infty$$

is fulfilled, then the statistic $\bar\omega_n^2$ converges in distribution to

$$\bar\omega^2 = \int_0^1 \xi^2(t)\, dt, \tag{2}$$

where $\xi(t)$, $t \in [0, 1]$, is the Gaussian process with zero mean and the covariance function

$$K_\psi(t, \tau) = \psi(t)\psi(\tau)(\min(t, \tau) - t\tau).$$

The Gaussian process $\xi(t)$ can be developed into the Karhunen–Loève series

$$\xi(t) = \sum_{i=1}^{\infty} \frac{\omega_k \varphi_k(t)}{\sqrt{\lambda_k}},$$

where, throughout this paper, $\omega_k \sim N(0, 1)$, $k = 1, 2, \ldots$, are independent normal random variables with mean zero and variance 1, and λ_k and $\varphi_k(t)$, $k = 1, 2, \ldots$, are the eigenvalues and eigenfunctions of the Fredholm integral equation

$$\varphi(t) = \lambda \int_0^1 \psi(t)\psi(\tau)(\min(t, \tau) - t\tau)\varphi(\tau)\, d\tau. \tag{3}$$

2.1.2 Cramér–von Mises Statistics with Power Weight Function

Deheuvels and Martynov (2003) describe the following result. Let $\{b(t) : 0 \le t \le 1\}$ be the Brownian bridge, i.e., the Gaussian process with mean zero and covariance function $K(t, \tau) = \min(t, \tau) - t\tau$. Then, for each $\beta > -1$, the Karhunen–Loève expansion of $\{t^\beta b(t) : 0 < t \le 1\}$ is given by

$$t^\beta b(t) = \sum_{k=1}^{\infty} \sqrt{\lambda_k}\, \omega_k e_k(t).$$

For $k = 1, 2, \ldots$, the eigenvalues and the corresponding eigenfunctions are $\lambda_k = (2\nu/z_{\nu,k})^2$ and

$$e_k(t) = \frac{t^{\frac{1}{2\nu}-\frac{1}{2}} J_\nu\left(z_{\nu,k}\, t^{\frac{1}{2\nu}}\right)}{\sqrt{\nu} J_{\nu-1}\left(z_{\nu,k}\right)}, \quad 0 < t \le 1,$$

respectively, where $z_{\nu,k}$, $k = 1, 2, \ldots$, are the zeros of the Bessel functions $J_\nu(z)$ for $\nu = 1/(2(\beta+1))$.

2.1.3 Classical Cramér–von Mises Statistic

The original Cramér–von Mises statistic has the weight function $\psi(t) := 1$ and has the form

$$\bar{\omega}_n^2 = n \int_0^1 (F_n(t) - t)^2 dt.$$

The eigenvalues and eigenfunctions of the covariance function $K(t, \tau)$ are

$$\lambda_i = (\pi i)^2 \quad \text{and} \quad \varphi_i(t) = \sqrt{2} \sin(\pi i t), \quad i = 1, 2, \ldots,$$

respectively. The limit in distribution $\bar{\omega}^2$ of $\bar{\omega}_n^2$ can be written as

$$\bar{\omega}^2 = \int_0^1 b^2(t) dt = \sum_{k=1}^{\infty} \frac{\omega_k^2}{(\pi k)^2}.$$

2.2 Multivariate Uniformity Test with Weight Function

We shall use the notation $s = (s_1, \ldots, s_d)^\top$ and $t = (t_1, \ldots, t_d)^\top$ for d-vectors.

Let $U = (U_1, \ldots, U_d)^\top$ be a random vector with the uniform distribution function on $[0, 1]^d$,

$$F(t) = P(U \le t) = \prod_{i=1}^{d} t_i,$$

and let

$$U_i = (U_{i1}, \ldots, U_{id}), \quad i = 1, \ldots, n,$$

be n independent observations of U. The empirical distribution function of this sample has the form

$$F_n(t) = \frac{1}{n} \sum_{i=1}^{n} 1_{\{U_i \le t\}}.$$

We can write the multivariate empirical process as

$$\alpha_n(t) = n^{1/2} (F_n(t) - F(t)).$$

The Cramér–von Mises statistic for testing that F is the uniform cdf on $[0, 1]^d$ is given by

$$\bar{\omega}_n^2 = \int_{[0,1]^d} \alpha_n^2(t) dt.$$

Deheuvels (2005) studied the weighted variant

$$\bar{\omega}_n^2 = \int_{[0,1]^d} t^{2B} \alpha_n^2(t) dt$$

of this statistic, where $t^B = t_1^{\beta_1} \cdot \ldots \cdot t_d^{\beta_d}$ and $B = (\beta_1, \ldots, \beta_d)$.

The weighted process $t^B \alpha_n(t)$ converges weakly in the Hilbert space $L^2([0, 1]^d)$ to the Gaussian process $\xi(t) = t^B b(t)$, where $b(t)$ is the standard multivariate Brownian bridge, with covariance function

$$K_b(s, t) = E(b(s)b(t)) = \prod_{j=1}^{d} \{s_j \wedge t_j\} - \prod_{i=1}^{d} \{s_j t_j\}.$$

The eigenvalues and eigenfunctions corresponding to the kernel $K_b(s, t)$ can be derived from the eigenvalues and eigenfunctions corresponding to the covariance function of the weighted multivariate Wiener process $w(t)$, which has covariance function

$$K_w(s, t) = \prod_{j=1}^{d} \{s_j^{\beta_j} t_j^{\beta_j}\}\{s_j \wedge t_j\}.$$

Using results from Deheuvels and Martynov (2003), the Karhunen–Loève expansion for $w(t)$ is given by

$$w(t) = \sum_{k_1=1}^{\infty} \cdots \sum_{k_d=1}^{\infty} \sqrt{\tilde{\lambda}_{k_1 \ldots k_d}} Y_{k_1 \ldots k_d} \tilde{e}_{k_1 \ldots k_d}(t), \tag{4}$$

where

$$\tilde{\lambda}_{k_1 \ldots k_d} = \prod_{j=1}^{d} \left\{ 2\nu_j / z_{\nu_j - 1, k_j} \right\}^2$$

and

$$\tilde{e}_{k_1 \ldots k_d}(t) = \prod_{j=1}^{d} t_j^{\frac{1}{2\nu_j} - \frac{1}{2}} J_{\nu_j} \left(z_{\nu_j - 1, k_j} t_j^{\frac{1}{2\nu_j}} \right) \Big/ \sqrt{\nu_j} J_{\nu_j} \left(z_{\nu_j - 1, k_j} \right).$$

Here, $0 < z_{\nu,1} < z_{\nu,2} < \ldots$ are the zeros of the Bessel function $J_\nu(\cdot)$ and $Y_{k_1 \ldots k_d}$: $k_1 \geq 1, \ldots, k_d \geq 1$, is an infinite array of i.i.d. $N(0, 1)$ random variables.

The multiplicities of $\tilde{\lambda}_{k_1 \ldots k_d}$ in (4) can be very different. If all β_j's are equal to 0, then they vary from 1 to ∞. This case is considered in Krivjakova et al. (1977). If all the β_j's are distinct, then the multiplicities of $\tilde{\lambda}_{k_1 \ldots k_d}$ can be larger than one, but this happens very rarely. For example, if $d = 2$, $\nu_1 = 2$ and $\nu_2 = 2,37927259 \ldots$ (or $\beta_1 = -1/2$ and $\beta_2 = -.637490078 \ldots$), then $\tilde{\lambda}_{1,15} = \tilde{\lambda}_{2,8}$.

To conclude this section, we note that the method described below is suitable also for testing uniformity on the finite-dimensional cube.

3 Cramér–von Mises Tests for Gaussian Distributions in a Hilbert Space

3.1 General Formulation of the Gaussianity Hypothesis

The problem considered in this subsection was first formulated in Martynov (1979 and 1992).

Let $(\mathcal{X}, \mathcal{B}, \nu)$ be a probability space, where $\mathcal{X}$ is a real separable Hilbert space of elementary events, $\mathcal{B}$ is the σ-algebra of Borel sets on $\mathcal{X}$, and ν is a probability measure over $(\mathcal{X}, \mathcal{B})$. Let $X^1, X^2, \dots, X^n$ be n independent observations of the random element X of $(\mathcal{X}, \mathcal{B}, \nu)$. Let μ be the Gaussian measure on $(\mathcal{X}, \mathcal{B})$ with mean zero and specified covariance operator K_X. If X is the Gaussian random element, then its characteristic function is

$$E e^{i(X,t)} = e^{-(K_X t, t)/2}, \ t \in \mathcal{X}. \tag{5}$$

We shall test the null hypothesis

$$H_{0\mu}: \ \nu = \mu$$

versus the alternative

$$H_{1\mu}: \ \nu \text{ is a probability measure different from } \mu.$$

Sometimes, it is convenient to apply a linear invertible transformation T to the random element X and to its observations. Denote the new random element $Y = TX$. Its mean is also equal to 0 and we denote its covariance operator by K_Y.

Let $e = (e_1, e_2, \dots)$ be the orthonormal basis of the eigenvectors of K_Y, and denote by $\sigma_1^2, \sigma_2^2, \dots$ the corresponding eigenvalues. Let $x = (x_1, x_2, \dots)$ be the representation of x in the basis e. The random element $X = (X_1, X_2, \dots)$ has independent components with distributions $N(0, \sigma_j^2)$, $j = 1, 2, \dots$. We can transform coordinate-wise the probability space $(\mathcal{X}, \mathcal{B}, \nu)$ to the probability space

$$([0, 1]^\infty, \ U^\infty, \ \Gamma)$$

with the transformations $t_j = \Phi(x_j; 0, \sigma_j^2)$, where, for $j = 1, 2, \dots$, $\Phi(\cdot; \mu, \sigma^2)$ is the normal distribution function with mean μ and variance σ^2, and $t = (t_1, t_2, \dots) \in$

$[0, 1]^\infty$. The σ-algebra U^∞ is the completion of the set

$$\{[0, t_1] \times [0, t_2] \times \ldots, \ t = (t_1, t_2, \ldots) \in [0, 1]^\infty\}$$

to the minimal σ-algebra U^∞, and Γ is the probability measure corresponding to ν.

We denote the transformed random element X as $T = (T_1, T_2, \ldots) \in [0, 1]^\infty$. Here the random variables T_j, $j = 1, 2, \ldots$, are independent and uniformly distributed on $[0, 1]$ under the null hypothesis. Let Υ be the "uniform" measure on $([0, 1]^\infty, U^\infty)$, which corresponds to the measure μ. We can see that

$$\Upsilon([0, t_1] \times [0, t_2] \times \ldots) = t_1 t_2 \ldots$$

For example,

$$\Upsilon\left\{[0, t] \times [0, t^{1/2}] \times [0, t^{1/4}] \times [0, t^{1/8}] \times \ldots, \right\} = t^2, \quad \text{for } 0 < t < 1.$$

The observations X^i of X are transformed into $T^i = (T^{i1}, T^{i2}, T^{i3}, \ldots)$, $i = 1, 2, \ldots$, where $T^{ij} = \Phi(X^{ij}; 0, \sigma_j^2)$, $i = 1, 2, \ldots, n$, $j = 1, 2, \ldots$. Each T^i belongs to $[0, 1]^\infty$. The transformed observations remain independent.

The null hypothesis $H_{0\mu}$ and the alternative $H_{1\mu}$ can be reformulated as

$$H_{0\Upsilon} : \ \Gamma = \Upsilon \quad \text{and} \quad H_{1\Upsilon} : \ \Gamma \neq \Upsilon,$$

respectively.

3.1.1 Example 1. Gaussian Process Testing

We consider here the problem of testing that an observed random process $x(t)$ on the interval $[0, 1]$ is Gaussian with specified covariance function. Our null hypothesis is

$$H_{0x} : x(t), \ t \in [0, 1] \ \text{is Gaussian with} \ Ex(t) = 0$$
$$\text{and} \ E(x(t)x(\tau)) = K_x(t, \tau), \ t, \tau \in [0, 1].$$

The alternative H_{1x} is the set of all other Gaussian processes z on the interval $[0, 1]$. On $K_z(t, \tau)$ we only make the assumption that

$$\int_0^1 K_z(t, t)dt < \infty, \ t, \tau \in [0, 1]. \tag{6}$$

Our test is performed with n independent observations $x_1(t), x_2(t), \ldots, x_n(t)$ of $x(t)$.

The process $x(t)$ can be transformed into another Gaussian process, of the form $y(t) = \psi_x(t)x(t)$, where $\psi_x(t)$ is a weight function. This is the simplest linear invertible transformation of the process $x(t)$. Similarly, all the observations of $x(t)$

are transformed to new observations $y_i(t) = \psi_x(t)x_i(t), \quad i = 1, 2, \ldots, n$, of the process $y(t)$. The Gaussian process $y(t)$ has zero mean and covariance function $K_y(t, \tau) = \psi_x(t)\psi_x(\tau)K_x(t, \tau)$. The null hypothesis H_{0x} is transformed to the equivalent null hypothesis

$$H_{0y} : y(t), \quad t \in [0, 1], \quad \text{is the Gaussian process}$$

$$\text{with } Ey(t) = 0 \text{ and } E(y(t)\, y(\tau)) = K_y(t, \tau), \quad t, \tau \in [0, 1].$$

The alternative is modified correspondingly.

Realizations of the process $y(t)$ belong with probability 1 to the separable Hilbert space $H = L^2([0, 1])$. As a basis for H, we choose the orthonormal basis formed by the eigenfunctions $g_1(t), g_2(t), \ldots$ of the integral equation

$$g(t) = \lambda \int_0^1 K_y(t, \tau)g(\tau)d\tau. \tag{7}$$

Denote by $\lambda_1, \lambda_2, \lambda_3, \ldots$ the eigenvalues of Eq. (7). Let $h = (h^1, h^2, h^3, \ldots)$ denote the coordinates of $y(t)$ in that basis, namely,

$$h^j = \int_0^1 y(t)g_j(t)dt, \quad j = 1, 2, \ldots \tag{8}$$

Analogously, the observations $y_i(t)$ of the random process $y(t)$ can be represented in H as

$$h_i = (h^{i1}, h^{i2}, h^{i3}, \ldots), i = 1, 2, \ldots, n,$$

where

$$h^{ij} = \int_0^1 y_i(t)g_j(t)dt, \, , i = 1, 2, \ldots, n, \ j = 1, 2, \ldots \tag{9}$$

The random variables h^{ij} are independent mean zero normal random variables.

3.1.2 Example 2. Wiener Process Testing

Consider testing the null hypothesis H_0 that the observed random process on [0,1] is Gaussian with mean zero and covariance function

$$K_0(t, \tau) = \min(t, \tau).$$

We transform this process (and all its observations) by dividing $x(t)$ by $\sqrt{t}$. This transformation is invertible. The resulting process has unit variance and covariance function

$$K(t, \tau) = \min(t, \tau)/\sqrt{t\tau}.$$

The corresponding covariance operator has eigenvalues $\lambda_k = (z_{0,k}/2)^2$ and eigenfunctions

$$\varphi_k(t) = J_0(z_{0,k}t)/\sqrt{t}, \ k = 1, 2, \ldots,$$

where $J_0(w)$ is the Bessel function of the first kind, and $z_{0,k}, \ k = 1, 2, \ldots,$ are its zeros.

The test with the weight function $1/\sqrt{t}$ was considered by Scott (1999).

3.2 Definition of the Test Statistic

Our test statistic is analogous to the one-dimensional Cramér–von Mises statistic (1). However, the commonly used definition of a distribution function is not suitable for infinite-dimensional spaces. A convenient alternative to the classical definition is the *generalized distribution function*

$$F^*(t) = t_1^{r_1} t_2^{r_2} t_3^{r_3} \ldots, \ t = (t_1, t_2, \ldots) \in [0, 1]^\infty, \tag{10}$$

where $1 = r_1 \geq r_2 \geq r_3 \geq \ldots$, and $r_j \to 0$. For example, we can take $r_i = i^{-a}, \ a > 1$. For this distribution function, the measure Υ can be uniquely recovered by associating to each $t \in [0, 1]^\infty$ the set

$$([0, t_1^{r_1}] \times [0, t_2^{r_2}] \times [0, t_3^{r_3}] \times \ldots).$$

Let $T^{(i)} = (T^{i1}, T^{i2}, \ldots)$ be the ith observation of T. The corresponding empirical distribution function is

$$F_n^*(t) = \frac{1}{n} \sum_{i=1}^\infty I_{\{T^{i1} \leq t_1^{r_1}, \ T^{i2} \leq t_2^{r_2}, \ldots\}} = \frac{1}{n} \sum_{i=1}^n \prod_{j=1}^\infty I_{\{T^{ij} \leq t_j^{r_j}\}}. \tag{11}$$

The Cramér–von Mises statistics can be written analogously to the one- and multi-dimensional cases as

$$\Omega_n^2 = n \int_{[0,1]^\infty} \left(F_n^*(t) - \prod_{j=1}^\infty t_j^{r_j} \right)^2 \Upsilon(dt). \tag{12}$$

In the sequel, it will be convenient to denote $\Upsilon(dt)$ by $dt_1 dt_2 \ldots$ The measure Υ plays here the role of the weight function. Note that any measure that is absolutely continuous with respect to Υ can be substituted for Υ itself.

The corresponding "empirical process" is

$$\xi_n^*(t) = \sqrt{n} \left(F_n^*(t) - \prod_{j=1}^\infty t_j^{r_j} \right), \ t \in [0, 1]^\infty,$$

or

$$\xi_n^*(t) = \frac{1}{\sqrt{n}} \sum_{i=1}^{n} \left(\prod_{j=1}^{\infty} I_{\{T^{ij} < t_j^{r_j}\}} - \prod_{j=1}^{\infty} t_j^{r_j} \right), \quad t \in [0,1]^{\infty}. \tag{13}$$

The covariance function of $\xi_n^*(t)$ is

$$K^*(t,\tau) = E\xi_n^*(t)\xi_n^*(\tau)$$

$$= \frac{1}{n} E \sum_{i=1}^{n} \sum_{k=1}^{n} \left(\prod_{j=1}^{\infty} I_{\{T^{ij} < t_j^{r_j}\}} - \prod_{j=1}^{\infty} t_j^{r_j} \right) \left(\prod_{m=1}^{\infty} I_{\{T^{km} < \tau_m^{r_m}\}} - \prod_{m=1}^{\infty} \tau_m^{r_m} \right).$$

Finally, we obtain

$$K^*(t,\tau) = \prod_{j=1}^{\infty} \min(t_j^{r_j}, \tau_j^{r_j}) - \prod_{j=1}^{\infty} t_j^{r_j} \tau_j^{r_j}, \quad t,\tau \in [0,1]^{\infty}. \tag{14}$$

This empirical process can be written as the normalized sum of independent identically distributed random functions in the form

$$\xi_n^*(t) = \frac{1}{\sqrt{n}} \sum_{i=1}^{n} v_i(t), \quad t \in [0,1]^{\infty}, \tag{15}$$

where

$$v_i(t) = \prod_{j=1}^{\infty} I_{\{T^{ij} < t_j^{r_j}\}} - \prod_{j=1}^{\infty} t_j^{r_j}.$$

Under the condition

$$\int_{[0,1]^{\infty}} K^*(t,t)dt < \infty \tag{16}$$

this empirical process converges weakly in $L_2([0,1]^{\infty})$ to a mean zero Gaussian process with covariance function $K^*(t,\tau)$. This follows from the corresponding theorem on weak convergence in a Hilbert space of normalized sums of independent identically distributed random functions to a Gaussian process (see for example van der Vaart and Wellner (1996)). Then, (16) implies that

$$\int_{[0,1]^{\infty}} K^*(t,t)dt = \prod_{j=1}^{\infty} \frac{1}{r_j + 1} - \prod_{j=1}^{\infty} \frac{1}{2r_j + 1} < \infty.$$

This holds for all positive sequences $\{r_j, \quad j = 1, 2, \dots\}$. We are interested in the case when $1 \geq r_j \downarrow 0$ as $j \to \infty$, and the empirical process is nondegenerate, i.e.,

$$\int_{[0,1]^\infty} K^*(t, t)dt > 0. \tag{17}$$

To fulfill this condition, it is sufficient to find a sequence which satisfies

$$\prod_{j=1}^{\infty} \frac{1}{r_j + 1} > 0 \tag{18}$$

(an example of such a sequence is $r_j = j^{-a}, \quad a > 1, \quad j = 1, 2, \dots$). It is not necessary for the sequence r_j to begin to decrease from $r_1 = 1$. We can write

$$K^*(t, \tau) = K_0^*(t, \tau) - w(t)w(\tau),$$

where

$$K_0^*(t, \tau) = \prod_{j=1}^{\infty} K_{0j}^*(t_j, \tau_j) = \prod_{j=1}^{\infty} \min\left(t_j^{r_j}, \tau_j^{r_j}\right),$$

$$w(t) = \prod_{j=1}^{\infty} w_j(t), \quad w_j(t) = t_j^{r_j}, \quad t, \tau \in [0, 1]^\infty$$

with

$$K_{0j}^*(t_j, \tau_j) = \min\left(t_j^{r_j}, \tau_j^{r_j}\right), \quad j = 1, 2 \dots$$

We will note that the distribution of the statistic Ω_n^2 converges in distribution to the quadratic form

$$Q = \sum_{k=1}^{\infty} \frac{z_i^2}{\lambda_i^*},$$

where z_i are independent identically distributed random variables with standard normal distribution and λ_i^* are the eigenvalues of the linear integral operator with kernel $K^*(t, \tau)$.

4 Limit Distribution of the Test Statistic Under the Null Hypothesis

4.1 Eigenvalues and Eigenfunctions of K_{0j}^*

First, we shall consider the elementary kernel $K_0(t, \tau, r)$ in the one-dimensional case with arbitrary $0 < r \leq 1$

$$K_0(t, \tau, r) = \min(t^r, \tau^r), \quad 0 \leq t, \ \tau \leq 1.$$

Then $K_{0j}^*(t_j, \tau_j) = K_0(t_j, \tau_j, r_j)$, for $j = 1, 2, \ldots$

The eigenfunctions and eigenvalues of $K_0(t, \tau, r)$ can be found from the Fredholm integral equation

$$\varphi_r(t) = \lambda_r \int_0^1 \min(t^r, \tau^r)\varphi_r(\tau)d\tau, \ t \in [0, 1]. \tag{19}$$

We make the following change of variable $\varphi_r(t) = h(t^r)$ and substitute $\varphi_r(t)$ into Eq. (19). Its eigenfunctions and eigenvalues can be determined from the integral equation

$$h_r(t^r) = \frac{\lambda_r}{r} \int_0^1 \min(t^r, \tau^r)h_r(\tau^r)\tau^{1-r}d\tau^r, \ t \in [0, 1],$$

or

$$h_r(x) = \rho_r \int_0^1 y^{\frac{1}{r}-1} \min(x, y)h_r(y)dy, \ x \in [0, 1], \tag{20}$$

where $\rho_r = \lambda_r/r$. We can represent this equation as

$$h_r(x) = \rho_r \int_0^x y^{\frac{1}{r}} h_r(y)dy + \rho_r x \int_x^1 y^{\frac{1}{r}-1}h_r(y)dy.$$

Then, by successive differentiation of this equation, we obtain

$$h_r'(x) = \rho_r \int_x^1 y^{\frac{1}{r}-1}h_r(y)dy \tag{21}$$

and

$$h_r''(x) + \rho_r x^{\frac{1}{r}-1}h_r(x) = 0. \tag{22}$$

The function h must satisfy the conditions

$$h_r(0) = 0 \ \text{ and } \ h_r'(1) = 0. \tag{23}$$

The first condition follows from (20), the second one from (21).

The next lemma follows from Theorem 1.3 and Lemma 3.5 in Deheuvels and Martynov (2003).

Lemma 1 *The equation*

$$y''(t) + \theta t^{2\beta} y(t) = 0 \tag{24}$$

with $\beta > -1$ and under the conditions $y(0) = 0$ and $y'(1) = 0$ has eigenvalues

$$\theta_k = \left(\frac{z_{\nu-1,k}}{2\nu} \right)^2, \quad k = 1, 2, \ldots, \tag{25}$$

and non-normalized eigenfunctions

$$y_k(t) = \sqrt{t} J_\nu \left(z_{\nu-1,k} t^{1/(2\nu)} \right), \quad k = 1, 2, \ldots,$$

where $\nu = 1/(2(\beta + 1))$, and $z_{a,k}$, $k = 1, 2, \ldots$ are the zeros of the Bessel function $J_a(x)$.

It can be concluded by comparing Eqs. (22) and (24) that β corresponds to $(1-r)/(2r)$. Hence, ν corresponds to $\mu = r/(1+r)$. Also, θ_k transforms to $\rho_{r,k}$. Then the eigenvalues of Eq. (19) are

$$\lambda_{r,k} = \rho_{r,k} r = r \left(\frac{z_{\mu-1,k}}{2\mu} \right)^2, \quad k = 1, 2, \ldots \tag{26}$$

In our study, r varies from 1 to zero while μ varies from 1/2 to zero.

On the other hand, the eigenfunctions of Eq. (22) with conditions (23) are

$$h_{r,k}(t) = \sqrt{t} J_\mu \left(z_{\mu-1,k} t^{1/(2\mu)} \right), \quad k = 1, 2, \ldots,$$

and the non-normalized eigenfunctions of Eq. (19) are, with the change of variable $\varphi_r(t) = h_r(t^r)$,

$$\widetilde{\varphi}_{r,k}(t) = t^{r/2} J_\mu \left(z_{\mu-1,k} t^{(1+r)/2} \right), \quad k = 1, 2, \ldots, \tag{27}$$

or

$$\widetilde{\varphi}_{r,k}(t) = t^{\mu/(2(1-\mu))} J_\mu \left(z_{\mu-1,k} t^{1/(2(1-\mu))} \right), \quad k = 1, 2, \ldots \tag{28}$$

The squared normalizing divisor for $\widetilde{\varphi}_{r,k}(t)$ is

$$
D_{r,k}^2 = \frac{(\mu-1)z_{\mu-1,k}^{2\mu}\Gamma(2+\mu)\Gamma(1+2\mu)}{\sqrt{\pi}}
$$
$$
\times \, {}_1F_2\left(\frac{1}{2}+\mu; 2+\mu, 1+2\mu; -z_{\mu-1,k}^2\right).
$$

Here, ${}_pF_q(a_1, \ldots, a_p; b_1, \ldots, b_q, z)$ is the generalized hypergeometric function (see Luke (1969)) defined by the series

$$
{}_pF_q(a_1, \ldots, a_p; b_1, \ldots, b_q, z) = 1 + \sum_{m=1}^{\infty} \frac{(a_1)_m \ldots (a_p)_m}{(b_1)_m \ldots (b_q)_m \, m!} t^m, \qquad (29)
$$

and $(a)_m = a(a+1)\cdots(a+m-1)$, where $(a)_0 = 1$, is the Pochhammer symbol. This series converges for all $|t| < 1$. We shall use the notation $\varphi_{r,k}(t) = \widetilde{\varphi}_{r,k}(t)/D_{r,k}$ for the normalized eigenfunctions of Eq. (19).

4.2 Eigenvalues and Eigenfunctions of K_0^* in the infinite-dimensional case

Further, we obtain the eigenvalues and eigenfunctions of the kernel $K_0^*(t, \tau)$, $t, \tau \in [0, 1]^{\infty}$. In this case, the Fredholm equation becomes

$$
\varphi_r(t) = \lambda \int_{[0,1]^{\infty}} \prod_{j=1}^{\infty} \min(t_j^{r_j}, \tau_j^{r_j})\varphi_r(\tau)d\tau, \quad t, \tau \in [0, 1]^{\infty}. \qquad (30)
$$

Here, r_j, $j = 1, 2, \ldots$, is a prescribed sequence of numbers monotonically decreasing from 1 to 0 and satisfying condition (17). Denote by $\alpha_{j,k} = 1/\lambda_{r_j,k}$, $k = 1, 2, 3, \ldots$, the characteristic numbers for Eq. (19). Let $A_j = \{\alpha_{j,1}, \alpha_{j,2}, \alpha_{j,3} \ldots\}$ be the set of characteristic numbers for Eq. (19) corresponding to the value $r = r_j$. Then the set of the characteristic numbers for Eq. (30) consists of all possible products

$$
\alpha_{k_1,k_2,k_3,\ldots} = \prod_{j=1}^{\infty} \alpha_{j,k_j}, \qquad (31)
$$

where $(k_1, k_2, k_3, \ldots) \in \mathbb{N}^{\mathbb{N}}$, is the set of all sequences of positive natural numbers. The normalized eigenfunction corresponding to $\alpha_{k_1,k_2,k_3,\ldots}$ is

$$
\varphi_{k_1,k_2,k_3,\ldots}(t) = \prod_{j=1}^{\infty} \varphi_{r_j,k_j}(t). \qquad (32)
$$

Among all the characteristic numbers $\alpha_{k_1,k_2,k_3,\ldots}$, groups may appear with equal values. The cardinalities of such groups are called the *multiplicities* of the corresponding characteristic numbers. But, if the sequence of integers $r_1, r_2, r_3, \ldots$ is chosen arbitrarily and without "malicious intent", all multiplicities will be equal to one. For the sake of simplicity, we consider this case only. We arrange all the characteristic numbers in descending order, by giving them the serial numbers $\alpha_1 > \alpha_2 > \alpha_3 > \ldots$.

The following theorem gives us an idea about the behavior of those characteristic numbers.

Theorem 1 *If $r \to 0$, then the limit form of the Fredholm equation (19) has a unique root, which is equal to one.*

Proof The limit form of Eq. (19) is

$$\varphi(t) = \lambda \int_0^1 \varphi(\tau)d\tau, \quad t \in [0, 1]. \tag{33}$$

Hence, $\varphi(t)$ is a constant, i.e., Eq. (19) in this case has the only eigenvalue $\lambda = 1$. This assertion follows from Parseval's equality because here, formally, $K(t, t) = 1$ and integrates to 1. $\qquad\square$

It follows from Theorem 1 that $\lambda_{r_j,1} \to 1$, but $\lambda_{r_j,k} \to \infty$ ($\alpha_{j,k} \to 0$), $k = 2, 3, \ldots$, as $j \to \infty$.

The behavior of the characteristic numbers when r_j tends to zero is shown in Table 1. Here and below, we take $r_i = i^{-a(1-i^{-b})}$.

We can compute the largest characteristic number. It is equal to

$$\alpha_1 = \prod_{j=1}^{\infty} \alpha_{j,1} \approx 0.0642. \tag{34}$$

It is also obvious that the second largest characteristic number is

$$\alpha_2 = \alpha_1 \alpha_{1,2}/\alpha_{1,1} \approx 0.007137. \tag{35}$$

The corresponding eigenfunctions are

$$\varphi_1(t) = \prod_{j=1}^{\infty} \varphi_{r_j,1}(t) \quad \text{and} \quad \varphi_2(t) = \varphi_1(t)\varphi_{r_1,2}(t)/\varphi_{r_1,1}(t).$$

Notice that the sequence $k_1, k_2, k_3, \ldots$ in (31) may contain only a finite number of terms different from 1. Denote $v_{j,k} = \alpha_{j,k}/\alpha_{j,1}$, $j \geq 2$. Then the mth characteristic number can be represented as

$$\alpha_m = \alpha_1 v_{m,k_{m,1}} v_{m,k_{m,2}} \cdots v_{m,k_{m,s_m}}, \tag{36}$$

Table 1 The behavior of $\alpha_{r_j,k}$ when r_i varies from one to zero

j	r_j	$\alpha_{j,1}$	$\alpha_{j,2}$	$\alpha_{j,3}$	$\alpha_{j,4}$	$\alpha_{j,5}$	
1	1	0.4053	0.0450	0.0162	0.0083	0.0050	...
2	0.60197	0.5298	0.0463	0.0160	0.0081	0.0048	...
3	0.31323	0.6829	0.0400	0.0132	0.0065	0.0039	...
4	0.17678	0.7917	0.0303	0.0097	0.0047	0.0028	...
5	0.10816	0.8610	0.0219	0.0068	0.0033	0.0019	...
10	0.01952	0.9716	0.0050	0.0015	0.0007	0.0004	...
25	0.00160	0.9976	0.0004	0.0001	0.0001	0.0000	...
50	0.00023	0.9997	0.0001	0.0000	0.0000	0.0000	...
99	0.00003	1.0000	0.0000	0.0000	0.0000	0.0000	...
$\vdots$	$\vdots$	$\vdots$	$\vdots$	$\vdots$	$\vdots$	$\vdots$	$\vdots$
∞	0	1	0	0	0	0	...

where $v_{m,k_{m,1}} v_{m,k_{m,2}} \cdots v_{m,k_{m,s_m}}$ is the mth value among the finite products

$$v_{j,k_1} v_{j,k_2} \cdots v_{j,k_s} \quad j \geq 2, \quad 1 \leq k_1 < k_2 < \cdots < k_s \leq \infty, \quad s \geq 1,$$

ranked in descending order.

Table 2 represents the realization of formula (36) for the considered sequence r_i. Denote its elements as s_{jk}. Column k in the table corresponds to the kth characteristic number α_k of K_0^* in descending order by the formula

$$\alpha_k = \prod_{j=1}^{\infty} \alpha_{j,s_{jk}}.$$

Table 3 shows some real values of the characteristic numbers in accordance with Table 2.

4.3 Eigenvalues of K^*

It follows from Darling (1955), Durbin (1973), and Krivjakova et al. (1977) that the characteristic numbers α_i^* of $K^*(s, t)$ are solutions of the equation

$$\sum_{k=1}^{\infty} \frac{C_k^2}{\alpha_k - \alpha} = 1,$$

Table 2 Symbolic table for computing the first 50 eigenfunctions of $K_0^*(s,t)$

j	k									
	1 → 5	6 → 10	11 → 15	16 → 20	21 → 25	26 → 30	31 → 35	36 → 40	41 → 45	46 → 50
	1 **2** 1 1 3	1 1 1 4 1	1 1 5 1 1	2 1 1 1 6	1 1 2 1 1	7 1 1 1 1	8 1 2 1 1	1 1 1 3 9	2 1 1 1 2	10 1 1 1 1
1	1 1 **2** 1 1	1 3 1 1 1	1 4 1 1 1	2 1 5 1 1	1 1 1 6 1	1 1 1 1 2	1 7 1 1 1	1 1 1 2 1	3 2 8 1 1	1 1 1 1 1
↓	1 1 1 **2** 1	1 1 1 1 3	1 1 1 1 1	1 4 1 1 1	1 1 2 1 1	1 5 1 1 2	1 1 1 1 1	1 6 1 1 1	1 1 1 1 1	1 1 7 1 1
5	1 1 1 1 1	**2** 1 1 1 1	1 1 1 1 3	1 1 1 1 1	1 1 1 1 4	1 1 1 1 1	1 1 2 1 1	1 1 5 1 1	1 2 1 1 1	1 1 1 1 1
	1 1 1 1 1	1 1 **2** 1 1	1 1 1 1 1	1 1 1 1 1	3 1 1 1 1	1 1 1 1 1	1 1 1 1 4	1 1 1 1 1	1 1 1 1 2	1 1 1 1 1
	1 1 1 1 1	1 1 1 1 1	**2** 1 1 1 1	1 1 1 1 1	1 1 1 1 1	1 1 3 1 1	1 1 1 1 1	1 1 1 1 1	1 1 1 1 1	1 1 1 1 4
6	1 1 1 1 1	1 1 1 1 1	1 1 1 **2** 1	1 1 1 1 1	1 1 1 1 1	1 1 1 1 1	1 1 1 1 1	3 1 1 1 1	1 1 1 1 1	1 1 1 1 1
↓	1 1 1 1 1	1 1 1 1 1	1 1 1 1 1	1 1 1 **2** 1	1 1 1 1 1	1 1 1 1 1	1 1 1 1 1	1 1 1 1 1	1 1 1 1 1	1 3 1 1 1
10	1 1 1 1 1	1 1 1 1 1	1 1 1 1 1	1 1 1 1 1	1 **2** 1 1 1	1 1 1 1 1	1 1 1 1 1	1 1 1 1 1	1 1 1 1 1	1 1 1 1 1
	1 1 1 1 1	1 1 1 1 1	1 1 1 1 1	1 1 1 1 1	1 1 1 1 1	1 1 1 **2** 1	1 1 1 1 1	1 1 1 1 1	1 1 1 1 1	1 1 1 1 1
	1 1 1 1 1	1 1 1 1 1	1 1 1 1 1	1 1 1 1 1	1 1 1 1 1	1 1 1 1 1	1 1 1 **2** 1	1 1 1 1 1	1 1 1 1 1	1 1 1 1 1
11	1 1 1 1 1	1 1 1 1 1	1 1 1 1 1	1 1 1 1 1	1 1 1 1 1	1 1 1 1 1	1 1 1 1 1	1 1 1 1 1	1 1 1 **2** 1	1 1 1 1 1
↓	1 1 1 1 1	1 1 1 1 1	1 1 1 1 1	1 1 1 1 1	1 1 1 1 1	1 1 1 1 1	1 1 1 1 1	1 1 1 1 1	1 1 1 1 1	1 1 1 **2** 1
15	1 1 1 1 1	1 1 1 1 1	1 1 1 1 1	1 1 1 1 1	1 1 1 1 1	1 1 1 1 1	1 1 1 1 1	1 1 1 1 1	1 1 1 1 1	1 1 1 1 1
⋮	⋮	⋮	⋮	⋮	⋮	⋮	⋮	⋮	⋮	⋮

Table 3 Upper left corner of Table 2 with $\alpha_{j,s_{j,k}}$

j	k	
	$1 \to 5$	$6 \to 10$
1 $\downarrow$ 5	$\alpha_{1,1}\ \mathbf{1,2}\ \alpha_{1,1}\ \alpha_{1,1}\ \mathbf{1,3}$	$\alpha_{1,1}\ \alpha_{1,1}\ \alpha_{1,1}\ \mathbf{1,4}\ \alpha_{1,1}$
	$\alpha_{2,1}\ \alpha_{2,1}\ \mathbf{2,2}\ \alpha_{2,1}\ \alpha_{2,1}$	$\alpha_{2,1}\ \mathbf{2,3}\ \alpha_{2,1}\ \alpha_{2,1}\ \alpha_{2,1}$
	$\alpha_{3,1}\ \alpha_{3,1}\ \alpha_{3,1}\ \alpha_{3,2}\ \alpha_{3,1}$	$\alpha_{3,1}\ \alpha_{3,1}\ \alpha_{3,1}\ \alpha_{3,1}\ \mathbf{3,3}$
	$\alpha_{4,1}\ \alpha_{4,1}\ \alpha_{4,1}\ \alpha_{4,1}\ \alpha_{4,1}$	$\mathbf{4,2}\ \alpha_{4,1}\ \alpha_{4,1}\ \alpha_{4,1}\ \alpha_{4,1}$
	$\alpha_{5,1}\ \alpha_{5,1}\ \alpha_{5,1}\ \alpha_{5,1}\ \alpha_{5,1}$	$\alpha_{5,1}\ \alpha_{5,1}\ \mathbf{5,2}\ \alpha_{5,1}\ \alpha_{5,1}$

where

$$C_k = \prod_{j=1}^{\infty} C_{j,k}, \quad k = 1, 2, \ldots$$

and

$$C_{j,k} = \int_0^1 w_j(t)\varphi_{j,k}(t)dt = \int_0^1 t^{\mu_j/1-\mu_j}\varphi_{j,k}(t)dt, \quad j = 1, 2, \ldots$$

From this, it can be derived that

$$C_{r,k}^2 = \left(2^{-\mu}(\mu - 1)\Gamma(2 + \mu)z_{\mu-1,k}^{\mu}\ {}_0F_1\left(; 2 + \mu; -\frac{1}{4}z_{\mu-1,k}^2\right)\right)^2 \Big/ D_{r,k}^2,$$

where ${}_0F_1$ is the *generalized hypergeometric function*

$$_0F_1(; a; z) = \sum_{k=0}^{\infty} \frac{z^k}{(a)_k k!}.$$

4.4 Limit Distribution

In the previous section, we obtained the eigenvalues $\lambda_k^* = 1/\alpha_k^*$ for the covariance function $K^*(s, t)$. For calculation of the limiting distribution function of Ω^2, we can use the *Smirnov formula*, namely, for $t > 0$,

$$P\left(\Omega^2 > t\right) = \frac{1}{\pi}\sum_{k=1}^{\infty}(-1)^{k+1}\int_{\lambda_{2k-1}^*}^{\lambda_{2k}^*} \frac{e^{-tu/2}du}{u\sqrt{\left|\prod_{k=1}^{\infty}\left\{1 - \frac{u}{\lambda_k^*}\right\}\right|}}.$$

The Smirnov formula is designed for distinct eigenvalues. This formula, and the formula permitting eigenvalues with multiplicities, was considered, for example, in Deheuvels and Martynov (1996), Martynov (1975, 1992).

4.5 The Limiting Distribution of Ω_n^2

The method described in Sects. 4.2–4.4, is implemented for the sequence $r_i = i^{-2.5(1-i^{-0.5})}$, $i = 1, 2, \ldots$. Some results of the calculations have been presented in these sections. We obtained 1000 values $\alpha_i^* = 1/\lambda_i^*$. The distribution of the statistic Ω_n^2 is approximated by a finite quadratic form

$$Q_{100} = \sum_{k=1}^{100} \frac{z_i^2}{\lambda_i^*},$$

where z_i are independent identically distributed random variables with standard normal distribution. The mathematical expectation of Q_{100} is

$$EQ_{100} = \sum_{k=1}^{100} \frac{1}{\lambda_i^*} = 0.0742689.$$

On the other hand, the limiting mathematical expectation of the statistic Ω_n^2 is equal to the mathematical expectation of the infinite quadratic form

$$EQ = \sum_{k=1}^{\infty} \frac{1}{\lambda_i^*} = \prod_{j=1}^{\infty} \frac{1}{r_j + 1} - \prod_{j=1}^{\infty} \frac{1}{2r_j + 1} = 0.10390.$$

The residue of the quadratic form Q_{100},

$$\sum_{k=101}^{\infty} \frac{z_k^2}{\lambda_i^*},$$

is replaced by its expectation 0.0296327. As a result, we obtain the following percent points of the statistic Ω_n^2 :

$$P\{\Omega^2 \le 0.90\} \approx 0.16450, \quad P\{\Omega^2 \le 0.95\} \approx 0.20371,$$

$$P\{\Omega_n^2 \le 0.99\} \approx 0.30039, \quad P\{\Omega_n^2 \le 0.995\} \approx 0.34350,$$

$$P\{\Omega_n^2 \le 0.999\} \approx 0.44563.$$

4.6 Monte Carlo Results

The distribution of the statistic Ω_n^2 was calculated also by the Monte Carlo method. The Cramér–von Mises statistic can be represented as

$$\Omega_n^2 = n \int_{[0,1]^\infty} \left(\frac{1}{n} \sum_{i=1}^n \prod_{j=1}^\infty I_{\{T^{ij} < t_j^{r_i}\}} - \prod_{i=1}^\infty t_i^{r_i} \right)^2 dt.$$

Here we used double-loop calculations by a Monte Carlo method. The value of the statistic Ω_n^2 was computed in the inner loop, while its distribution was modeled in the outer loop. The number of summands in the integral was 100, the number of Monte Carlo iterations to calculate the statistic Ω_n^2 was 500, and the number of Monte Carlo iterations for calculating the percentage points was 10,000.

Here are a few estimated quantiles of the limiting distribution of Ω_n^2, with $r_i = i^{-a(1-i^{-b})}$. When $a = 2.5$ and $b = 0.5$, the exact expectation is $E\Omega^2 \approx 0.1039$ and the estimated expectation is $\hat{E}\omega^2 \approx 0.104$. The corresponding percentage points are

$$P\{\Omega_n^2 \le 0.90\} \approx 0.17 \quad \text{and} \quad P\{\Omega_n^2 \le 0.95\} \approx 0.21.$$

When $a = 3$ and $b = 0.5$, the exact expectation is $E\Omega^2 \approx 0.1306$ and the estimated expectation is $\hat{E}\Omega^2 \approx 0.132$. The corresponding percentage points are

$$P\{\Omega_n^2 \le 0.90\} \approx 0.22 \quad \text{and} \quad P\{\Omega_n^2 \le 0.95\} \approx 0.28.$$

The simulation results and exact calculations agree with each other within the expected precision. Simulation results confirm the possibility of calculating the statistics Ω_n^2 values by a Monte Carlo method. The results of both sections, however, should be considered as preliminary.

Acknowledgments The author would like to express his appreciation and thanks to Professor Paul Deheuvels for their long cooperation. The author is also grateful to the editors of this Festschrift for their constructive comments.

The work was partly supported by the Russian Foundation for Basic Research (RFBR), grant N 13-01-12447.

References

Darling, D. A. (1955). The Cramèr-Smirnov test in the parametric case. *Annals of Mathematical Statistics, 26*, 1–20.

Deheuvels, P. (2005). Weighted multivariate Cramér-von Mises-type statistics. *Africa Statistica, 1*, 1–14.

Deheuvels, P., & Martynov, G. V. (1996). Cramér-Von Mises-type tests with applications to tests of independence for multivariate extreme-value distributions. *Communications in Statistics—Theory and Methods, 25*, 871–908.

Deheuvels, P., & Martynov, G. (2003). Karhunen-Loève expansions for weighted Wiener processes and Brownian bridges via Bessel functions. *Progress in Probability, Birkhäuser, Basel, Switzerland, 55*, 57–93.

Durbin, J. (1973). Distribution theory for tests based upon the sample distribution function. *Regional Conference Series in Applied Mathematics* (Vol. 9). Philadelphia: SIAM.

Krivjakova, E. N., Martynov, G. V., & Tjurin, J. N. (1977). The distribution of the omega square statistics in the multivariate case. *Theory of Probability and Its Applications, 22,* 415–420.

Luke, Y. L. (1969). *The special functions and their approximations* (Vol. 1). New York: Academic Press Inc.

Martynov, G. V. (1975). Computation of distribution function of quadratic forms of normally distributed random variables. *Theory of Probability and Its Applications, 20,* 782–793.

Martynov, G. V. (1979). *The omega square tests.* Moscow (in Russian): Nauka.

Martynov, G. V. (1992). Statistical tests based on empirical processes and related questions. *Journal of Soviet Mathematics, 61,* 2195–2271.

Scott, W. F. (1999). A weighted Cramér-von Mises statistic, with some applications to clinical trials. *Communications in Statistics—Theory and Methods, 28,* 3001–3008.

Van der Vaart, A. W., & Wellner, J. A. (1996). *Weak convergence and empirical processes.* New York: Springer.

New U-empirical Tests of Symmetry Based on Extremal Order Statistics, and Their Efficiencies

Ya. Yu. Nikitin and M. Ahsanullah

Abstract We use a characterization of symmetry in terms of extremal order statistics which enables to build several new nonparametric tests of symmetry. We discuss their limiting distributions and calculate their local exact Bahadur efficiency under location alternative which is mostly high.

1 Introduction

The idea of building statistical tests based on characterizations belongs to Linnik (1953). Suppose we have a sample $X_1, \ldots, X_n$ of i.i.d. observations with distribution function F, and consider testing the hypothesis $\mathcal{H} : F \in \mathcal{F}$, where $\mathcal{F}$ is some family of distributions, against the alternative $\mathcal{A} : F \notin \mathcal{F}$. Common examples of $\mathcal{F}$ are the families of exponential or normal distributions with unknown parameters or the class of symmetric distributions with known or unknown center of symmetry.

Assume that the family $\mathcal{F}$ is characterized by the fact that two statistics $g_1(X_1, \ldots, X_r)$ and $g_2(X_1, \ldots, X_s)$ under $F \in \mathcal{F}$ have the same distribution. We introduce the two U-empirical distribution functions

Y.Y. Nikitin—Research supported by RFBR grant No. 13-01-00172, and by SPbGU grant No.6.38.672.2013.

Y.Y. Nikitin (✉)
Department of Mathematics and Mechanics, Saint-Petersburg State University,
Universitetsky Pr. 28, Stary Peterhof 198504, St. Petersburg, Russia
e-mail: yanikit47@gmail.com

Y.Y. Nikitin
National Research University - Higher School of Economics,
Souza Pechatnikov, 16, 190008 St. Petersburg, Russia

M. Ahsanullah
Department of Management Sciences, Rider University,
Lawrenceville, NJ 08648, USA
e-mail: ahsan@rider.edu

© Springer International Publishing Switzerland 2015
M. Hallin et al. (eds.), *Mathematical Statistics and Limit Theorems*,
DOI 10.1007/978-3-319-12442-1_13

$$G_{1n}(t) = \binom{n}{r}^{-1} \sum_{1 \le i_1 < \cdots < i_r \le n} \mathbf{1}\{g_1(X_{i_1}, \ldots, X_{i_r}) < t\}, \quad t \in R^1, \quad r \ge 1,$$

$$G_{2n}(t) = \binom{n}{s}^{-1} \sum_{1 \le i_1 < \cdots < i_s \le n} \mathbf{1}\{g_2(X_{i_1}, \ldots, X_{i_s}) < t\}, \quad t \in R^1, \quad s \ge 1.$$

According to the Glivenko–Cantelli theorem for U-empirical distribution functions, see Helmers et al. (1988), $G_{1n}(t)$ and $G_{2n}(t)$ converge uniformly and a.s. to the distribution functions $G_1(t) = P(g_1 < t)$ and $G_2(t) = P(g_2 < t)$ as $n \to \infty$. As under $\mathscr{H}$ one has $G_1(t) \equiv G_2(t)$, it follows that, a.s. under $\mathscr{H}$,

$$D_n := \sup_{t \in R^1} \mid G_{1n}(t) - G_{2n}(t) \mid \longrightarrow 0, \ n \to \infty.$$

Hence the Kolmogorov-type statistic D_n can be used for testing $\mathscr{H}$ against $\mathscr{A}$. We can also use some U-empirical integral statistics, e.g.,

$$I_n = \int_R (G_{1n}(t) - G_{2n}(t)) \, dF_n(t), \tag{1}$$

where F_n is the usual empirical distribution function, *in case they are consistent.* The use of ω^2-type statistics of the type

$$\Omega_n^2 = \int_R (G_{1n}(t) - G_{2n}(t))^2 \, dF_n(t)$$

is likely to be unjustified because of their complexity and the considerable difficulty of establishing their limit behavior.

The examples of such *goodness-of-fit tests* together with their asymptotic analysis and related calculation of efficiencies can be found in Baringhaus and Henze (1992), Henze and Meintanis (2002), Morris and Szynal (2001), Muliere and Nikitin (2002), Nikitin (1996b), Nikitin (2010), Nikitin and Volkova (2010), and some other related papers.

Testing of *symmetry* based on characterizations has been much less explored. Consider the classical hypothesis

$$H_0 : 1 - F(x) - F(-x) = 0, \ \forall x \in R^1, \tag{2}$$

against the alternative H_1 under which the equality (2) is violated at least in one point. The first step in construction of such tests was made by Baringhaus and Henze (1992).

Suppose that X and Y are i.i.d. rv's with continuous distribution function F. Baringhaus and Henze proved that the distributions of $|X|$ and $|\max(X, Y)|$ coincide iff F is symmetric with respect to zero, that is, iff (2) holds. They also proposed suitable Kolmogorov-type and omega-square type tests of symmetry. Some efficiency calculations were then performed in Nikitin (1996a), see also Nikitin (2010).

An integral test of symmetry similar to (1) was proposed by Litvinova (2001). Below we reconsider, inter alia the Litvinova's test.

In the present paper we are interested in new tests of symmetry with respect to zero based on the following characterization by Ahsanullah (1992):

Let $X_1, \ldots, X_k$, $k \geq 2$, be i.i.d. rv's with absolutely continuous distribution function $F(x)$. Denote $X_{1,k} := \min(X_1, \ldots, X_k)$ and $X_{k,k} := \max(X_1, \ldots, X_k)$. Then $|X_{1,k}|$ and $|X_{k,k}|$ are identically distributed iff F is symmetric about zero, i.e. iff (2) holds.

Subsequently, we refer to this result as *Ahsanullah's characterization of order k*.

In the sequel, we construct new tests of symmetry using this characterization and explore their asymptotic properties with emphasis on their local Bahadur efficiency. We shall see that corresponding tests of symmetry for $k = 2$ and $k = 3$ are asymptotically equivalent to the test of Litvinova and to the Kolmogorov-type test of Baringhaus and Henze. In case of location alternative they are competitive and exhibit rather high Bahadur and Pitman efficiency in comparison to many other tests of symmetry. At the same time, higher values of k, $k > 3$, lead us to different tests, presumably with lower efficiency values in case of common alternatives.

In the rest of the Introduction we briefly review some results on the asymptotic normality of U-statistics and the calculation of Bahadur efficiencies which are repeatedly used later on and might be helpful for the reader.

Currently, U-statistics play an important role in Statistics and Probability. They appeared in the middle of the 1940s in problems of unbiased estimation Halmos (1946). After the seminal paper of Hoeffding (1948), it became clear that many valuable statistics were just U-statistics (or von Mises functionals, with very similar asymptotic theory). We refer to the monographs Korolyuk and Borovskikh (1994) and Lee (1990) for a complete exposition of the theory.

We consider U-statistics of the form

$$U_n = \binom{n}{m}^{-1} \sum_{1 \leq i_1 < \cdots < i_m \leq n} \Psi(X_{i_1}, \ldots, X_{i_m}), \qquad n \geq m,$$

where $X_1, X_2, \ldots$ is a sequence of i.i.d. rv's with common distribution P, while the kernel $\Psi : R^m \to R^1$ is a measurable symmetric function of m variables. The number m is called the *degree* of the kernel. We assume that the kernel Ψ is integrable on R^m and denote

$$\theta(P) = \int \ldots \int_{R^m} \Psi(x_1, \ldots, x_m) dP(x_1) \ldots dP(x_m).$$

In the sequel we need the notation

$$\psi(x) := \mathbb{E}_P\{\Psi(X_1, \ldots, X_m)|X_1 = x\}, \quad \Delta^2 := \mathbb{E}_P \psi^2(X_1) - (\theta(P))^2.$$

The function ψ is called the one-dimensional *projection* of the kernel Ψ and plays an important role in asymptotic theory. If $\Delta^2 > 0$ (the so-called nondegenerate case), the limiting distribution of $U-$statistics is normal as discovered by Hoeffding (1948). He proved that if $\mathbb{E}_P \Psi^2(X_1, \ldots, X_m) < \infty$ and $\Delta^2 > 0$, then as $n \to \infty$ one has convergence in distribution

$$\sqrt{\frac{n}{m^2 \Delta^2}} \, (U_n - \theta(P)) \xrightarrow{\ d\ } N(0, 1). \tag{3}$$

Bahadur efficiency is one of several possible approaches evaluating the asymptotic relative efficiency (ARE) of two statistical tests. The Bahadur approach, proposed in Bahadur (1967) and (1971), consists in fixing the power of concurrent tests, then comparing the exponential rates of decrease of their sizes for increasing number of observations under some fixed alternative. This exponential rate for a sequence of statistics $\{T_n\}$ is usually proportional to some nonrandom function $c_T(\theta)$ depending on the alternative parameter θ which is called the *exact slope* of the sequence $\{T_n\}$. The Bahadur ARE $e^B_{V,T}(\theta)$ of two sequences of statistics $\{V_n\}$ and $\{T_n\}$ is defined by means of the formula

$$e^B_{V,T}(\theta) = c_V(\theta) \big/ c_T(\theta).$$

The Bahadur exact slope of the sequence of test statistics $\{T_n\}$ can be evaluated as $c_T(\theta) = 2f(b_T(\theta))$, where $b_T(\theta)$ is the limit in probability of T_n under the alternative, while the continuous function $f(t)$ describes the logarithmic large deviation asymptotics of this sequence under the null hypothesis, see details in Bahadur (1971) or Nikitin (1995).

It is important to note that there exists an upper bound for exact slopes Bahadur (1967) and (1971)

$$c_T(\theta) \le 2K(\theta), \tag{4}$$

where the Kullback–Leibler information number $K(\theta)$ measures the "statistical distance" between the alternative and the null hypothesis. It is sometimes compared in the literature with the Cramér–Rao inequality in the estimation theory. Therefore, the absolute (nonrelative) Bahadur efficiency of the sequence $\{T_n\}$ can be defined as $e^B_T(\theta) := c_T(\theta)/2K(\theta)$.

Computing the exact Bahadur ARE for arbitrary alternatives (depending on θ) is often infeasible; but it is possible to calculate the local Bahadur ARE as θ approaches the null hypothesis. Then one speaks about *local* efficiency and *local* Bahadur slopes: see Nikitin (1995).

The indisputable merit of Bahadur efficiency is its ability to handle statistics with nonnormal asymptotic distributions. This is the primary reason for using it in the present paper, as the Kolmogorov-type statistics have nonnormal limiting distribution.

2 Integral Test of Symmetry for $k = 2$ and Its Asymptotic Theory

In this section we study the simplest integral test. Consider two V-empirical distribution functions

$$G_n(t) = n^{-2}\sum_{1\leq i,j\leq n}\mathbf{1}\{|\min(X_i, X_j)| < t\}, t \in R^1,$$

$$H_n(t) = n^{-2}\sum_{1\leq i,j\leq n}\mathbf{1}\{|\max(X_i, X_j)| < t\}, t \in R^1,$$

and let Q_n be the empirical distribution function corresponding to the sample $|X_i|, i = 1, \ldots, n$.

We introduce the integral statistic

$$J_n = \int_{R^1} [G_n(t) - H_n(t)]dQ_n(t).$$

Let us show that this statistic is distribution free under the hypothesis of symmetry. Denote by F^{-1} the inverse distribution function of the sample assuming for simplicity that it is strictly monotone. Then

$$J_n = \int_0^1 [H_n(F^{-1}(u)) - G_n(F^{-1}(u))]dQ_n(F^{-1}(u)).$$

By symmetry of F,

$$-F^{-1}(u) = F^{-1}(1 - u), \qquad \forall u \in [0, 1].$$

Hence, for any u,

$$\begin{aligned}
G_n(F^{-1}(u)) &= n^{-2} \sum_{1\leq i,j\leq n} \mathbf{1}\{-F^{-1}(u) < \min(X_i, X_j) < F^{-1}(u)\} \\
&= n^{-2} \sum_{1\leq i,j\leq n} \mathbf{1}\{F^{-1}(1 - u) < \min(X_i, X_j) < F^{-1}(u)\} \\
&= n^{-2} \sum_{1\leq i,j\leq n} \mathbf{1}\{(1 - u) < \min(F(X_i), F(X_j)) < u\} \\
&= n^{-2} \sum_{1\leq i,j\leq n} \mathbf{1}\{1 - u < \min(U_i, U_j) < u\},
\end{aligned}$$

where $U_1, \ldots, U_n$ are independent standard uniform rv's, and we see that $G_n(F^{-1}(u))$ does not depend on F. Similar arguments can be invoked for $H_n(F^{-1}(u))$ and $Q_n(F^{-1}(u))$. Hence J_n is distribution free. Thus we may assume in the sequel that F is the (symmetric) uniform distribution on $[-1, 1]$. Now we see that

$$J_n = n^{-3} \sum_{1 \le i,j,k \le n} \left(\mathbf{1}\{|\min(X_i, X_j)| < |X_k|\} - \mathbf{1}\{|\max(X_i, X_j)| < |X_k|\} \right)$$

$$= n^{-3} \sum_{1 \le i,j,k \le n} \Psi_3(X_i, X_j, X_k),$$

where the kernel Ψ_3 of degree 3 of the last V-statistic is given after symmetrization by

$$3\Psi_3(X, Y, Z) = \mathbf{1}\{|\min(X, Y)| < |Z|\} + \mathbf{1}\{|\min(X, Z)|$$
$$< |Y|\} + \mathbf{1}\{|\min(Y, Z)| < |X|\}$$
$$- \mathbf{1}\{|\max(X, Y)| < |Z|\} - \mathbf{1}\{|\max(X, Z)|$$
$$< |Y|\} - \mathbf{1}\{|\max(Y, Z)| < |X|\}.$$

As U- and V-statistics with the same kernel have the same asymptotic distribution (see Korolyuk and Borovskikh (1994)), we can replace the V-statistic J_n by the asymptotically equivalent U-statistic $I_n^{(3)}$ of degree 3

$$I_n^{(3)} = \binom{n}{3}^{-1} \sum_{1 \le i < j < k \le n} \Psi_3(X_i, X_j, X_k),$$

which is simpler to calculate.

In what follows we use a system of notations for statistics $I_n^{(k)}$ and $D_n^{(k)}$ in such a way that the index k always corresponds to the *degree* of the associated U-statistic or to the *degree* of the corresponding *family* of U-statistics in case of supremum-type tests. At the same time these statistics correspond to Ahsanullah's characterization *of order $k - 1$*.

Let us calculate the projection of the kernel Ψ_3. We should find

$$\psi_3(s) := \mathbb{E}[\Psi_3(X, Y, Z)|Z = s].$$

Due to the underlying characterization, we have

$$\mathbb{E}\left(\mathbf{1}\{|\min(X, Y)| < |s|\} - \mathbf{1}\{|\max(X, Y)| < |s|\}\right) = 0.$$

It is clear that

$$\mathbb{E}\, \mathbf{1}\{|\min(X, s)| < |Y|\} = \mathbb{E}\, \mathbf{1}\{|\min(Y, s)| < |X|\} = \mathbb{P}\{|\min(X, s)| < |Y|\}.$$

The simplest way to calculate this probability is to use geometric considerations, evaluating

$$\frac{1}{4}\lambda\{(x, y) : -1 \le x, y \le 1, |\min(x, s)| < |y|\} = \begin{cases} (s^2 - 2s + 3)/4, & \text{if } s > 0, \\ (-s^2 + 2s + 3)/4, & \text{if } s \le 0. \end{cases}$$

where λ stands for the Lebesgue measure on R^2.

The values of the expectations

$$\mathbb{E}\mathbf{1}\{|\max(X, s)| < |Y|\} = \mathbb{E}\mathbf{1}\{|\max(Y, s)| < |X|\} = \mathbb{P}\{|\max(X, s)| < |Y|\}$$

are slightly different, and are given by

$$\frac{1}{4}\lambda\{(x, y) : -1 \le x, y \le 1, |\max(x, s)| < |y|\} = \begin{cases} (-s^2 - 2s + 3)/4, & \text{if } s > 0, \\ (s^2 + 2s + 3)/4, & \text{if } s \le 0. \end{cases}$$

Hence

$$\mathbb{E}\mathbf{1}\{|\min(X, s)| < |Y|\} - \mathbb{E}\mathbf{1}\{|\max(X, s)| < |Y|\} = \begin{cases} s^2/2, & \text{if } s > 0, \\ -\,s^2/2, & \text{if } s \le 0. \end{cases}$$

Taking in account the same value for $\mathbb{E}\mathbf{1}\{|\min(Y, s)| < |X|\} - \mathbb{E}\mathbf{1}\{|\max(Y, s)| < |X|\}$, we conclude that the required projection is given by

$$\psi_3(s) = \begin{cases} s^2/3, & \text{if } s > 0, \\ -\,s^2/3, & \text{if } s \le 0. \end{cases}$$

Consequently, the projection's variance equals

$$\sigma_3^2 := \mathbb{E}\psi_3^2(X_1) = \frac{1}{18} \int_{-1}^{1} x^4 \, dx = \frac{1}{45} > 0,$$

so that our kernel Ψ_3 is nondegenerate. According to Hoeffding's theorem, see (3), we have the weak convergence

$$\sqrt{n} I_n^{(3)} \xrightarrow{\text{d}} N(0, \tfrac{1}{5}).$$

Now we can describe the rough large deviation asymptotics under H_0. The following result can be derived from more general theorems proved in Nikitin and Ponikarov (1999):

For $a > 0$, it holds true under H_0 that

$$\lim_{n \to \infty} n^{-1} \ln P(I_n^{(3)} > a) = -f_3(a),$$

where the function f_3 is analytic for sufficiently small $a > 0$, and such that

$$f_3(a) \sim \frac{a^2}{18\sigma^2} = \frac{5}{2} a^2, \quad \text{as } a \to 0.$$

Now we apply Bahadur's theory Bahadur (1971), Nikitin (1995) to evaluate the local Bahadur efficiency of this test. By the Law of Large Numbers for U- and V-statistics, see Korolyuk and Borovskikh (1994), we have a.s. convergence under the parametric alternative P_θ:

$$I_n^{(3)} \longrightarrow b_I^{(3)}(\theta) = \mathbb{E}_\theta \Psi_3(X, Y, Z), \quad n \to \infty.$$

In efficiency calculations we shall not go beyond location alternatives, but for a few remarks on common parametric alternatives. Let P_θ denote the alternative distribution function $F(x, \theta) = F(x - \theta)$ with some symmetric distribution function F. Under these notations we obtain

$$\begin{aligned}
b_I^{(3)}(\theta) &= P_\theta\{|\min(Y, Z)| < |X|\} - P_\theta\{|\max(Y, Z)| < |X|\} \\
&= \int_0^\infty \left((1 - F(-x - \theta))^2 - (1 - F(x - \theta))^2\right) d(F(x - \theta) - F(-x - \theta)) \\
&\quad - \int_0^\infty \left(F^2(x - \theta) - F^2(-x - \theta)\right) d(F(x - \theta) - F(-x - \theta)) \\
&= 2 \int_0^\infty (F(x - \theta) - F(-x - \theta)) \left(1 - F(x - \theta) - F(-x - \theta)\right) \\
&\quad \times d(F(x - \theta) - F(-x - \theta)).
\end{aligned}$$

Assuming that F is differentiable with the density f, we have for any x and $\theta \to 0$

$$F(x - \theta) - F(-x - \theta) = 2F(x) - 1 + O(\theta^2)$$

and

$$\begin{aligned}
1 - F(x - \theta) - F(-x - \theta) &= 1 - F(x) - F(-x) + 2\theta f(x) + O(\theta^2) \\
&= 2f(x)\theta + O(\theta^2).
\end{aligned}$$

Consequently, under weak regularity conditions imposed on F, we have

$$b_I^{(3)}(\theta) \sim 8 \int_0^\infty (2F(x) - 1) f^2(x) dx \cdot \theta, \quad \theta \to 0.$$

It follows that the local exact Bahadur slope (Bahadur (1971, Sect. 7) and Nikitin (1995)) is equivalent as $\theta \to 0$ to

$$c_I^{(3)}(\theta) \sim 320 \left(\int_0^\infty (2F(x) - 1) f^2(x) dx\right)^2 \theta^2.$$

Same result can be obtained using general considerations from Nikitin and Peaucelle (2004).

This local exact slope is equivalent to that of Litvinova's test studied in Litvinova (2001) and (2004). Her test was based on the Baringhaus-Henze characterization, and the test statistic appeared as a U-statistic with centered kernel

$$\Phi(x, y, z) = \frac{1}{2} - \frac{1}{3}\left(\mathbf{1}\{|\max(x, y)| < |z|\} + \mathbf{1}\{|\max(x, z)| < |y|\}\right.$$
$$\left. + \mathbf{1}\{|\max(y, z)| < |x|\}\right).$$

The calculations are similar. While the limiting distributions have distinct variances and hence large deviation asymptotics are also different, the local exact slope is the same. Hence both tests are statistically equivalent for large samples, at least from the point of view of local Bahadur efficiency (and also limiting Pitman efficiency).

According to the inequality (4), in our case of a location parameter, we have under mild regularity conditions (see Bahadur (1971) and Nikitin (1995), Sect. 4.4),

$$320\left(\int_0^\infty (2F(x) - 1)f^2(x)dx\right)^2 \leq I(f), \tag{5}$$

where

$$I(f) = \int_{-\infty}^\infty \frac{\left(f'(x)\right)^2}{f(x)}dx$$

is the Fisher information for location. The local Bahadur efficiency is equal to the ratio of the left- and right-hand sides in (5). Litvinova also found rather high values of this efficiency for some concrete distributions. For instance, she found an efficiency as high as 0.977 for the normal distribution and 0.938 for the logistic distribution. At the same time this efficiency is only 0.488 for the Cauchy distribution.

Similar local efficiencies hold for skew alternatives with densities of the form $2f(x)F(\theta x)$ (see Azzalini (2014)). Litvinova (2004) also explored contamination and Lehmann alternatives, against which local efficiencies also are rather high.

It follows that our test is also quite efficient with respect to those alternatives. Which test is better is to ascertain and can be explored either by power simulation or by the calculation of variances for corresponding P-values in the spirit of Lambert and Hall (1982).

3 Kolmogorov-Type Test of Symmetry for $k = 2$

In this section, we consider tests based on supremum-type test statistics of the form

$$D_n^{(2)} = \sup_t |G_n(t) - H_n(t)|. \tag{6}$$

As this statistic is also distribution free under H_0, we may assume that the rv's X_i are again uniformly distributed on $[-1, 1]$ and that the supremum can be taken over $[-1, 1]$.

Limiting distributions and critical values for this statistic are unknown, but can be obtained via simulation. Therefore, we will focus on large deviations. Statistic (6) is the supremum of a *family* of U-statistics with kernel of degree 2 depending on t, namely

$$\Xi_2(X, Y, t) = \mathbf{1}\{|\min(X, Y)| < t\} - \mathbf{1}\{|\max(X, Y)| < t\}, \ 0 \le t \le 1. \quad (7)$$

In the sequel, we need the projection function of the family of kernels (7) (see Nikitin (2010))

$$\xi_2(z, t) = \mathbb{E}[\Xi_2(X, Y, t)|Y = z] = \mathbb{P}\{|\min(z, X)| < t\} - \mathbb{P}\{|\max(z, X)| < t\}.$$

This function depends on the relationship between z and t, and after some calculations we get

$$\xi_2(z; t) = \begin{cases} -t, & -1 \le z < -t, \\ 0, & -t \le z \le t, \\ t, & t < z \le 1. \end{cases}$$

Therefore, we can calculate the so-called *variance function* Nikitin (2010) of the family of kernels (7). We get

$$\xi_2(t) := \mathbb{E}\xi_2^2(Y; t) = t^2(1 - t), \ 0 \le t \le 1.$$

The maximum of this function is attained for $t = \frac{2}{3}$ and is equal to $\frac{4}{27}$. We note that the variance function is nondegenerate in the sense of Nikitin (2010), and hence we get, due to Nikitin (2010), the large deviation asymptotics

$$\lim_{n \to \infty} n^{-1} \ln \mathbb{P}(D_n^{(2)} > a) = -h_2(a) \sim -\frac{27}{32}a^2, \text{ as } a \to 0,$$

where h_2 is some analytic function in the neighborhood of zero.

Hence the exact slope of our statistic $D_n^{(2)}$ is $2h_2(b_D^{(2)}(\theta))$, where

$$b_D^{(2)}(\theta) = \lim_{n \to \infty} D_n^{(2)}$$

a.s. under the alternative. Under location alternatives, we can use the same calculations as above, and get, under minimal regularity assumptions,

$$b_D^{(2)}(\theta) = \sup_t |P_\theta\{|\min(X, Y)| < t\} - P_\theta\{|\max(X, Y)| < t\}$$

$$= \sup_t |(1 - F(-t - \theta))^2 - (1 - F(t - \theta))^2 - F^2(t - \theta) + F^2(-t - \theta)|$$

$$= 2 \sup_t |(F(t - \theta) - F(-t - \theta))(1 - F(t - \theta) - F(-t - \theta))|$$

$$\sim 4 \sup_t |(2F(t) - 1)| f(t)\theta, \quad \theta \to 0.$$

Thus the local exact slope of the sequence $D_n^{(2)}$ satisfies the relation

$$c_D^{(2)}(\theta) \sim 27 \sup_t (2F(t) - 1)^2 f^2(t)\theta^2, \quad \theta \to 0.$$

This local exact slope coincides with that of the Kolmogorov-type tests from Baringhaus and Henze (1992) as evaluated in Nikitin (1996a). The latter test is formally different, being based on the difference of the U-empirical distribution functions F_n and H_n, but turns out to be asymptotically equivalent to our statistic $D_n^{(2)}$.

In any case, in Nikitin (1996a) the local Bahadur efficiency of both tests is calculated for location alternatives. It is 0.764 for the normal, 0.750 for the logistic, and 0.376 for the Cauchy distribution. For the Kolmogorov-type tests it is an adoptable result as such tests usually are less efficient than the integral ones Nikitin (1995).

4 Integral Tests in the General Case

We see that the characterization of symmetry we used for $k = 2$ leads to tests which are asymptotically equivalent and equally as efficient as well-known ones. Let us consider the general case when the tests are built on the characterization by the fact that

$$|\min(X_1, \ldots, X_k)| \overset{\mathrm{d}}{=} |\max(X_1, \ldots, X_k)|, \quad k \geq 3. \tag{8}$$

In the sequel, the index k or $k + 1$ corresponds again to the degree of the kernel of a U-statistic. As in previous sections we associate with the condition (8) of order k the U-statistic of degree $k + 1$

$$I_n^{(k+1)} = \binom{n}{k+1}^{-1} \sum_{1 \leq i_1 < \cdots < i_{k+1} \leq n} \Psi_{k+1}(X_{i_1}, \ldots, X_{i_{k+1}}),$$

where the kernel Ψ_{k+1} of degree $k+1$ is given after symmetrization by

$$
\begin{aligned}
(k+1)\Psi_{k+1}(X_1,\ldots,X_{k+1}) = \mathbf{1}\{|\min(X_1,\ldots,X_k)| \\
< |X_{k+1}|\} + \cdots + \mathbf{1}\{|\min(X_2,\ldots,X_{k+1})| < |X_1|\} \\
- \mathbf{1}\{|\max(X_1,\ldots,X_k)| < |X_{k+1}|\} \\
- \cdots - \mathbf{1}\{|\max(X_2,\ldots,X_{k+1})| < |X_1|\}.
\end{aligned}
$$

In Sect. 2 we studied the special case of this kernel for $k = 2$.

When calculating the projection ψ_{k+1} of this kernel, we are first interested in

$$
\mathbb{P}(|\min(X_1,\ldots,X_{k-1},s)| < t) - \mathbb{P}(|\max(X_1,\ldots,X_{k-1},s)| < t).
$$

Reasoning as above, we have for $s > 0$ and $t \in [0,1]$

$$
\mathbb{P}(|\min(X_1,\ldots,X_{k-1},s)| < t) = \begin{cases} (1+t)^{k-1}/2^{k-1}, & \text{if } s \le t, \\ (1+t)^{k-1}/2^{k-1} - (1-t)^{k-1}/2^{k-1}, & \text{if } t < s \le 1. \end{cases}
$$

Therefore, integrating, we get for $s > 0$

$$
\mathbb{P}(|\min(X_1,\ldots,X_{k-1},s)| < |Z|) = \left(2^k - 2 + (1-s)^k\right)/k2^{k-1}.
$$

In the same manner for $s \le 0$ we obtain

$$
\mathbb{P}(|\min(X_1,\ldots,X_{k-1},s)| < |Z|) = \left(2^k - (1-s)^k\right)/k2^{k-1}.
$$

Quite analogously we find the probabilities related to the maximum, namely

$$
\mathbb{P}(|\max(X_1,\ldots,X_{k-1},s)| < |Z|) = \begin{cases} \left(2^k - (1+s)^k\right)/k2^{k-1}, & s > 0, \\ \left(2^k - 2 + (1+s)^k\right)/k2^{k-1}, & s \le 0. \end{cases}
$$

Taking together our calculations, we obtain the projection of our kernel as

$$
\psi_{k+1}(s) = \begin{cases} \dfrac{(1+s)^k + (1-s)^k - 2}{(k+1)2^{k-1}}, & \text{if } s > 0; \\[2ex] \dfrac{2 - (1+s)^k - (1-s)^k}{(k+1)2^{k-1}}, & \text{if } s \le 0. \end{cases}
$$

Now we can calculate the variance $\sigma_{k+1}^2 = \mathbb{E}\psi_{k+1}^2(X_1)$. It is given, for any $k \ge 2$, by

$$
\sigma_{k+1}^2 = \frac{1}{2^{2k-2}(k+1)^2} \int_0^1 \left((1+s)^k + (1-s)^k - 2\right)^2 \, ds > 0.
$$

Table 1 Some exact values of the variance σ^2_{k+1}

k	Variance σ^2_{k+1}
$k = 2$	1/45
$k = 3$	9/320
$k = 4$	2843/126000
$k = 5$	2335/145152
$k = 6$	421691/37669632

In Table 1 we give some values of this variance which apparently has no nice explicit form.

By Hoeffding's theorem, see (3), the limiting distribution of $\sqrt{n}\,I_n^{(k+1)}$ is $N(0, (k+1)^2\sigma^2_{k+1})$. The large deviation asymptotics under H_0, see Sect. 2, is given by

$$\lim_{n\to\infty} n^{-1}\ln P(I_n^{(k+1)} > a) = -f_{k+1}(a),$$

where the function f_{k+1}, $k \geq 2$, is analytic for sufficiently small $a > 0$, and such that

$$f_{k+1}(a) \sim \frac{a^2}{2(k+1)^2\sigma^2_{k+1}}, \quad \text{as } a \to 0.$$

Thus the local exact slope of the sequence of statistics $I_n^{(k+1)}$, $k \geq 2$, is equivalent to

$$c_I^{(k+1)}(\theta) \sim (b_I^{(k+1)}(\theta))^2/(k+1)^2\sigma^2_{k+1}, \text{ as } \theta \to 0.$$

We see that

$$b_I^{(k+1)}(\theta) = P_\theta\{|\min(X_1,\ldots,X_k| < |Z|\} - P_\theta\{|\max(X_1,\ldots,X_k| < |Z|\}$$

$$= \int_0^\infty \left((1 - F(-x - \theta))^k - (1 - F(x - \theta))^k\right)$$

$$\times d(F(x - \theta) - F(-x - \theta))$$

$$- \int_0^\infty \left(F^k(x - \theta) - F^k(-x - \theta)\right) d(F(x - \theta) - F(-x - \theta))$$

$$\sim 4k \int_0^\infty \left(F^{k-1}(x) - F^{k-1}(-x)\right) f^2(x)dx \cdot \theta.$$

Hence, the local exact slope of the statistic of order k is equal to

$$c_I^{(k+1)}(\theta) \sim \frac{16k^2}{(k+1)^2\sigma^2_{k+1}} \left(\int_0^\infty \left(F^{k-1}(x) - F^{k-1}(-x)\right) f^2(x)dx\right)^2 \theta^2. \quad (9)$$

It is somewhat surprising to see that, for $k = 3$, we get from (9), as $\theta \to 0$,

$$c_I^{(4)}(\theta) \sim c_I^{(3)}(\theta) \sim 320 \left(\int_0^\infty (2F(x) - 1) f^2(x) dx \right)^2 \theta^2.$$

But this equivalence is not long. Already for $k = 4$ we get the variance $\sigma_5^2 = 2843/126000$, hence the local exact slope is equivalent as $\theta \to 0$ to the expression

$$c_I^{(5)}(\theta) \sim \frac{1290240}{2843} \left(\int_0^\infty \left(F^3(x) - F^3(-x) \right) f^2(x) dx \right)^2 \theta^2,$$

which is different from the case $k = 2$ and $k = 3$.

For instance, in case of logistic distribution we have as $\theta \to 0$

$$c_I^{(5)}(\theta) \sim \frac{1290240}{2843} \left(\int_0^\infty \frac{(e^{3x} - 1)e^{2x}}{(e^x + 1)^7} dx \right)^2 \theta^2 = \frac{1290240}{2843} \cdot \left(\frac{5}{192} \right)^2 \theta^2 \approx 0.308\,\theta^2.$$

As the Fisher information in this case is $\frac{1}{3}$, the efficiency of our test is 0.925. This is high value comparable with the value 0.938 in case of lower dimensions $k = 2$ and $k = 3$.

In the case of normal law we get

$$c_I^{(5)}(\theta) \sim \frac{1290240}{2843 \cdot 4\pi^2} \left(\int_0^\infty \left(\Phi^3(x) - \Phi^3(-x) \right) \exp(-x^2)\, dx \right)^2 \theta^2 \approx 0.975\,\theta^2.$$

Note that 0.975 is just the value of local efficiency as the Fisher information is equal to 1. This is also high value. On the contrary, in the Cauchy case we get again much lower value of local efficiency 0.332.

It is interesting to compare the calculations of efficiencies for other common symmetric distributions and for other alternatives.

5 Local Efficiency of Kolmogorov-Type Test in the General Case

Using the condition (8) for any $k > 2$, we can construct the Kolmogorov-type statistic $D_n^{(k)}$ according to (6). We concentrate here on large deviations and local efficiencies of such statistics for location alternatives. It is necessary to consider the family of kernels, depending on $t \in [0, 1]$ in a following way:

$$\Psi_k(X_1, \ldots, X_k, t) = \mathbb{P}(|\min(X_1, \ldots, X_k)| < t) - \mathbb{P}(|\max(X_1, \ldots, X_k)| < t).$$

Let us calculate the projection of this family. We have

$$\xi_k(z, t) := \mathbb{E}\left(\Psi(X_1, \ldots, X_k, t) | X_k = z\right)$$
$$= \mathbb{P}(|\min(X_1, \ldots, X_{k-1}, z)| < t) - \mathbb{P}(|\max(X_1, \ldots, X_{k-1}, z)| < t).$$

Using the calculations performed above, we obtain

$$\xi_k(z, t) = \begin{cases} (1-t)^{k-1}/2^{k-1} - (1+t)^{k-1}/2^{k-1}, & -1 \le z < -t, \\ 0, & -t \le z \le t, \\ (1+t)^{k-1}/2^{k-1} - (1-t)^{k-1}/2^{k-1}, & t < z \le 1. \end{cases}$$

Consequently, the variance function is equal to

$$\xi_k(t) = \mathbb{E}(\xi_k(Z, t))^2 = \frac{1}{2} \int_{-1}^{-t} ((1-t)^{k-1}/2^{k-1} - (1+t)^{k-1}/2^{k-1})^2 dx$$
$$+ \frac{1}{2} \int_t^1 ((1-t)^{k-1}/2^{k-1} - (1+t)^{k-1}/2^{k-1})^2 dx$$
$$= (1-t)((1+t)^{k-1}/2^{k-1} - (1-t)^{k-1}/2^{k-1})^2.$$

For $k = 3$ we have again, as in the case $k = 2$, the variance function

$$\xi_3(t) = (1-t)t^2, \, -1 \le t \le 1,$$

with the same maximum $\frac{4}{27}$, so that the large deviation asymptotics, see Nikitin (2010), is given by the formula

$$\lim_{n \to \infty} n^{-1} \ln \mathbb{P}(D_n^{(3)} > a) = -h_3(a) = -\frac{3}{8} a^2 (1 + o(1)), \text{ as } a \to 0,$$

where h_3 is some analytic function in the vicinity of zero.

It is easy to see that the a.s. limit under the alternative of statistics $D_n^{(k)}$ admits the representation

$$b_D^{(k)}(\theta) \sim 2k \sup_x f(x)[F^{k-1}(x) - F^{k-1}(-x)] \cdot \theta, \, \theta \to 0.$$

It follows that for $k = 3$ the local exact slope has the form

$$c_D^{(3)}(\theta) \sim 27 \left(\sup_x [f(x)(2F(x) - 1)] \right)^2 \theta^2, \, \theta \to 0,$$

and the test is again equivalent to that of the case $k = 2$ as in the instance of integral tests.

But in the case $k = 4$ the situation changes as the variance function is

$$\xi_4(t) = \frac{1}{16}(1 - t)(3t + t^3)^2, \ 0 \le t \le 1.$$

We find numerically that the maximum of the variance function is equal to $0.1123\ldots$. Hence the large deviation result is different and reads

$$\lim_{n \to \infty} n^{-1} \ln \mathbb{P}(D_n^{(4)} > a) = -h_4(a) = -0.2783 \cdots a^2(1 + o(1)), \text{ as } a \to 0.$$

Therefore, the exact slope admits the representation

$$c_D^{(4)}(\theta) \sim 35.622\ldots \sup_x \left[f(x) \left(F^3(x) - F^3(-x) \right) \right]^2 \theta^2, \ \theta \to 0.$$

In case of logistic distribution and $k = 4$ we have in the right-hand side

$$35.622\ldots \sup_x \left(\frac{e^x(e^{3x} - 1)}{(1 + e^x)^5} \right)^2 \approx 0.232,$$

that gives for local efficiency lower result 0.696 than in previous cases.

For the normal law we find that

$$\frac{1}{2\pi} \sup_x e^{-x^2} \left(\Phi^3(x) - \Phi^3(-x) \right)^2 \approx 0.0206.$$

Consequently, the efficiency is approximately 0.733. Similar calculations show that for the Cauchy law the local efficiency equals 0.313. All these efficiencies are reasonable but moderate.

6 Discussion

We can resume the calculations of efficiencies in Table 2. One sees that for logistic and normal distributions the values of efficiencies of integral tests for location alternative are rather high in comparison with other nonparametric tests of symmetry, see Nikitin (1995, Chap. 4).

At the same time the results for the Cauchy law are mediocre. It would be of interest to study other alternatives and to compare the efficiency values with the power simulations for moderate sample size.

The efficiencies of Kolmogorov-type tests are lower but have tolerable values. One should keep in mind that these tests are always consistent while the integral tests of structure (1) have mostly one-sided character, and their consistency depends on the alternative.

Table 2 Local Bahadur efficiencies in location case

Statistic/Density	Logistic	Normal	Cauchy
$I_n^{(3)}$, $I_n^{(4)}$	0.938	0.977	0.488
$I_n^{(5)}$	0.925	0.975	0.332
$D_n^{(2)}$, $D_n^{(3)}$	0.750	0.764	0.376
$D_n^{(4)}$	0.696	0.733	0.313

We can also presume the deterioration of efficiency properties for our tests with the growth of their order and degree of complexity, at least for location alternative. Hence the simplest test statistics $I_n^{(3)}$ and $D_n^{(2)}$ and their equivalents described above seem to be most suitable for practical use.

Acknowledgments We are thankful to the referee and the editors for their comments on the paper leading to a substantial improvement of the presentation.

References

Ahsanullah, M. (1992). On some characteristic property of symmetric distributions. *Pakistan Journal of Statistics, 8*, 19–22.

Azzalini A. with the collaboration of A. Capitanio (2014). *The Skew-normal and related families*. New York: Cambridge University Press.

Bahadur, R. R. (1967). Rates of convergence of estimates and test statistics. *Annals of Mathematical Statistics, 38*, 303–324.

Bahadur, R. R. (1971). *Some Limit Theorems in Statistics*. Philadelphia: SIAM.

Baringhaus, L., & Henze, N. (1992). A characterization of and new consistent tests for symmetry. *Communications in Statistics - Theory, 21*, 1555–1566.

Halmos, P. R. (1946). The theory of unbiased estimation. *Annals of Mathematical Statistics, 17*, 34–43.

Helmers, R., Janssen, P., & Serfling, R. (1988). Glivenko-Cantelli properties of some generalized empirical dfs and strong convergence of generalized L-statistics. *Probability Theory and Related Fields, 79*, 75–93.

Henze, N., & Meintanis, S. (2002). Goodness-of-fit tests based on a new characterization of the exponential distribution. *Communication in Statistics- Theory, 31*, 1479–1497.

Hoeffding, W. (1948). A class of statistics with asymptotically normal distribution. *Annals of Mathematical Statistics, 19*, 293–325.

Korolyuk, V. S., & Borovskikh, Y. V. (1994). *Theory of U-statistics*. Dordrecht: Kluwer.

Lambert, D., & Hall, W. J. (1982). Asymptotic lognormality of P-values. *Annals of Statistics, 10*, 44–64.

Lee, A. J. (1990). *U-statistics: Theory and practice*. New York: Dekker.

Linnik, Yu. V. (1953). Linear forms and statistical criteria. I, II. *Ukrainian Mathematical Journal, 5*(207–243), 247–290.

Litvinova, V. V. (2001). New nonparametric test for symmetry and its asymptotic efficiency. *Vestnik of St.Petersburg University. Mathematics, 34*, 12–14.

Litvinova, V. V. (2004) Asymptotic properties of goodness-of-fit and symmetry tests based on characterizations. Ph.D. thesis. Saint-Petersburg University

Morris, K., & Szynal, D. (2001). Goodness-of-fit tests using characterizations of continuous distributions. *Applied Mathematics (Warsaw), 28*, 151–168.

Muliere, P., & Nikitin, Y. Y. (2002). Scale-invariant test of normality based on Polya's characterization. *Metron, 60*, 21–33.

Nikitin, Y. (1995). *Asymptotic efficiency of nonparametric tests.* New York: Cambridge University Press.

Nikitin, Y. Y. (1996a). On Baringhaus-Henze test for symmetry: Bahadur efficiency and local optimality for shift alternatives. *Mathematical Methods of Statistics, 5*, 214–226.

Nikitin, Y. Y. (1996b). Bahadur efficiency of a test of exponentiality based on a loss-of-memory type functional equation. *Journal of Nonparametric Statistic, 6*, 13–26.

Nikitin, Y. Y. (2010). Large deviations of U-empirical Kolmogorov-Smirnov tests, and their efficiency. *Journal of Nonparametric Statistics, 22*, 649–668.

Nikitin, Y. Y., & Peaucelle, I. (2004). Efficiency and local optimality of distribution-free tests based on U- and V-statistics. *Metron, LXII*, 185–200.

Nikitin, Y.Y, & Ponikarov, E.V. (1999) Rough large deviation asymptotics of Chernoff type for von Mises functionals and U-statistics. *Proceedings of the St. Petersburg Mathematical Society* (pp. 124–167) vol 7. (Y. Y. Nikitin, E. V. Ponikarov, AMS Transl, series 2, 203, pp. 107–146, Trans.)

Nikitin, Y. Y., & Volkova, K. Y. (2010). Asymptotic efficiency of exponentiality tests based on order statistics characterization. *Georgian Mathematical Journal, 17*, 749–763.

Optimal Rank-Based Tests for the Location Parameter of a Rotationally Symmetric Distribution on the Hypersphere

Davy Paindaveine and Thomas Verdebout

Abstract Rotationally symmetric distributions on the unit hyperpshere are among the most commonly met in directional statistics. These distributions involve a finite-dimensional parameter θ and an infinite-dimensional parameter g, that play the role of "location" and "angular density" parameters, respectively. In this paper, we focus on hypothesis testing on θ, under unspecified g. We consider (i) the problem of testing that θ is equal to some given θ_0, and (ii) the problem of testing that θ belongs to some given great "circle". Using the uniform local and asymptotic normality result from Ley et al. (Statistica Sinica 23:305–333, 2013), we define parametric tests that achieve Le Cam optimality at a target angular density f. To improve on the poor robustness of these parametric procedures, we then introduce a class of rank tests for these problems. Parallel to parametric tests, the proposed rank tests achieve Le Cam optimality under correctly specified angular densities. We derive the asymptotic properties of the various tests and investigate their finite-sample behavior in a Monte Carlo study.

1 Introduction

Spherical or directional data naturally arise in a plethora of earth sciences such as geology (see, e.g., Fisher 1989), seismology (Storetvedt and Scheidegger 1992), astrophysics (Briggs 1993), oceanography (Bowers et al. 2000), or meteorology (Fisher 1987), as well as in studies of animal behavior (Fisher et al. 1987) or even in

Paindaveine—Research supported by an A.R.C. contract from the Communauté Française de Belgique and by the IAP research network grant P7/06 of the Belgian government (Belgian Science Policy).

D. Paindaveine (✉)
Université libre de Bruxelles, Avenue F.D. Roosevelt, 50, ECARES - CP 114/04, 1050 Brussels, Belgium
e-mail: dpaindav@ulb.ac.be

T. Verdebout
Université Lille 3, Domaine universitaire du "pont de bois", Maison de la recherche, Rue du barreau, BP 60149, 59653 Villeneuve d'ascq Cedex, France
e-mail: thomas.verdebout@univ-lille3.fr

© Springer International Publishing Switzerland 2015

M. Hallin et al. (eds.), *Mathematical Statistics and Limit Theorems*,
DOI 10.1007/978-3-319-12442-1_14

neuroscience (Leong and Carlile 1998). For decades, spherical data were explored through linear approximations trying to circumvent the "curved" nature of the data. Then the seminal paper Fisher (1953) showed that linearization hampers a correct study of several phenomena (such as, e.g., the remanent magnetism found in igneous or sedimentary rocks) and that it was therefore crucial to take into account the non-linear, spherical nature of the data. Since then, a huge literature has been dedicated to a more appropriate study of spherical data; we refer to Mardia (1975), Jupp and Mardia (1989), Mardia and Jupp (2000) or to the second chapter of Merrifield (2006) for a detailed overview.

Spherical data are commonly viewed as realizations of a random vector $\mathbf{X}$ taking values in the unit hypersphere $\mathscr{S}^{k-1} := \left\{ \mathbf{x} \in \mathbb{R}^k \,|\, \|\mathbf{x}\| := \sqrt{\mathbf{x}'\mathbf{x}} = 1 \right\}$ $(k \geq 2)$. In the last decades, numerous (classes of) distributions on $\mathscr{S}^{k-1}$ have been proposed and investigated. In this paper, we focus on the class of *rotationally symmetric distributions* on $\mathscr{S}^{k-1}$, that were introduced in Saw (1978). This class is characterized by the fact that the probability mass at $\mathbf{x}$ is a monotone nondecreasing function of the "spherical distance" $\mathbf{x}'\boldsymbol{\theta}$ between $\mathbf{x}$ and a given $\boldsymbol{\theta} \in \mathscr{S}^{k-1}$. This implies that the resulting equiprobability contours are the $(k-2)$-hyperspheres $\mathbf{x}'\boldsymbol{\theta} = c$ $(c \in [-1, 1])$, and that this "north pole" $\boldsymbol{\theta}$ may be considered as a "spherical mode," hence may be interpreted as a *location* parameter.

Of course, this assumption of rotational symmetry may seem very restrictive. Yet, the latter is often used to model real phenomena. Indeed, according to Jupp and Mardia (1989), rotationally symmetric spherical data appear *inter alia* in situations where the observation process imposes such symmetrization (e.g., the rotation of the earth; see Mardia and Edwards 1982). Another instance where rotational symmetry is appropriate is obtained when the observation scheme does not allow to make a distinction between the measurements $\mathbf{x}$ and $\mathbf{O}_{\boldsymbol{\theta}}\mathbf{x}$ for any rotation matrix $\mathbf{O}_{\boldsymbol{\theta}}$ such that $\mathbf{O}_{\boldsymbol{\theta}}\boldsymbol{\theta} = \boldsymbol{\theta}$. In such a case, indeed, only the projection of $\mathbf{x}$ onto the modal axis $\boldsymbol{\theta}$ can be observed; see Clark (1983).

In the absolutely continuous case (with the dominating measure being the uniform distribution on $\mathscr{S}^{k-1}$), rotationally symmetric distributions have a probability density function (pdf) of the form $\mathbf{x} \mapsto cg(\mathbf{x}'\boldsymbol{\theta})$, for some nondecreasing function $g : [-1, 1] \to \mathbb{R}^+$. Hence, this model is intrinsically of a semiparametric nature. While inference about $\boldsymbol{\theta}$ has been considered in many papers (see, among others, Chang 2004; Tsai and Sen 2007), semiparametrically efficient inference procedures in the rotationally symmetric case have not been developed in the literature. The only exception is the very recent contribution by Ley et al. (2013), where rank-based estimators of $\boldsymbol{\theta}$ that achieve semiparametric efficiency at a target angular density are defined. Their methodology, that builds on Hallin and Werker (2003), relies on invariance arguments and on the uniform local and asymptotic normality—with respect to $\boldsymbol{\theta}$, at a fixed g—of the model considered.

Ley et al. (2013), however, considers point estimation only, hence does not address situations where one would like to test the null hypothesis that the location parameter $\boldsymbol{\theta}$ is equal to a given $\boldsymbol{\theta}_0$. In this paper, we therefore extend the results from Ley et al. (2013) to hypothesis testing. This leads to a class of rank tests for the aforementioned testing problem, that when based on correctly specified scores, are semiparametrically optimal. The proposed tests are invariant both with respect to the group of continuous monotone increasing transformations (of spherical distances) and with respect to the group of orthogonal transformations fixing the null value θ_0. Their main advantage over "studentized" parametric tests is that they are not only validity-robust but are also efficiency-robust. We also treat a more involved testing problem, in which one needs to test the hypothesis that θ belongs to some given great "circle"—more precisely, to the intersection of $\mathscr{S}^{k-1}$ with a given vectorial subspace of $\mathbb{R}^k$.

The outline of the paper is as follows. In Sect. 2, we carefully define the class of rotationally symmetric distributions considered, introduce the main assumptions needed, and state the uniform local and asymptotic normality result that will be the main technical tool for this work. In Sect. 3, we focus on the problem of testing that θ is equal to some given θ_0, derive optimal parametric tests and study their asymptotic behaviour. In Sect. 4, we discuss the group invariance structure of this testing problem, propose a class of (invariant) rank tests, and study their asymptotic properties. In Sect. 5, we treat the problem of testing that θ belongs to a given great circle. We conduct in Sect. 6 a Monte Carlo study to investigate the finite-sample behaviour of the proposed tests. Finally, an Appendix collects technical proofs.

2 Rotationally Symmetric Distributions and ULAN

The random vector $\mathbf{X}$, with values in the unit sphere $\mathscr{S}^{k-1}$ of $\mathbb{R}^k$, is said to be *rotationally symmetric* about $\theta (\in \mathscr{S}^{k-1})$ if and only if, for all orthogonal $k \times k$ matrices $\mathbf{O}$ satisfying $\mathbf{O}\theta = \theta$, the random vectors $\mathbf{OX}$ and $\mathbf{X}$ are equal in distribution. If $\mathbf{X}$ is further absolutely continuous (with respect to the usual surface area measure on $\mathscr{S}^{k-1}$), then the corresponding density is of the form

$$f_{\theta,g} : \mathscr{S}^{k-1} \to \mathbb{R}^+ \tag{1}$$

$$\mathbf{x} \mapsto c_{k,g}\, g(\mathbf{x}'\theta),$$

where $c_{k,g} (>0)$ is a normalization constant and $g : [-1, 1] \to \mathbb{R}$ is some nonnegative function—called an *angular function* in the sequel. Throughout the paper, we then (tacitly) adopt the following assumption on the data generating process.

ASSUMPTION (A). The observations $\mathbf{X}_1, \ldots, \mathbf{X}_n$ are mutually independent and admit a common density of the form (1), for some $\theta \in \mathscr{S}^{k-1}$ and some angular function g in the collection $\mathscr{F}$ of functions from $[-1, 1]$ to $\mathbb{R}^+$ that are positive and monotone nondecreasing.

The notation f (instead of g) will be used when considering a fixed angular density. An angular function that plays a fundamental role in directional statistics is then

$$t \mapsto f_{\exp,\kappa}(t) = \exp(\kappa t), \tag{2}$$

for some "concentration" parameter $\kappa\,(>0)$. Clearly, $f_{\exp,\kappa}$ satisfies the conditions in Assumption (A). The resulting rotationally symmetric distribution was introduced in Fisher (1953) and is known as the Fisher-von Mises-Langevin (FvML(κ)) distribution. Other examples are the so-called "linear" rotationally symmetric distributions (LIN(a)), that are obtained for angular densities defined by $f(t) = t+a$, with $a > 1$.

In the sequel, the joint distribution of $\mathbf{X}_1, \ldots, \mathbf{X}_n$ under Assumption (A) will be denoted as $\mathrm{P}_{\boldsymbol{\theta},g}^{(n)}$. Note that under $\mathrm{P}_{\boldsymbol{\theta},g}^{(n)}$, the random variables $\mathbf{X}_1'\boldsymbol{\theta}, \ldots, \mathbf{X}_n'\boldsymbol{\theta}$ are mutually independent and admit the common density (with respect to the Lebesgue measure over the real line)

$$
t \mapsto \tilde{g}(t) := \frac{\omega_k\, c_{k,g}}{B\left(\frac{1}{2}, \frac{1}{2}(k-1)\right)}\, g(t)(1 - t^2)^{(k-3)/2}\, \mathbb{I}_{[-1,1]}(t), \tag{3}
$$

where $B(\cdot, \cdot)$ is the beta function, $\omega_k = 2\pi^{k/2}/\Gamma(k/2)$ is the surface area of $\mathscr{S}^{k-1}$, and $\mathbb{I}_A(\cdot)$ stands for the indicator function of the set A. The corresponding cdf will be denoted by $t \mapsto \tilde{G}(t) = \int_{-1}^{t} \tilde{g}(s)ds$. Still under $\mathrm{P}_{\boldsymbol{\theta},g}^{(n)}$, the random vectors $\mathbf{S}_1(\boldsymbol{\theta}), \ldots, \mathbf{S}_n(\boldsymbol{\theta})$, where

$$
\mathbf{S}_i(\boldsymbol{\theta}) := \frac{\mathbf{X}_i - (\mathbf{X}_i'\boldsymbol{\theta})\boldsymbol{\theta}}{\|\mathbf{X}_i - (\mathbf{X}_i'\boldsymbol{\theta})\boldsymbol{\theta}\|}, \qquad i = 1, \ldots, n, \tag{4}
$$

are well-defined with probability one are independent of the $\mathbf{X}_i'\boldsymbol{\theta}$'s, and are i.i.d., with a common distribution that is uniform over the unit $(k-2)$-sphere $\mathscr{S}^{k-1}(\boldsymbol{\theta}^\perp) := \{\mathbf{x} \in \mathscr{S}^{k-1} : \mathbf{x}'\boldsymbol{\theta} = 0\}$. It is easy to check that the common mean vector and covariance matrix of the $\mathbf{S}_i(\boldsymbol{\theta})$'s are given by $\mathbf{0}$ and $(\mathbf{I}_k - \boldsymbol{\theta}\boldsymbol{\theta}')/(k-1)$, respectively.

Fix now an angular density f and consider the parametric family of probability measures $\mathscr{P}_f^{(n)} := \{\mathrm{P}_{\boldsymbol{\theta},f}^{(n)} \,|\, \boldsymbol{\theta} \in \mathscr{S}^{k-1}\}$. For $\mathscr{P}_f^{(n)}$ to be *uniformly locally and asymptotically normal (ULAN)*, the angular density f needs to satisfy some mild regularity conditions; more precisely, as we will state in Proposition 1 below, ULAN holds if f belongs to the collection $\mathscr{F}_{\mathrm{ULAN}}$ of angular densities in $\mathscr{F}$ (see Assumption (A) above) that (i) are absolutely continuous (with a.e. derivative f', say) and such that (ii), letting $\varphi_f := f'/f$, the quantity

$$
\mathscr{J}_k(f) := \int_{-1}^{1} \varphi_f^2(t)(1 - t^2)\tilde{f}(t)\, dt = \int_{0}^{1} \varphi_f^2(\tilde{F}^{-1}(u))(1 - (\tilde{F}^{-1}(u))^2)\, du
$$

is finite.

As usual, uniform local and asymptotic normality describes the asymptotic behaviour of local likelihood ratios of the form

$$
\frac{\mathrm{P}_{\boldsymbol{\theta}+n^{-1/2}\boldsymbol{\tau}^{(n)},f}^{(n)}}{\mathrm{P}_{\boldsymbol{\theta},f}^{(n)}},
$$

where the sequence $(\boldsymbol{\tau}^{(n)})$ is bounded. In the present *curved* setup, $(\boldsymbol{\tau}^{(n)})$ should be such that $\boldsymbol{\theta} + n^{-1/2}\boldsymbol{\tau}^{(n)} \in \mathcal{S}^{k-1}$ for all n, which imposes that $\boldsymbol{\theta}'\boldsymbol{\tau}^{(n)} = O(n^{-1/2})$. For the sake of simplicity, we will assume throughout that $\boldsymbol{\tau}^{(n)} = \boldsymbol{\tau} + O(n^{-1/2})$, with $\boldsymbol{\theta}'\boldsymbol{\tau} = 0$.

We then have the following result (see Ley et al. 2013 for the proof).

Proposition 1 *Fix* $f \in \mathcal{F}_{\mathrm{ULAN}}$. *Then the family* $\mathcal{P}_f^{(n)} = \{\mathrm{P}_{\boldsymbol{\theta},f}^{(n)} \mid \boldsymbol{\theta} \in \mathcal{S}^{k-1}\}$ *is* ULAN, *with* central sequence

$$\boldsymbol{\Delta}_{\boldsymbol{\theta},f}^{(n)} := \frac{1}{\sqrt{n}} \sum_{i=1}^{n} \varphi_f(\mathbf{X}_i'\boldsymbol{\theta})\sqrt{1 - (\mathbf{X}_i'\boldsymbol{\theta})^2}\, \mathbf{S}_i(\boldsymbol{\theta}) \tag{5}$$

and Fisher information matrix

$$\boldsymbol{\Gamma}_{\boldsymbol{\theta},f} := \frac{\mathcal{J}_k(f)}{k-1}\,(\mathbf{I}_k - \boldsymbol{\theta}\boldsymbol{\theta}'). \tag{6}$$

More precisely, (i) *for any sequence* $(\boldsymbol{\tau}^{(n)})$ *as above,*

$$\log\left(\frac{\mathrm{P}_{\boldsymbol{\theta}+n^{-1/2}\boldsymbol{\tau}^{(n)},f}^{(n)}}{\mathrm{P}_{\boldsymbol{\theta},f}^{(n)}}\right) = (\boldsymbol{\tau}^{(n)})'\boldsymbol{\Delta}_{\boldsymbol{\theta},f}^{(n)} - \frac{1}{2}(\boldsymbol{\tau}^{(n)})'\boldsymbol{\Gamma}_{\boldsymbol{\theta},f}(\boldsymbol{\tau}^{(n)}) + o_{\mathrm{P}}(1)$$

as $n \to \infty$ *under* $\mathrm{P}_{\boldsymbol{\theta},f}^{(n)}$, *and* (ii) $\boldsymbol{\Delta}_{\boldsymbol{\theta},f}^{(n)}$, *still under* $\mathrm{P}_{\boldsymbol{\theta},f}^{(n)}$, *is asymptotically normal with mean zero and covariance matrix* $\boldsymbol{\Gamma}_{\boldsymbol{\theta},f}$.

As we show in the next section, this ULAN result allows to define Le Cam optimal tests for $\mathcal{H}_0 : \boldsymbol{\theta} = \boldsymbol{\theta}_0$ under specified angular density f.

3 Optimal Parametric Tests for $\mathcal{H}_0 : \boldsymbol{\theta} = \boldsymbol{\theta}_0$

For some fixed $\boldsymbol{\theta}_0 \in \mathcal{S}^{k-1}$ and $f \in \mathcal{F}_{\mathrm{ULAN}}$, consider the problem of testing $\mathcal{H}_0 :$ $\boldsymbol{\theta} = \boldsymbol{\theta}_0$ versus $\mathcal{H}_1 : \boldsymbol{\theta} \neq \boldsymbol{\theta}_0$ in $\mathcal{P}_f^{(n)}$, that is, consider the testing problem

$$\begin{cases} \mathcal{H}_0 : \{\mathrm{P}_{\boldsymbol{\theta}_0,f}^{(n)}\} \\ \mathcal{H}_1 : \bigcup_{\boldsymbol{\theta} \neq \boldsymbol{\theta}_0} \{\mathrm{P}_{\boldsymbol{\theta},f}^{(n)}\}. \end{cases} \tag{7}$$

For this problem, we define the test $\phi_f^{(n)}$ that at asymptotic level α, rejects the null of (7) whenever

$$Q_f^{(n)} := (\Delta_{\theta_0,f}^{(n)})' \, \Gamma_{\theta_0,f}^{-} \, \Delta_{\theta_0,f}^{(n)} \tag{8}$$

$$= \frac{k-1}{n\, \mathscr{J}_k(f)} \sum_{i,j=1}^{n} \varphi_f(\mathbf{X}_i'\theta_0)\varphi_f(\mathbf{X}_j'\theta_0)\sqrt{1-(\mathbf{X}_i'\theta_0)^2} \tag{9}$$

$$\times \sqrt{1-(\mathbf{X}_j'\theta_0)^2}\,(\mathbf{S}_i(\theta_0))'\mathbf{S}_j(\theta_0) > \chi^2_{k-1,1-\alpha},$$

where $\mathbf{A}^-$ denotes the Moore–Penrose inverse of $\mathbf{A}$ and $\chi^2_{k-1,1-\alpha}$ stands for the α-upper quantile of a chi-square distribution with $k-1$ degrees of freedom. Applying in the present context the general results in Hallin et al. (2010) about hypothesis testing in curved ULAN families, yields that $\phi_f^{(n)}$ is Le Cam optimal—more precisely, locally and asymptotically maximin—at asymptotic level α for the problem (7). The asymptotic properties of this test are stated in the following result.

Theorem 1 *Fix* $\theta_0 \in \mathscr{S}^{k-1}$ *and* $f \in \mathscr{F}_{\mathrm{ULAN}}$. *Then, (i) under* $\mathrm{P}_{\theta_0,f}^{(n)}$, $Q_f^{(n)}$ *is asymptotically chi-square with* $k-1$ *degrees of freedom; (ii) under* $\mathrm{P}_{\theta_0+n^{-1/2}\tau^{(n)},f}^{(n)}$, *where the sequence* $(\tau^{(n)})$ *in* $\mathbb{R}^k$ *satisfies* $\tau^{(n)} = \tau + O(n^{-1/2})$, *with* $\theta_0'\tau = 0$, $Q_f^{(n)}$ *is asymptotically non-central chi-square, still with* $k-1$ *degrees of freedom, and non-centrality parameter*

$$\tau'\,\Gamma_{\theta_0,f}\tau = \frac{\mathscr{J}_k(f)}{k-1}\,\|\tau\|^2; \tag{10}$$

(iii) the sequence of tests $\phi_f^{(n)}$ *has asymptotic size* α *under* $\mathrm{P}_{\theta_0,f}^{(n)}$; *(iv)* $\phi_f^{(n)}$ *is locally asymptotically maximin, at asymptotic level* α, *when testing* $\{\mathrm{P}_{\theta_0,f}^{(n)}\}$ *against alternatives of the form* $\bigcup_{\theta\neq\theta_0}\{\mathrm{P}_{\theta,f}^{(n)}\}$.

For the particular case of the fixed-κ Fisher-von Mises-Langevin (FvML(κ)) distribution (obtained for $f_{\exp,\kappa}$; see (2)), we obtain

$$Q_{f_{\exp,\kappa}}^{(n)} = \frac{\kappa^2(k-1)}{n\,\mathscr{J}_k(f_{\exp,\kappa})} \sum_{i,j=1}^{n} (\mathbf{X}_i - (\mathbf{X}_i'\theta_0)\theta_0)'(\mathbf{X}_j - (\mathbf{X}_j'\theta_0)\theta_0)$$

$$= \frac{\kappa^2(k-1)}{n\,\mathscr{J}_k(f_{\exp,\kappa})} \sum_{i,j=1}^{n} \mathbf{X}_i'(\mathbf{I}_k - \theta_0\theta_0')\mathbf{X}_j$$

$$=: \frac{\kappa^2(k-1)n}{\mathscr{J}_k(f_{\exp,\kappa})} \bar{\mathbf{X}}'(\mathbf{I}_k - \theta_0\theta_0')\bar{\mathbf{X}}. \tag{11}$$

The main drawback of the parametric tests $\phi_f^{(n)}$ is their lack of (validity-)robustness: under angular density $g \neq f$, there is no guarantee that $\phi_f^{(n)}$ asymptotically meets the nominal level constraint. Indeed $Q_f^{(n)}$ is, in general, not asymptotically

χ^2_{k-1} under $P^{(n)}_{\theta_0,g}$. In practice, however, the underlying angular density may hardly be assumed to be known, and it is therefore needed to define robustified versions of $\phi^{(n)}_f$ that will combine (a) Le Cam optimality at f (Theorem 1(iv)) and (b) validity under a broad collection of angular densities $\{g\}$.

A first way to perform such a robustification is to rely on "studentization". This simply consists in considering test statistics of the form

$$Q^{(n)}_{f;\text{Stud}} := (\Delta^{(n)}_{\theta_0,f})' (\hat{\Gamma}^g_{\theta_0,f})^- \Delta^{(n)}_{\theta_0,f},$$

where $\hat{\Gamma}^g_{\theta_0,f}$ is an arbitrary consistent estimator of the covariance matrix in the asymptotic multinormal distribution of $\Delta^{(n)}_{\theta_0,f}$ under $P^{(n)}_{\theta_0,g}$. The resulting tests, that reject the null $\mathcal{H}_0 : \theta = \theta_0$ (with unspecified angular density) whenever $Q^{(n)}_{f;\text{Stud}} > \chi^2_{k-1,1-\alpha}$, are *validity-robust*—that is, they asymptotically meet the level constraint under a broad range of angular densities—and remain Le Cam optimal at f.

Of special interest is the FvML studentized test—$\phi^{(n)}_{f_{\exp};\text{Stud}}$, say—that rejects the null hypothesis whenever

$$Q^{(n)}_{f_{\exp};\text{Stud}} = \frac{k-1}{n\hat{\mathcal{L}}_k} \sum_{i,j=1}^{n} \mathbf{X}'_i (\mathbf{I}_k - \theta_0\theta'_0)\mathbf{X}_j, \tag{12}$$

where $\hat{\mathcal{L}}_k := 1 - \frac{1}{n}\sum_{i=1}^{n}(\mathbf{X}'_i\theta_0)^2$ is a consistent estimator of $\mathcal{L}_k(g) := 1 - E^{(n)}_{\theta_0,g}[(\mathbf{X}'_i\theta_0)^2]$ (this quantity does not depend on θ_0, which justifies the notation); this test was studied in Watson (1983). From studentization, this test is valid under any rotationally symmetric distribution; moreover, since $Q^{(n)}_{f_{\exp};\text{Stud}} = Q^{(n)}_{f_{\exp,\kappa}} + o_P(1)$ as $n \to \infty$ under $P^{(n)}_{\theta_0,f_{\exp,\kappa}}$ for any κ, this test is also optimal in the Le Cam sense under any FvML distribution.

Studentization, however, typically leads to tests that fail to be *efficiency-robust*, in the sense that the resulting type 2 risk may dramatically increase when the underlying angular density g much deviates from the target density—or target densities, in the case of the FvML studentized test $\phi^{(n)}_{f_{\exp};\text{Stud}}$—at which they are optimal. That is why studentization will not be considered in this paper. Instead, we will take advantage of the group invariance structure of the testing problem considered, in order to introduce invariant tests that are both validity- and efficiency-robust. As we will see, invariant tests in the present context are *rank* tests.

4 Optimal Rank Tests for $\mathcal{H}_0 : \boldsymbol{\theta} = \boldsymbol{\theta}_0$

We start by describing the group invariance structure of the testing problem considered above (Sect. 4.1). Then we introduce (and study the properties of) rank-based versions of the central sequences from Proposition 1 (Sect. 4.2). This will allow us to develop the resulting (optimal) rank tests and to derive their asymptotic properties (Sect. 4.3).

4.1 Group Invariance Structure

Still for some given $\boldsymbol{\theta}_0 \in \mathcal{S}^{k-1}$, consider the problem of testing $\mathcal{H}_0 : \boldsymbol{\theta} = \boldsymbol{\theta}_0$ against $\mathcal{H}_1 : \boldsymbol{\theta} \neq \boldsymbol{\theta}_0$ *under unspecified angular density* g, that is, consider the testing problem

$$\begin{cases} \mathcal{H}_0 : \bigcup_{g \in \mathcal{F}} \{ P_{\boldsymbol{\theta}_0, g}^{(n)} \} \\ \mathcal{H}_1 : \bigcup_{\boldsymbol{\theta} \neq \boldsymbol{\theta}_0} \bigcup_{g \in \mathcal{F}} \{ P_{\boldsymbol{\theta}, g}^{(n)} \}. \end{cases} \tag{13}$$

This testing problem is invariant under two groups of transformations, which we now quickly describe.

(i) To define the first group, we introduce the *tangent-normal decomposition*

$$\mathbf{X}_i = (\mathbf{X}_i' \boldsymbol{\theta}_0) \boldsymbol{\theta}_0 + \| \mathbf{X}_i - (\mathbf{X}_i' \boldsymbol{\theta}_0) \boldsymbol{\theta}_0 \| \, \mathbf{S}_i(\boldsymbol{\theta}_0), \qquad i = 1, \dots, n$$

of the observations $\mathbf{X}_i, i = 1, \dots, n$. The first group of transformations we consider is then $\mathcal{G} = \{ g_h : h \in \mathcal{H} \}$, $\circ$, with

$$g_h : \qquad (\mathcal{S}^{k-1})^n \quad \to (\mathcal{S}^{k-1})^n$$

$$(\mathbf{X}_1, \dots, \mathbf{X}_n) \mapsto (h(\mathbf{X}_1' \boldsymbol{\theta}_0) \boldsymbol{\theta}_0 + \| \mathbf{X}_1 - h(\mathbf{X}_1' \boldsymbol{\theta}_0) \boldsymbol{\theta}_0 \| \, \mathbf{S}_1(\boldsymbol{\theta}_0), \dots,$$

$$h(\mathbf{X}_n' \boldsymbol{\theta}_0) \boldsymbol{\theta}_0 + \| \mathbf{X}_n - h(\mathbf{X}_n' \boldsymbol{\theta}_0) \boldsymbol{\theta}_0 \| \, \mathbf{S}_n(\boldsymbol{\theta}_0)),$$

where $\mathcal{H}$ is the collection of mappings $h : [-1, 1] \mapsto [-1, 1]$ that are continuous, monotone increasing, and satisfy $h(\pm 1) = \pm 1$.

The null hypothesis of (13) is clearly invariant under the group $\mathcal{G}, \circ$. The invariance principle therefore suggests restricting to tests that are invariant with respect to this group. As it was shown in Ley et al. (2013), the maximal invariant $\mathbf{I}^{(n)}(\boldsymbol{\theta}_0)$ associated with $\mathcal{G}, \circ$ is the sign-and-rank statistic $(\mathbf{S}_1(\boldsymbol{\theta}_0), \dots, \mathbf{S}_n(\boldsymbol{\theta}_0), R_1(\boldsymbol{\theta}_0), \dots, R_n(\boldsymbol{\theta}_0))$, where $R_i(\boldsymbol{\theta}_0)$ denotes the rank of $\mathbf{X}_i' \boldsymbol{\theta}_0$ among $\mathbf{X}_1' \boldsymbol{\theta}_0, \dots, \mathbf{X}_n' \boldsymbol{\theta}_0$. Consequently, the class of invariant tests coincides with the collection of tests that are measurable with respect to $\mathbf{I}^{(n)}(\boldsymbol{\theta}_0)$, in short, with the class of *(sign-and-)rank*—or, simply, *rank*—tests.

It is easy to check that $\mathscr{G}, \circ$ is actually a generating group for the null hypothesis $\bigcup_{g \in \mathscr{F}} \{ P_{\theta_0, g}^{(n)} \}$ in (13). As a direct corollary, rank tests are distribution-free under the whole null hypothesis. This explains why rank tests will be validity-robust.

(ii) Of course, the null hypothesis in (13) is also invariant under orthogonal transformations fixing the null location value θ_0. More precisely, it is invariant under the group $\mathscr{G}_{\mathrm{rot}} = \{ g_{\mathbf{O}} : \mathbf{O} \in \mathscr{O}_{\theta_0} \}, \circ$, with

$$g_{\mathbf{O}} : (\mathscr{S}^{k-1})^n \quad \to (\mathscr{S}^{k-1})^n$$
$$(\mathbf{X}_1, \ldots, \mathbf{X}_n) \mapsto (\mathbf{O}\mathbf{X}_1, \ldots, \mathbf{O}\mathbf{X}_n),$$

where $\mathscr{O}_{\theta_0}$ is the collection of all $k \times k$ orthogonal matrices $\mathbf{O}$ satisfying $\mathbf{O}\theta_0 = \theta_0$. Clearly, the vectors of signs and ranks above is not invariant under $\mathscr{G}_{\mathrm{rot}}, \circ$, but the statistic

$$\left((\mathbf{S}_1(\theta_0))' \mathbf{S}_2(\theta_0), (\mathbf{S}_1(\theta_0))' \mathbf{S}_3(\theta_0), \ldots, (\mathbf{S}_{n-1}(\theta_0))' \mathbf{S}_n(\theta_0), R_1(\theta_0), \ldots, R_n(\theta_0) \right)$$

$$(14)$$

is. Tests that are measurable with respect to the statistic in (14) will therefore be invariant with respect to both groups considered above.

4.2 Rank-Based Central Sequences

To combine validity-robustness/invariance with Le Cam optimality at a target angular density f, we introduce rank-based versions of the central sequences that appear in the ULAN property above (Proposition 1). More precisely, we consider rank statistics of the form

$$\underset{\sim}{\Delta}_{\theta, K}^{(n)} = \frac{1}{\sqrt{n}} \sum_{i=1}^{n} K \left(\frac{R_i(\theta)}{n+1} \right) \mathbf{S}_i(\theta),$$

where the *score function* $K : [0, 1] \to \mathbb{R}$ is throughout assumed to be continuous (which implies that it is bounded and square-integrable over $[0, 1]$).

In order to state the asymptotic properties of the rank-based random vector $\underset{\sim}{\Delta}_{\theta, K}^{(n)}$, we introduce the following notation. For any $g \in \mathscr{F}$, write

$$\Delta_{\theta, K, g}^{(n)} := \frac{1}{\sqrt{n}} \sum_{i=1}^{n} K \left(\tilde{G}(\mathbf{X}_i' \theta) \right) \mathbf{S}_i(\theta),$$

where $\tilde{G}$ denotes the cdf of $\mathbf{X}_i'\boldsymbol{\theta}$ under $P_{\boldsymbol{\theta},g}^{(n)}$. For any $g \in \mathscr{F}_{\mathrm{ULAN}}$, define further

$$\boldsymbol{\Gamma}_{\boldsymbol{\theta},K} := \frac{\mathscr{J}_k(K)}{k-1}\,(\mathbf{I}_k - \boldsymbol{\theta}\boldsymbol{\theta}') \quad \text{and} \quad \boldsymbol{\Gamma}_{\boldsymbol{\theta},K,g} := \frac{\mathscr{J}_k(K,g)}{k-1}\,(\mathbf{I}_k - \boldsymbol{\theta}\boldsymbol{\theta}'),$$

with $\mathscr{J}_k(K) := \int_0^1 K^2(u)\,du$ and $\mathscr{J}_k(K,g) := \int_0^1 K(u)K_g(u)\,du$, where we wrote

$$K_g(u) := \varphi_g(\tilde{G}^{-1}(u))\sqrt{1 - (\tilde{G}^{-1}(u))^2}$$

for any $u \in [0, 1]$. We then have the following result (see Ley et al. 2013).

Proposition 2 *Fix $\boldsymbol{\theta} \in \mathscr{S}^{k-1}$ and let $(\boldsymbol{\tau}^{(n)})$ be a sequence in $\mathbb{R}^k$ that satisfies $\boldsymbol{\tau}^{(n)} = \boldsymbol{\tau} + O(n^{-1/2})$, with $\boldsymbol{\theta}'\boldsymbol{\tau} = 0$. Then (i) under $P_{\boldsymbol{\theta},g}^{(n)}$, with $g \in \mathscr{F}$, $\underset{\sim}{\boldsymbol{\Delta}}_{\boldsymbol{\theta},K}^{(n)} = \boldsymbol{\Delta}_{\boldsymbol{\theta},K,g}^{(n)} + o_{L^2}(1)$ as $n \to \infty$; (ii) under $P_{\boldsymbol{\theta},g}^{(n)}$, with $g \in \mathscr{F}$, $\underset{\sim}{\boldsymbol{\Delta}}_{\boldsymbol{\theta},K}^{(n)}$ is asymptotically multinormal with mean zero and covariance matrix $\boldsymbol{\Gamma}_{\boldsymbol{\theta},K}$; (iii) under $P_{\boldsymbol{\theta}+n^{-1/2}\boldsymbol{\tau}^{(n)},g}^{(n)}$, with $g \in \mathscr{F}_{\mathrm{ULAN}}$, $\underset{\sim}{\boldsymbol{\Delta}}_{\boldsymbol{\theta},K}^{(n)}$ is asymptotically multinormal with mean $\boldsymbol{\Gamma}_{\boldsymbol{\theta},K,g}\boldsymbol{\tau}$ and covariance matrix $\boldsymbol{\Gamma}_{\boldsymbol{\theta},K}$; (iv) under $P_{\boldsymbol{\theta},g}^{(n)}$, with $g \in \mathscr{F}_{\mathrm{ULAN}}$, $\underset{\sim}{\boldsymbol{\Delta}}_{\boldsymbol{\theta}+n^{-1/2}\boldsymbol{\tau}^{(n)},K}^{(n)} = \underset{\sim}{\boldsymbol{\Delta}}_{\boldsymbol{\theta},K}^{(n)} - \boldsymbol{\Gamma}_{\boldsymbol{\theta},K,g}\boldsymbol{\tau}^{(n)} + o_{\mathrm{P}}(1)$ as $n \to \infty$.*

For any $f \in \mathscr{F}_{\mathrm{ULAN}}^{C^1}$, where $\mathscr{F}_{\mathrm{ULAN}}^{C^1}$ denotes the collection of angular densities in $\mathscr{F}_{\mathrm{ULAN}}$ that are continuously differentiable over $[-1, 1]$, the function $u \mapsto K_f(u) := \varphi_f(\tilde{F}^{-1}(u))\sqrt{1 - (\tilde{F}^{-1}(u))^2}$ is a valid score function to be used in rank-based central sequences. Proposition 2(i) then entails that under $P_{\boldsymbol{\theta},f}^{(n)}$, $\underset{\sim}{\boldsymbol{\Delta}}_{\boldsymbol{\vartheta},K_f}^{(n)}$ is asymptotically equivalent—in L^2, hence also in probability—to the parametric f-central sequence $\boldsymbol{\Delta}_{\boldsymbol{\theta},f}^{(n)} = \boldsymbol{\Delta}_{\boldsymbol{\theta},K_f,f}^{(n)}$. This provides the key to develop rank tests that are Le Cam optimal at any given $f \in \mathscr{F}_{\mathrm{ULAN}}^{C^1}$. As for Proposition 2(ii)–(iii), they allow to derive the asymptotic properties of the resulting optimal rank tests.

4.3 Rank Tests for $\mathscr{H}_0 : \boldsymbol{\theta} = \boldsymbol{\theta}_0$

Fix a score function K as above. The previous sections then make it natural to consider the rank test—$\phi_K^{(n)}$, say—that at asymptotic level α, rejects the null hypothesis $\mathscr{H}_0 : \boldsymbol{\theta} = \boldsymbol{\theta}_0$ (with unspecified angular density g) whenever

$$Q_K^{(n)} = (\underset{\sim}{\Delta}_{\theta_0,K}^{(n)})' \Gamma_{\theta_0,K}^{-} \underset{\sim}{\Delta}_{\theta_0,K}^{(n)}$$

$$= \frac{k-1}{n \mathscr{J}_k(K)} \sum_{i,j=1}^{n} K\left(\frac{R_i(\theta_0)}{n+1}\right) K\left(\frac{R_j(\theta_0)}{n+1}\right) (\mathbf{S}_i(\theta_0))' \mathbf{S}_j(\theta_0)$$

$$> \chi_{k-1,1-\alpha}^2.$$

Clearly, this test is invariant with respect to both groups introduced in Sect. 4.1, since it is measurable with respect to the statistic in (14). The following result, that summarizes the asymptotic properties of $\phi_K^{(n)}$, easily follows from Proposition 2.

Theorem 2 *Let $(\tau^{(n)})$ be a sequence in $\mathbb{R}^k$ that satisfies $\tau^{(n)} = \tau + O(n^{-1/2})$, with $\theta_0'\tau = 0$. Then, (i) under $\bigcup_{g \in \mathscr{F}}\{\mathrm{P}_{\theta_0,g}^{(n)}\}$, $Q_K^{(n)}$ is asymptotically chi-square with $k-1$ degrees of freedom; (ii) under $\mathrm{P}_{\theta_0+n^{-1/2}\tau^{(n)},g}^{(n)}$, with $g \in \mathscr{F}_{\mathrm{ULAN}}$, is asymptotically non-central chi-square, still with $k-1$ degrees of freedom, and non-centrality parameter*

$$\tau' \Gamma_{\theta,K,g} \Gamma_{\theta_0,K}^{-} \Gamma_{\theta,K,g} \tau = \frac{\mathscr{J}_k^2(K,g)}{(k-1)\mathscr{J}_k(K)} \|\tau\|^2; \tag{15}$$

(iii) the sequence of tests $\phi_K^{(n)}$ has asymptotic size α under $\bigcup_{g \in \mathscr{F}}\{\mathrm{P}_{\theta_0,g}^{(n)}\}$; (iv) $\phi_{K_f}^{(n)}$, with $f \in \mathscr{F}_{\mathrm{ULAN}}^{C^1}$, is locally and asymptotically maximin, at asymptotic level α, when testing $\bigcup_{g \in \mathscr{F}}\{\mathrm{P}_{\theta_0,g}^{(n)}\}$ against alternatives of the form $\bigcup_{\theta \neq \theta_0}\{\mathrm{P}_{\theta,f}^{(n)}\}$.

Note that for $K = K_f$ and $g = f$ (with $f \in \mathscr{F}_{\mathrm{ULAN}}^{C^1}$), the non-centrality parameter in (15) above rewrites

$$\frac{\mathscr{J}_k^2(K_f, f)}{(k-1)\mathscr{J}_k(K_f)} \|\tau\|^2 = \frac{\mathscr{J}_k(f)}{k-1} \|\tau\|^2,$$

hence coincides with the non-centrality parameter in (10). This establishes the optimality statement in Theorem 2(iv).

5 Testing Great Circle Hypotheses

In this section, we turn to another classical testing problem involving rotationally symmetric distributions, namely to the problem of testing that θ belongs to some given "great circle" of $\mathscr{S}^{k-1}$, where the term *great circle* here refers to the intersection of $\mathscr{S}^{k-1}$ with a vectorial subspace of $\mathbb{R}^k$. In other words, letting Υ be some given full-rank $k \times s$ ($s < k$) matrix, we consider the problem of testing $\mathscr{H}_0^{\Upsilon} : \theta \in \mathscr{S}^{k-1} \cap \mathscr{M}(\Upsilon)$ against $\mathscr{H}_1^{\Upsilon} : \theta \notin \mathscr{S}^{k-1} \cap \mathscr{M}(\Upsilon)$, where $\mathscr{M}(\Upsilon)$ denotes the s-dimensional subspace of $\mathbb{R}^k$ that is spanned by the columns of Υ. More precisely, the testing problem is

$$\begin{cases} \mathcal{H}_0^{\boldsymbol{\Upsilon}} : \bigcup_{\boldsymbol{\theta} \in \mathcal{S}^{k-1} \cap \mathcal{M}(\boldsymbol{\Upsilon})} \bigcup_g \{P_{\boldsymbol{\theta},g}^{(n)}\} \\ \mathcal{H}_1^{\boldsymbol{\Upsilon}} : \bigcup_{\boldsymbol{\theta} \notin \mathcal{S}^{k-1} \cap \mathcal{M}(\boldsymbol{\Upsilon})} \bigcup_g \{P_{\boldsymbol{\theta},g}^{(n)}\}. \end{cases} \tag{16}$$

This problem has been studied by Watson (1983), that provided an FvML score test, and in Fujikoshi and Watamori (1992) and Watamori (1992), that investigated the properties of the FvML likelihood ratio test.

For any $f \in \mathcal{F}_{\mathrm{ULAN}}$, the construction of f-optimal parametric tests for this problem proceeds as follows. Fix $\boldsymbol{\theta} \in \mathcal{S}^{k-1} \cap \mathcal{M}(\boldsymbol{\Upsilon})$ (the unspecification of $\boldsymbol{\theta}$ under the null will be taken care of later on) and consider a local perturbation of the form $\boldsymbol{\theta} + n^{-1/2} \boldsymbol{\tau}^{(n)}$ where $\boldsymbol{\tau}^{(n)}$ is such that $\boldsymbol{\tau}^{(n)} = \boldsymbol{\tau} + O(n^{-1/2})$, with $\boldsymbol{\theta}'\boldsymbol{\tau} = 0$ (see the comment just before Proposition 1). It directly follows from Hallin et al. (2010) that a locally (in the vicinity of $\boldsymbol{\theta}$) and asymptotically most stringent test can be obtained by considering the most stringent test for the linear constraint $\boldsymbol{\tau}^{(n)} \in \mathcal{M}(\mathbf{I}_k - \boldsymbol{\theta}\boldsymbol{\theta}') \cap \mathcal{M}(\boldsymbol{\Upsilon})$. Letting $\boldsymbol{\Upsilon}_{\boldsymbol{\theta}}$ be such that $\mathcal{M}(\boldsymbol{\Upsilon}_{\boldsymbol{\theta}}) = \mathcal{M}(\mathbf{I}_k - \boldsymbol{\theta}\boldsymbol{\theta}') \cap \mathcal{M}(\boldsymbol{\Upsilon})$, the resulting f-optimal test therefore rejects the null hypothesis $\mathcal{H}_{0,f}^{\boldsymbol{\Upsilon}} : \bigcup_{\boldsymbol{\theta} \in \mathcal{S}^{k-1} \cap \mathcal{M}(\boldsymbol{\Upsilon})} \{P_{\boldsymbol{\theta},f}^{(n)}\}$ for large values of

$$Q_{\boldsymbol{\theta},f}^{(n)} = (\boldsymbol{\Delta}_{\boldsymbol{\theta},f}^{(n)})' \left(\boldsymbol{\Gamma}_{\boldsymbol{\theta},f}^{-} - \boldsymbol{\Upsilon}_{\boldsymbol{\theta}} (\boldsymbol{\Upsilon}_{\boldsymbol{\theta}}' \boldsymbol{\Gamma}_{\boldsymbol{\theta},f} \boldsymbol{\Upsilon}_{\boldsymbol{\theta}})^{-} \boldsymbol{\Upsilon}_{\boldsymbol{\theta}}' \right) \boldsymbol{\Delta}_{\boldsymbol{\theta},f}^{(n)}.$$

Using the identity $(\mathbf{I}_k - \boldsymbol{\theta}\boldsymbol{\theta}')\boldsymbol{\Upsilon}_{\boldsymbol{\theta}} = \boldsymbol{\Upsilon}_{\boldsymbol{\theta}}$ (which follows from the fact that $\mathbf{I}_k - \boldsymbol{\theta}\boldsymbol{\theta}'$ is the projection matrix onto $\mathcal{M}(\mathbf{I}_k - \boldsymbol{\theta}\boldsymbol{\theta}')$, that by definition, contains every column vector of $\boldsymbol{\Upsilon}_{\boldsymbol{\theta}}$), then the fact that $\boldsymbol{\Upsilon}_{\boldsymbol{\theta}} (\boldsymbol{\Upsilon}_{\boldsymbol{\theta}}' \boldsymbol{\Upsilon}_{\boldsymbol{\theta}})^{-} \boldsymbol{\Upsilon}_{\boldsymbol{\theta}}' (\mathbf{I}_k - \boldsymbol{\theta}\boldsymbol{\theta}') = \boldsymbol{\Upsilon} (\boldsymbol{\Upsilon}'\boldsymbol{\Upsilon})^{-1} \boldsymbol{\Upsilon}' (\mathbf{I}_k - \boldsymbol{\theta}\boldsymbol{\theta}')$, we obtain

$$Q_{\boldsymbol{\theta},f}^{(n)} = \frac{k-1}{\mathcal{J}_k(f)} (\boldsymbol{\Delta}_{\boldsymbol{\theta},f}^{(n)})' \left(\mathbf{I}_k - \boldsymbol{\Upsilon}_{\boldsymbol{\theta}} (\boldsymbol{\Upsilon}_{\boldsymbol{\theta}}' \boldsymbol{\Upsilon}_{\boldsymbol{\theta}})^{-} \boldsymbol{\Upsilon}_{\boldsymbol{\theta}}' \right) \boldsymbol{\Delta}_{\boldsymbol{\theta},f}^{(n)}$$

$$= \frac{k-1}{\mathcal{J}_k(f)} (\boldsymbol{\Delta}_{\boldsymbol{\theta},f}^{(n)})' (\mathbf{I}_k - \boldsymbol{\Upsilon} (\boldsymbol{\Upsilon}'\boldsymbol{\Upsilon})^{-} \boldsymbol{\Upsilon}') \boldsymbol{\Delta}_{\boldsymbol{\theta},f}^{(n)}. \tag{17}$$

We will show below that under $P_{\boldsymbol{\theta},f}^{(n)}$, $Q_{\boldsymbol{\theta},f}^{(n)}$ is asymptotically chi-square with $k - s$ degrees of freedom, so that the resulting test rejects the null, at asymptotic level α, whenever $Q_{\boldsymbol{\theta},f}^{(n)}$ exceeds the corresponding upper α-quantile $\chi^2_{k-s,1-\alpha}$.

Since $\boldsymbol{\theta} \in \mathcal{M}(\boldsymbol{\Upsilon})$ implies that $\boldsymbol{\Upsilon} (\boldsymbol{\Upsilon}'\boldsymbol{\Upsilon})^{-1} \boldsymbol{\Upsilon}' \boldsymbol{\theta} = \boldsymbol{\theta}$ (or equivalently, that $(\mathbf{I}_k - \boldsymbol{\Upsilon} (\boldsymbol{\Upsilon}'\boldsymbol{\Upsilon})^{-1} \boldsymbol{\Upsilon}') \boldsymbol{\theta} = \mathbf{0}$), the FvML($\kappa$) version of $Q_{\boldsymbol{\theta},f}^{(n)}$ is given by

$$Q_{\boldsymbol{\theta},f_{\mathrm{exp},\kappa}}^{(n)} = \frac{n\kappa^2(k-1)}{\mathcal{J}_k(f_{\mathrm{exp},\kappa})} \bar{\mathbf{X}}' (\mathbf{I}_k - \boldsymbol{\Upsilon} (\boldsymbol{\Upsilon}'\boldsymbol{\Upsilon})^{-1} \boldsymbol{\Upsilon}') \bar{\mathbf{X}}. \tag{18}$$

As in Sect. 3, this leads to defining the FvML studentized test—$\phi_{f_{\mathrm{exp}};\mathrm{Stud}}^{(n)}$, say—that rejects $\mathcal{H}_0^{\boldsymbol{\Upsilon}}$ whenever

$$Q^{(n)}_{f_{\exp};\mathrm{Stud}} = \frac{n(k-1)}{\hat{\mathscr{L}}_k(\hat{\boldsymbol{\theta}})} \bar{\mathbf{X}}'(\mathbf{I}_k - \boldsymbol{\Upsilon}(\boldsymbol{\Upsilon}'\boldsymbol{\Upsilon})^{-1}\boldsymbol{\Upsilon}')\bar{\mathbf{X}} > \chi^2_{k-s,1-\alpha},$$

where $\hat{\mathscr{L}}_k(\boldsymbol{\theta}) = 1 - \frac{1}{n}\sum_{i=1}^{n}(\mathbf{X}_i'\boldsymbol{\theta})^2$ is evaluated at an arbitrary consistent estimator $\hat{\boldsymbol{\theta}}$ of $\boldsymbol{\theta}$. When based on $\hat{\boldsymbol{\theta}} = \bar{\mathbf{X}}/|\bar{\mathbf{X}}|$, this test is actually the Watson (1983) score test. Consequently, the following result, that states the asymptotic and optimality properties of $\phi^{(n)}_{f_{\exp};\mathrm{Stud}}$, clarifies the exact optimality properties of this classical test.

Theorem 3 *Fix $\boldsymbol{\theta} \in \mathscr{S}^{k-1} \cap \mathscr{M}(\boldsymbol{\Upsilon})$ and let $(\boldsymbol{\tau}^{(n)})$ be a sequence in $\mathbb{R}^k$ that satisfies $\boldsymbol{\tau}^{(n)} = \boldsymbol{\tau} + O(n^{-1/2})$, with $\boldsymbol{\theta}'\boldsymbol{\tau} = 0$. Then, (i) under $\mathrm{P}^{(n)}_{\boldsymbol{\theta},g}$, with $g \in \mathscr{F}$, $Q^{(n)}_{f_{\exp};\mathrm{Stud}}$ is asymptotically chi-square with $k-s$ degrees of freedom; (ii) under $\mathrm{P}^{(n)}_{\boldsymbol{\theta}+n^{-1/2}\boldsymbol{\tau}^{(n)},g}$, with $g \in \mathscr{F}_{\mathrm{ULAN}}$, $Q^{(n)}_{f_{\exp};\mathrm{Stud}}$ is asymptotically non-central chi-square, still with $k-s$ degrees of freedom, and non-centrality parameter*

$$\frac{k-1}{\mathscr{L}_k(g)} \boldsymbol{\tau}'(\mathbf{I}_k - \boldsymbol{\Upsilon}(\boldsymbol{\Upsilon}'\boldsymbol{\Upsilon})^{-1}\boldsymbol{\Upsilon}')\boldsymbol{\tau}; \tag{19}$$

(iii) the sequence of tests $\phi^{(n)}_{f_{\exp};\mathrm{Stud}} = \mathbb{I}[Q^{(n)}_{f_{\exp};\mathrm{Stud}} > \chi^2_{k-s,1-\alpha}]$ has asymptotic size α under $\cup_{\boldsymbol{\theta} \in \mathscr{S}^{k-1} \cap \mathscr{M}(\boldsymbol{\Upsilon})} \cup_{g \in \mathscr{F}} \{\mathrm{P}^{(n)}_{\boldsymbol{\theta},g}\}$; (iv) $\phi^{(n)}_{f_{\exp};\mathrm{Stud}}$ is locally asymptotically most stringent, at asymptotic level α, when testing $\cup_{\boldsymbol{\theta} \in \mathscr{S}^{k-1} \cap \mathscr{M}(\boldsymbol{\Upsilon})} \cup_{g \in \mathscr{F}} \{\mathrm{P}^{(n)}_{\boldsymbol{\theta},g}\}$ against alternatives of the form $\cup_{\boldsymbol{\theta} \in \mathscr{S}^{k-1} \setminus \mathscr{M}(\boldsymbol{\Upsilon})} \cup_{\kappa > 0} \{\mathrm{P}^{(n)}_{\boldsymbol{\theta},f_{\exp,\kappa}}\}$.

This test is therefore valid under any rotationally symmetric distribution, hence is validity-robust. It is optimal under any FvML distribution, but is not efficiency-robust outside the class of FvML distributions. As for the first testing problem we considered in the previous sections, this motivates building rank-based tests that combine validity- and efficiency-robustness.

The appropriate rank test statistics are obtained by replacing in (17) the parametric central sequence $\boldsymbol{\Delta}^{(n)}_{\boldsymbol{\theta},f}$ with its rank-based counterpart $\underset{\sim}{\boldsymbol{\Delta}}^{(n)}_{\boldsymbol{\theta},f} = \underset{\sim}{\boldsymbol{\Delta}}^{(n)}_{\boldsymbol{\theta},K_f}$. More generally, we will consider general-score rank statistics of the form

$$\underset{\sim}{Q}^{(n)}_{\boldsymbol{\theta},K} = \frac{k-1}{\mathscr{J}_k(K)} \left(\underset{\sim}{\boldsymbol{\Delta}}^{(n)}_{\boldsymbol{\theta},K}\right)' \left(\mathbf{I}_k - \boldsymbol{\Upsilon}(\boldsymbol{\Upsilon}'\boldsymbol{\Upsilon})^{-1}\boldsymbol{\Upsilon}'\right) \underset{\sim}{\boldsymbol{\Delta}}^{(n)}_{\boldsymbol{\theta},K}.$$

As already mentioned, $\boldsymbol{\theta}$ is not specified under the null hypothesis. We will therefore rather consider the test—$\underset{\sim}{\phi}^{(n)}_K$, say—that rejects the null at asymptotic level α whenever $\underset{\sim}{Q}^{(n)}_K := \underset{\sim}{Q}^{(n)}_{\hat{\boldsymbol{\theta}},K} > \chi^2_{k-s,1-\alpha}$. The estimator $\hat{\boldsymbol{\theta}}$ to be used needs to be part of a sequence of estimators that satisfies the following assumption.

ASSUMPTION (B). The sequence of estimators $\hat{\boldsymbol{\theta}} = \hat{\boldsymbol{\theta}}^{(n)}$ is (i) *root-n consistent:* $\hat{\boldsymbol{\theta}} - \boldsymbol{\theta} = O_{\mathrm{P}}(n^{-1/2})$ under $\bigcup_{\boldsymbol{\theta} \in \mathscr{S}^{k-1} \cap \mathscr{M}(\boldsymbol{\Upsilon})} \bigcup_{g \in \mathscr{F}} \mathrm{P}^{(n)}_{\boldsymbol{\theta};g}$; (ii) *locally and asymptotically discrete:* for all $\boldsymbol{\theta}$ and for all $C > 0$, there exists a positive integer $M = M(C)$

such that the number of possible values of $\hat{\boldsymbol{\theta}}^{(n)}$ in balls of the form $\{\boldsymbol{\theta}' \in \mathbb{R}^k : \sqrt{n}\|\boldsymbol{\theta}' - \boldsymbol{\theta}\| \leq C\}$ is bounded by M, uniformly as $n \to \infty$; (iii) *constrained:* $\hat{\boldsymbol{\theta}}$ takes its values in $\mathcal{M}(\boldsymbol{\Upsilon})(\cap \mathscr{S}^{k-1})$.

The following result then states the asymptotic properties of the rank tests $\phi_K^{(n)}$.

Theorem 4 *Let Assumption (B) hold, fix* $g \in \mathscr{F}_{\mathrm{ULAN}}$, $\boldsymbol{\theta} \in \mathscr{S}^{k-1} \cap \mathcal{M}(\boldsymbol{\Upsilon})$, *and let* $(\boldsymbol{\tau}^{(n)})$ *be a sequence in* $\mathbb{R}^k$ *that satisfies* $\boldsymbol{\tau}^{(n)} = \boldsymbol{\tau} + O(n^{-1/2})$, *with* $\boldsymbol{\theta}'\boldsymbol{\tau} = 0$. *Then,*
(i) under $\mathrm{P}_{\boldsymbol{\theta},g}^{(n)}$, $Q_K^{(n)}$ *is asymptotically chi-square with* $k - s$ *degrees of freedom;*
(ii) under $\mathrm{P}_{\boldsymbol{\theta}+n^{-1/2}\boldsymbol{\tau}^{(n)},g}^{(n)}$, $Q_K^{(n)}$ *is asymptotically non-central chi-square, still with* $k - s$ *degrees of freedom, and non-centrality parameter*

$$\frac{\mathscr{J}_k^2(K,g)}{(k-1)\,\mathscr{J}_k(K)}\,\boldsymbol{\tau}'(\mathbf{I}_k - \boldsymbol{\Upsilon}(\boldsymbol{\Upsilon}'\boldsymbol{\Upsilon})^{-1}\boldsymbol{\Upsilon}')\boldsymbol{\tau};\tag{20}$$

(iii) the sequence of tests $\phi_K^{(n)} = \mathbb{I}[Q_K^{(n)} > \chi_{k-s,1-\alpha}^2]$ *has asymptotic size* α *under* $\cup_{\boldsymbol{\theta}\in\mathscr{S}^{k-1}\cap\mathcal{M}(\boldsymbol{\Upsilon})}\cup_{g\in\mathscr{F}}\{\mathrm{P}_{\boldsymbol{\theta},g}^{(n)}\}$; *(iv)* $\phi_{K_f}^{(n)}$, *with* $f \in \mathscr{F}_{\mathrm{ULAN}}^{C^1}$, *is locally asymptotically most stringent, at asymptotic level* α, *when testing* $\cup_{\boldsymbol{\theta}\in\mathscr{S}^{k-1}\cap\mathcal{M}(\boldsymbol{\Upsilon})}\cup_{g\in\mathscr{F}_{\mathrm{ULAN}}}\{\mathrm{P}_{\boldsymbol{\theta},g}^{(n)}\}$ *against alternatives of the form* $\cup_{\boldsymbol{\theta}\in\mathscr{S}^{k-1}\setminus\mathcal{M}(\boldsymbol{\Upsilon})}\{\mathrm{P}_{\boldsymbol{\theta},f}^{(n)}\}$.

Theorems 3–4 allow to compute in a straightforward way (as usual, as the ratios of the non-centrality parameters in the asymptotic non-null distributions of the corresponding statistics) the asymptotic relative efficiencies (AREs) of the proposed rank tests with respect to their FvML-score competitors from Watson (1983). These AREs are given by

$$\mathrm{ARE}_g\big[\,\phi_K^{(n)}/\phi_{f_{\mathrm{exp}};\mathrm{Stud}}^{(n)}\,\big] = \frac{\mathscr{L}_k^2(g)\,\mathscr{J}_k^2(K,g)}{(k-1)^2\,\mathscr{J}_k(K)},$$

and do not depend on $\boldsymbol{\theta}$ nor on $\boldsymbol{\tau}$. It is easy to check that these AREs, that coincides with the ones obtained in Ley et al. (2013) for point estimation, also hold for the testing problem considered in the previous sections.

6 Simulations

In this final section, we conduct a Monte Carlo study to investigate the finite-sample behaviour of the rank tests proposed in Sects. 4.3 and 5. Letting

$$\boldsymbol{\theta}_0 = \begin{pmatrix} 1 \\ 0 \\ 0 \end{pmatrix} \in \mathbb{R}^k = \mathbb{R}^3 \quad \text{and} \quad \boldsymbol{\Upsilon} = \begin{pmatrix} 1 & 0 \\ 0 & 0 \\ 0 & 1 \end{pmatrix},$$

we considered the problems of testing $\mathcal{H}_0 : \boldsymbol{\theta} = \boldsymbol{\theta}_0$ and $\mathcal{H}_0 : \boldsymbol{\theta} \in \mathcal{S}^{k-1} \cap \mathcal{M}(\boldsymbol{\Upsilon})$, respectively. We generated $M = 10{,}000$ (mutually independent) random samples of rotationally symmetric random vectors

$$\boldsymbol{\varepsilon}_i^{(\ell)}, \quad i = 1, \ldots, n = 250, \quad \ell = 1, 2, 3, 4,$$

with location $\boldsymbol{\theta}_0$ and with angular densities that are FvML(1), FvML(3), LIN(1.1), and LIN(2), for $\ell = 1, 2, 3$, and 4, respectively; see Sect. 2. Each random vector $\boldsymbol{\varepsilon}_i^{(\ell)}$ was then transformed into

$$\mathbf{X}_{i;\omega}^{(\ell)} := \mathbf{O}_\omega \boldsymbol{\varepsilon}_i^{(\ell)}, \quad i = 1, \ldots, n, \quad \ell = 1, 2, 3, 4, \quad \omega = 0, 1, 2, 3,$$

with

$$\mathbf{O}_\omega = \begin{pmatrix} \cos(\pi\omega/50) & -\sin(\pi\omega/50) & 0 \\ \sin(\pi\omega/50) & \cos(\pi\omega/50) & 0 \\ 0 & 0 & 1 \end{pmatrix}.$$

For both testing problems considered, the value $\omega = 0$ corresponds to the null hypothesis, whereas $\omega = 1, 2, 3$ correspond to increasingly severe alternatives.

On each replication of the samples $(\mathbf{X}_{1;\omega}^{(\ell)}, \ldots, \mathbf{X}_{n;\omega}^{(\ell)})$, $\ell = 1, 2, 3, 4$, $\omega = 0, 1, 2, 3$, we performed the following tests for $\mathcal{H}_0 : \boldsymbol{\theta} = \boldsymbol{\theta}_0$ and for $\mathcal{H}_0 : \boldsymbol{\theta} \in \mathcal{S}^{k-1} \cap \mathcal{M}(\boldsymbol{\Upsilon})$, all at asymptotic level $\alpha = 5\,\%$: (i) the tests $\phi_{f_{\exp,3}}^{(n)}$, that is, the parametric tests $\phi_f^{(n)}$ using a FvML(3) angular target density, (ii) the tests $\phi_{f_{\exp};\mathrm{Stud}}^{(n)}$ that are based on the FvML studentized statistics $Q_{f_{\exp};\mathrm{Stud}}^{(n)}$, and (iii) the rank-based tests $\not\!\phi_{K_{\mathrm{FvML}(1)}}^{(n)}$, $\not\!\phi_{K_{\mathrm{FvML}(3)}}^{(n)}$, $\not\!\phi_{K_{\mathrm{LIN}(2)}}^{(n)}$, and $\not\!\phi_{K_{\mathrm{LIN}(2)}}^{(n)}$ that are Le Cam optimal at FvML(1), FvML(3), Lin(1.1), and Lin(2) distributions, respectively. The resulting rejection frequencies are provided in Tables 1 and 2, for $\mathcal{H}_0 : \boldsymbol{\theta} = \boldsymbol{\theta}_0$ and for $\mathcal{H}_0 : \boldsymbol{\theta} \in \mathcal{S}^{k-1} \cap \mathcal{M}(\boldsymbol{\Upsilon})$, respectively.

These empirical results perfectly agree with the asymptotic theory : the parametric tests $\phi_{f_{\exp,3}}^{(n)}$ are the most powerful ones at the FvML(3) distribution, but their null size deviates quite much from the target size $\alpha = 5\,\%$ away from the FvML(3). The studentized parametric tests $\phi_{f_{\exp};\mathrm{Stud}}^{(n)}$, on the contrary, show a null behaviour that is satisfactory under all distributions considered. They also dominate their rank-based competitors under FvML densities, which is in line with the fact that they are optimal in the class of FvML densities. Outside this class, however, the proposed rank tests are more powerful than the studentized tests, which translates their better efficiency-robustness. The null behaviour of the proposed rank tests is very satisfactory, and their optimality under correctly specified angular densities is confirmed.

Table 1 Rejection frequencies (out of $M = 10{,}000$ replications), under the null $\mathscr{H}_0 : \theta = \theta_0$ ($\omega = 0$) and increasingly severe alternatives ($\omega = 1, 2, 3$), of the parametric FvML(3)-test ($\phi^{(n)}_{f_{\exp,3}}$), the studentized FvML-test ($\phi^{(n)}_{f_{\exp};\mathrm{Stud}}$), and of the rank tests achieving optimality at FvML(1), FvML(3), LIN(1.1), and LIN(2) densities ($\phi^{(n)}_{K_{\mathrm{FvML}(1)}}$, $\phi^{(n)}_{K_{\mathrm{FvML}(3)}}$, $\phi^{(n)}_{K_{\mathrm{LIN}(1.1)}}$, and $\phi^{(n)}_{K_{\mathrm{LIN}(2)}}$, respectively)

Underlying density	Test	ω			
		0	1	2	3
FvML(1)	$\phi^{(n)}_{f_{\exp,3}}$	0.1162	0.1572	0.2673	0.4493
	$\phi^{(n)}_{f_{\exp};\mathrm{Stud}}$	0.0486	0.0733	0.1502	0.2867
	$\phi^{(n)}_{K_{\mathrm{FvML}(1)}}$	0.0494	0.0732	0.1516	0.2891
	$\phi^{(n)}_{K_{\mathrm{FvML}(3)}}$	0.0525	0.0720	0.1467	0.2743
	$\phi^{(n)}_{K_{\mathrm{LIN}(1.1)}}$	0.0504	0.0662	0.1237	0.2248
	$\phi^{(n)}_{K_{\mathrm{LIN}(2)}}$	0.0527	0.0738	0.1505	0.2890
FvML(3)	$\phi^{(n)}_{f_{\exp,3}}$	0.0528	0.2250	0.7104	0.9719
	$\phi^{(n)}_{f_{\exp};\mathrm{Stud}}$	0.0536	0.2239	0.7062	0.9701
	$\phi^{(n)}_{K_{\mathrm{FvML}(1)}}$	0.0530	0.2177	0.6828	0.9615
	$\phi^{(n)}_{K_{\mathrm{FvML}(3)}}$	0.0541	0.2244	0.7056	0.9695
	$\phi^{(n)}_{K_{\mathrm{LIN}(1.1)}}$	0.0509	0.2066	0.6688	0.9598
	$\phi^{(n)}_{K_{\mathrm{LIN}(2)}}$	0.0538	0.2220	0.7025	0.9674
LIN(1.1)	$\phi^{(n)}_{f_{\exp,3}}$	0.1290	0.1661	0.2729	0.4556
	$\phi^{(n)}_{f_{\exp};\mathrm{Stud}}$	0.0498	0.0668	0.1372	0.2749
	$\phi^{(n)}_{K_{\mathrm{FvML}(1)}}$	0.0493	0.0676	0.1396	0.2747
	$\phi^{(n)}_{K_{\mathrm{FvML}(3)}}$	0.0496	0.0723	0.1588	0.3245
	$\phi^{(n)}_{K_{\mathrm{LIN}(1.1)}}$	0.0496	0.0711	0.1608	0.3359
	$\phi^{(n)}_{K_{\mathrm{LIN}(2)}}$	0.0501	0.0698	0.1499	0.3001
LIN(2)	$\phi^{(n)}_{f_{\exp,3}}$	0.1384	0.1496	0.1727	0.2284
	$\phi^{(n)}_{f_{\exp};\mathrm{Stud}}$	0.0526	0.0579	0.0741	0.1074
	$\phi^{(n)}_{K_{\mathrm{FvML}(1)}}$	0.0545	0.0585	0.0769	0.1100
	$\phi^{(n)}_{K_{\mathrm{FvML}(3)}}$	0.0542	0.0600	0.0773	0.1069
	$\phi^{(n)}_{K_{\mathrm{LIN}(1.1)}}$	0.0540	0.0600	0.0737	0.0968
	$\phi^{(n)}_{K_{\mathrm{LIN}(2)}}$	0.0540	0.0610	0.0781	0.1110

The sample size is $n = 250$ and the nominal level is 5 %; see Sect. 6 for further details

Table 2 Rejection frequencies (out of $M = 10{,}000$ replications), under the null $\mathscr{H}_0 : \theta \in \mathscr{S}^{k-1} \cap \mathscr{M}(\boldsymbol{\Upsilon})$ ($\omega = 0$) and increasingly severe alternatives ($\omega = 1, 2, 3$), of the parametric FvML(3)-test ($\phi^{(n)}_{f_{\exp,3}}$), the studentized FvML-test ($\phi^{(n)}_{f_{\exp};\mathrm{Stud}}$), and of the rank tests achieving optimality at FvML(1), FvML(3), LIN(1.1), and LIN(2) densities ($\underline{\phi}^{(n)}_{K_{\mathrm{FvML}(1)}}$, $\underline{\phi}^{(n)}_{K_{\mathrm{FvML}(3)}}$, $\underline{\phi}^{(n)}_{K_{\mathrm{LIN}(1.1)}}$, and $\underline{\phi}^{(n)}_{K_{\mathrm{LIN}(2)}}$, respectively)

Underlying density	Test	ω			
		0	1	2	3
FvML(1)	$\phi^{(n)}_{f_{\exp,3}}$	0.1027	0.1391	0.2902	0.4998
	$\phi^{(n)}_{f_{\exp};\mathrm{Stud}}$	0.0543	0.0794	0.1922	0.3783
	$\underline{\phi}^{(n)}_{K_{\mathrm{FvML}(1)}}$	0.0541	0.1943	0.1516	0.3786
	$\underline{\phi}^{(n)}_{K_{\mathrm{FvML}(3)}}$	0.0537	0.0789	0.1835	0.3568
	$\underline{\phi}^{(n)}_{K_{\mathrm{LIN}(1.1)}}$	0.0524	0.0712	0.1560	0.2997
	$\underline{\phi}^{(n)}_{K_{\mathrm{LIN}(2)}}$	0.0532	0.0792	0.1914	0.3695
FvML(3)	$\phi^{(n)}_{f_{\exp,3}}$	0.0464	0.2888	0.7955	0.9889
	$\phi^{(n)}_{f_{\exp};\mathrm{Stud}}$	0.0460	0.2884	0.7937	0.9891
	$\underline{\phi}^{(n)}_{K_{\mathrm{FvML}(1)}}$	0.0462	0.2725	0.7686	0.9839
	$\underline{\phi}^{(n)}_{K_{\mathrm{FvML}(3)}}$	0.0467	0.2873	0.7930	0.9897
	$\underline{\phi}^{(n)}_{K_{\mathrm{LIN}(1.1)}}$	0.0486	0.2632	0.7635	0.9849
	$\underline{\phi}^{(n)}_{K_{\mathrm{LIN}(2)}}$	0.0454	0.2831	0.7862	0.9883
LIN(1.1)	$\phi^{(n)}_{f_{\exp,3}}$	0.1080	0.1598	0.2922	0.4763
	$\phi^{(n)}_{f_{\exp};\mathrm{Stud}}$	0.0474	0.0798	0.1808	0.3363
	$\underline{\phi}^{(n)}_{K_{\mathrm{FvML}(1)}}$	0.0490	0.0793	0.1801	0.3345
	$\underline{\phi}^{(n)}_{K_{\mathrm{FvML}(3)}}$	0.0480	0.0876	0.2114	0.3920
	$\underline{\phi}^{(n)}_{K_{\mathrm{LIN}(1.1)}}$	0.0476	0.0896	0.2173	0.4075
	$\underline{\phi}^{(n)}_{K_{\mathrm{LIN}(2)}}$	0.0498	0.0841	0.1968	0.3653
LIN(2)	$\phi^{(n)}_{f_{\exp,3}}$	0.1119	0.1220	0.1659	0.2328
	$\phi^{(n)}_{f_{\exp};\mathrm{Stud}}$	0.0488	0.0558	0.0875	0.1366
	$\underline{\phi}^{(n)}_{K_{\mathrm{FvML}(1)}}$	0.0502	0.0567	0.0901	0.1391
	$\underline{\phi}^{(n)}_{K_{\mathrm{FvML}(3)}}$	0.0523	0.0613	0.0895	0.1386
	$\underline{\phi}^{(n)}_{K_{\mathrm{LIN}(1.1)}}$	0.0521	0.0594	0.0838	0.1224
	$\underline{\phi}^{(n)}_{K_{\mathrm{LIN}(2)}}$	0.0516	0.0599	0.0915	0.1422

The sample size is $n = 250$ and the nominal level is 5 %; see Sect. 6 for further details

Acknowledgments The authors are grateful to two anonymous referees and the Editor for their careful reading and insightful comments that allowed to improve the original manuscript.

Appendix

In this Appendix, we prove Theorems 3 and 4.

Proof of Theorem 3. (i) Recalling that $\hat{\theta}$ is an arbitrary consistent estimator of θ, we have that under $\mathrm{P}_{\theta,g}^{(n)}$,

$$\hat{\mathscr{L}}_k(\hat{\theta}) - \hat{\mathscr{L}}_k(\theta) = \frac{1}{n} \sum_{i=1}^{n} \left\{ (\mathbf{X}_i'\theta)^2 - (\mathbf{X}_i'\hat{\theta})^2 \right\}$$

$$= \frac{1}{n} \sum_{i=1}^{n} \left\{ \mathbf{X}_i'(\theta - \hat{\theta})\mathbf{X}_i'(\theta + \hat{\theta}) \right\} = (\theta - \hat{\theta})\left\{ \frac{1}{n} \sum_{i=1}^{n} \mathbf{X}_i \mathbf{X}_i' \right\}(\theta + \hat{\theta}) = o_{\mathrm{P}}(1)$$

as $n \to \infty$, so that $\hat{\mathscr{L}}_k(\hat{\theta})$ is a consistent estimator of $\mathscr{L}_k(g) = 1 - \mathrm{E}_{\theta,g}^{(n)}[(\mathbf{X}_i'\theta)^2]$. Consequently, $Q_{f_{\exp};\mathrm{Stud}}^{(n)} - Q_{\theta,g}^{(n)}$ is $o_{\mathrm{P}}(1)$ as $n \to \infty$ under $\mathrm{P}_{\theta,g}^{(n)}$, with

$$Q_{\theta,g}^{(n)} := \frac{n(k-1)}{\mathscr{L}_k(g)} \bar{\mathbf{X}}'(\mathbf{I}_k - \boldsymbol{\Upsilon}(\boldsymbol{\Upsilon}'\boldsymbol{\Upsilon})^{-1}\boldsymbol{\Upsilon}')\bar{\mathbf{X}}$$

$$= \frac{n(k-1)}{\mathscr{L}_k(g)} \bar{\mathbf{X}}'(\mathbf{I}_k - \boldsymbol{\Upsilon}(\boldsymbol{\Upsilon}'\boldsymbol{\Upsilon})^{-1}\boldsymbol{\Upsilon}')(\mathbf{I}_k - \theta\theta')\bar{\mathbf{X}} = \mathbf{Y}'(\mathbf{I}_k - \boldsymbol{\Upsilon}(\boldsymbol{\Upsilon}'\boldsymbol{\Upsilon})^{-1}\boldsymbol{\Upsilon}')\mathbf{Y},$$

where we used the fact that $\left(\mathbf{I}_k - \boldsymbol{\Upsilon}(\boldsymbol{\Upsilon}'\boldsymbol{\Upsilon})^{-1}\boldsymbol{\Upsilon}'\right)(\mathbf{I} - \theta\theta') = \mathbf{I}_k - \boldsymbol{\Upsilon}(\boldsymbol{\Upsilon}'\boldsymbol{\Upsilon})^{-1}\boldsymbol{\Upsilon}'$ and where we let $\mathbf{Y} := \sqrt{n(k-1)/\mathscr{L}_k(g)}(\mathbf{I}_k - \theta\theta')\bar{\mathbf{X}}$. Under $\mathrm{P}_{\theta,g}^{(n)}$,

$$\sqrt{n}(\mathbf{I} - \theta\theta')\bar{\mathbf{X}} = \frac{1}{\sqrt{n}} \sum_{i=1}^{n} \left\{ \sqrt{1 - (\mathbf{X}_i'\theta)^2} \, \mathbf{S}_i(\theta) \right\} \tag{21}$$

is asymptotically normal with mean zero and covariance matrix $\mathscr{L}_k(g)(\mathbf{I}_k - \theta\theta')/(k-1)$, so that $\mathbf{Y}$, under the same, is asymptotically normal with mean zero and covariance matrix $\mathbf{I}_k - \theta\theta'$. By using again the fact that $\left(\mathbf{I}_k - \boldsymbol{\Upsilon}(\boldsymbol{\Upsilon}'\boldsymbol{\Upsilon})^{-1}\boldsymbol{\Upsilon}'\right)(\mathbf{I} - \theta\theta') = \mathbf{I}_k - \boldsymbol{\Upsilon}(\boldsymbol{\Upsilon}'\boldsymbol{\Upsilon})^{-1}\boldsymbol{\Upsilon}'$ and by noting that $\mathrm{tr}[\mathbf{I}_k - \boldsymbol{\Upsilon}(\boldsymbol{\Upsilon}'\boldsymbol{\Upsilon})^{-1}\boldsymbol{\Upsilon}'] = k - s$, it is easy to check that Theorem 9.2.1 in Rao and Mitra (1971) provides the result.

(ii) Le Cam's third lemma implies that under $\mathrm{P}_{\theta+n^{-1/2}\boldsymbol{\tau}^{(n)},g}^{(n)}$, the random vector in (21) is asymptotically normal with mean

$$\lim_{n\to\infty} \mathrm{Cov}_{\theta,g}\left[\sqrt{n}(\mathbf{I} - \theta\theta')\bar{\mathbf{X}}, \Delta_{\theta,g}^{(n)} \right]\boldsymbol{\tau}^{(n)} \tag{22}$$

and covariance matrix $\mathscr{L}_k(g)(\mathbf{I}_k - \boldsymbol{\theta\theta}')/(k-1)$. Using (3) and integrating by parts yields

$$E_{\boldsymbol{\theta},g}\left[(1 - (\mathbf{X}_1'\boldsymbol{\theta})^2)\varphi_g(\mathbf{X}_1'\boldsymbol{\theta})\right] = \int_{-1}^{1}(1 - t^2)\varphi_g(t)\tilde{g}(t)\,dt = k - 1,$$

so that (22) can be rewritten as $E_{\boldsymbol{\theta},g}\left[(1 - (\mathbf{X}_1'\boldsymbol{\theta})^2)\varphi_g(\mathbf{X}_1'\boldsymbol{\theta})\right]E_{\boldsymbol{\theta},g}\left[S_1(\boldsymbol{\theta})(S_1(\boldsymbol{\theta}))'\right]\boldsymbol{\tau} = (\mathbf{I}_k - \boldsymbol{\theta\theta}')\boldsymbol{\tau} = \boldsymbol{\tau}$. Therefore, $\mathbf{Y}$, under $P^{(n)}_{\boldsymbol{\theta}+n^{-1/2}\boldsymbol{\tau}^{(n)},g}$, is asymptotically normal with mean $\boldsymbol{\mu} := \sqrt{(k-1)/\mathscr{L}_k(g)}\,\boldsymbol{\tau}$ and covariance matrix $\mathbf{I}_k - \boldsymbol{\theta\theta}'$. From contiguity, we still have that $Q^{(n)}_{f_{\exp};\mathrm{Stud}} - \mathbf{Y}'(\mathbf{I}_k - \boldsymbol{\Upsilon}(\boldsymbol{\Upsilon}'\boldsymbol{\Upsilon})^{-1}\boldsymbol{\Upsilon}')\mathbf{Y}$ is $o_{\mathrm{P}}(1)$ under $P^{(n)}_{\boldsymbol{\theta}+n^{-1/2}\boldsymbol{\tau}^{(n)},g}$. Theorem 9.2.1 in Rao and Mitra (1971) then shows that under this sequence of probability measures, $Q^{(n)}_{f_{\exp};\mathrm{Stud}}$ is asymptotically χ^2_{k-s} with non-centrality parameter $\boldsymbol{\mu}'(\mathbf{I}_k - \boldsymbol{\Upsilon}(\boldsymbol{\Upsilon}'\boldsymbol{\Upsilon})^{-1}\boldsymbol{\Upsilon}')\boldsymbol{\mu}$, which establishes the result.

(iii) This directly follows from the asymptotic null distribution given in (i) and the classical Helly–Bray theorem.

(iv) Fix $\kappa > 0$. Then, it follows from Part (i) of the proof that under $P^{(n)}_{\boldsymbol{\theta},f_{\exp,\kappa}}$, with $\boldsymbol{\theta} \in \mathscr{S}^{k-1} \cap \mathscr{M}(\boldsymbol{\Upsilon})$, $Q^{(n)}_{f_{\exp};\mathrm{Stud}}$ is asymptotically equivalent in probability to

$$\begin{aligned}
Q^{(n)}_{\boldsymbol{\theta};f_{\exp,\kappa}} &= \frac{n(k-1)}{\mathscr{L}_k(f_{\exp,\kappa})}\bar{\mathbf{X}}'(\mathbf{I}_k - \boldsymbol{\Upsilon}(\boldsymbol{\Upsilon}'\boldsymbol{\Upsilon})^{-1}\boldsymbol{\Upsilon}')\bar{\mathbf{X}} \\
&= \frac{n\kappa^2(k-1)}{\mathscr{J}_k(f_{\exp,\kappa})}\bar{\mathbf{X}}'(\mathbf{I}_k - \boldsymbol{\Upsilon}(\boldsymbol{\Upsilon}'\boldsymbol{\Upsilon})^{-1}\boldsymbol{\Upsilon}')\bar{\mathbf{X}},
\end{aligned}$$

which is the FvML(κ)-most stringent statistic we derived in (18). $\qquad\square$

In Theorem 3(ii), we assumed that $g \in \mathscr{F}_{\mathrm{ULAN}}$ to show, through Le Cam's third lemma, that $\sqrt{n}(\mathbf{I} - \boldsymbol{\theta\theta}')\bar{\mathbf{X}}$, under $P^{(n)}_{\boldsymbol{\theta}+n^{-1/2}\boldsymbol{\tau}^{(n)},g}$, is asymptotically normal with mean $\boldsymbol{\tau}$ and covariance matrix $\mathscr{L}_k(g)(\mathbf{I}_k - \boldsymbol{\theta\theta}')/(k-1)$. Actually, the result still holds for $g \in \mathscr{F}$, as it can be shown that as $n \to \infty$ under $P^{(n)}_{\boldsymbol{\theta}+n^{-1/2}\boldsymbol{\tau}^{(n)},g}$,

$$\sqrt{n}(\mathbf{I} - \boldsymbol{\theta\theta}')\bar{\mathbf{X}} = \mathbf{M}^{(n)} + (\mathbf{I}_k + \boldsymbol{\theta\theta}')\boldsymbol{\tau} + o_{\mathrm{P}}(1) = \mathbf{M}^{(n)} + \boldsymbol{\tau} + o_{\mathrm{P}}(1),$$

where $\mathbf{M}^{(n)} := \sqrt{n}(\mathbf{I} - (\boldsymbol{\theta}+n^{-1/2}\boldsymbol{\tau}^{(n)})(\boldsymbol{\theta}+n^{-1/2}\boldsymbol{\tau}^{(n)})')\bar{\mathbf{X}}$, under the same, is clearly asymptotically normal with mean zero and covariance matrix $\mathscr{L}_k(g)(\mathbf{I}_k - \boldsymbol{\theta\theta}')/(k-1)$.

Proof of Theorem 4. (i)–(ii) First note that since $\boldsymbol{\theta}'\boldsymbol{\tau}^{(n)} = O(n^{-1/2})$, Proposition 2(iv) rewrites

$$\underset{\sim}{\boldsymbol{\Delta}}^{(n)}_{\boldsymbol{\theta}+n^{-1/2}\boldsymbol{\tau}^{(n)},K} = \underset{\sim}{\boldsymbol{\Delta}}^{(n)}_{\boldsymbol{\theta},K} - \frac{\mathscr{J}(K,g)}{k-1}\boldsymbol{\tau}^{(n)} + o_{\mathrm{P}}(1) \qquad (23)$$

as $n \to \infty$ under $P_{\theta,g}^{(n)}$. Since Assumption (B) holds, Lemma 4.4 in Kreiss (1987) allows to replace in (23) the deterministic quantity $\tau^{(n)}$ with the random one $\sqrt{n}(\hat{\theta} - \theta)$, which yields

$$\underset{\sim}{\Delta}_{\hat{\theta},K}^{(n)} = \underset{\sim}{\Delta}_{\theta,K}^{(n)} - \frac{\mathcal{J}(K,g)}{k-1}\sqrt{n}(\hat{\theta} - \theta) + o_{\mathrm{P}}(1),$$

as $n \to \infty$, under $P_{\theta,g}^{(n)}$. This, jointly with Assumption (B)(iii) (which implies that $(\mathbf{I}_k - \boldsymbol{\Upsilon}(\boldsymbol{\Upsilon}'\boldsymbol{\Upsilon})^{-1}\boldsymbol{\Upsilon}')\hat{\theta} = 0$ almost surely), entails that under $P_{\theta,g}^{(n)}$, with $\theta \in \mathcal{M}(\boldsymbol{\Upsilon})$,

$$\left(\mathbf{I}_k - \boldsymbol{\Upsilon}(\boldsymbol{\Upsilon}'\boldsymbol{\Upsilon})^{-1}\boldsymbol{\Upsilon}'\right)\underset{\sim}{\Delta}_{\hat{\theta},K}^{(n)} = \left(\mathbf{I}_k - \boldsymbol{\Upsilon}(\boldsymbol{\Upsilon}'\boldsymbol{\Upsilon})^{-1}\boldsymbol{\Upsilon}'\right)\underset{\sim}{\Delta}_{\theta,K}^{(n)} + o_{\mathrm{P}}(1),$$

as $n \to \infty$. It follows that $\underset{\sim}{Q}_{K}^{(n)} = \underset{\sim}{Q}_{\theta,K}^{(n)} + o_{\mathrm{P}}(1)$ as $n \to \infty$ under $P_{\theta,g}^{(n)}$, with $\theta \in \mathcal{M}(\boldsymbol{\Upsilon})$, hence also under sequences of local alternatives. The results in (i)–(ii) then follow, as in the proof of Theorem 3(i)–(ii), from Theorem 9.2.1 in Rao and Mitra (1971) and Proposition 2(ii)–(iii) (recall that $(\mathbf{I}_k - \boldsymbol{\Upsilon}(\boldsymbol{\Upsilon}'\boldsymbol{\Upsilon})^{-1}\boldsymbol{\Upsilon}')(\mathbf{I} - \theta\theta') = \mathbf{I}_k - \boldsymbol{\Upsilon}(\boldsymbol{\Upsilon}'\boldsymbol{\Upsilon})^{-1}\boldsymbol{\Upsilon}')$.

(iii) As in the proof of Theorem 3(iii), this is a direct consequence of Part (i) of the result and the classical Helly–Bray theorem.

(iv) Then, under $P_{\theta,f}^{(n)}$, with $\theta \in \mathcal{S}^{k-1} \cap \mathcal{M}(\boldsymbol{\Upsilon})$, $\underset{\sim}{Q}_{K_f}^{(n)} = \underset{\sim}{Q}_{\theta,K_f}^{(n)} + o_{\mathrm{P}}(1)$ as $n \to \infty$. Now, Proposition 2(i) entails that under the same sequence of hypotheses, $\underset{\sim}{Q}_{\theta,K_f}^{(n)}$ is asymptotically equivalent in probability to

$$Q_{\theta,K_f}^{(n)} = \frac{k-1}{\mathcal{J}_k(K_f)}(\Delta_{\theta,K_f}^{(n)})'\left(\mathbf{I}_k - \boldsymbol{\Upsilon}(\boldsymbol{\Upsilon}'\boldsymbol{\Upsilon})^{-1}\boldsymbol{\Upsilon}'\right)\Delta_{\theta,K_f}^{(n)},$$

which coincides with the f-most stringent statistic in (17). The result follows. $\square$

References

Bowers, J. A., Morton, I. D., & Mould, G. I. (2000). Directional statistics of the wind and waves. *Applied Ocean Research, 22,* 401–412.

Briggs, M. S. (1993). Dipole and quadrupole tests of the isotropy of gamma-ray burst locations. *Astrophysical Journal, 407,* 126–134.

Chang, T. (2004). Spatial statistics. *Statistical Science, 19,* 624–635.

Clark, R. M. (1983). Estimation of parameters in the marginal Fisher distribution. *Australian Journal of Statistics, 25,* 227–237.

Fisher, N. I. (1953). Dispersion on a sphere. *Proceedings of the Royal Society of London A, 217,* 295–305.

Fisher, N. I. (1987). Problems with the current definitions of the standard deviation of wind direction. *Journal of Climate and Applied Meteorology, 26,* 1522–1529.

Fisher, N. I. (1989). Smoothing a sample of circular data. *Journal of Structural Geology, 11*, 775–778.

Fisher, N. I., Lewis, T., & Embleton, B. J. J. (1987). *Statistical analysis of spherical data.* Cambridge: Cambridge University Press.

Fujikoshi, Y., & Watamori, Y. (1992). Tests for the mean direction of the Langevin distribution with large concentration parameter. *Journal of Multivariate Analysis, 42*, 210–225.

Hallin, M., & Werker, B. J. (2003). Semi-parametric efficiency, distribution-freeness and invariance. *Bernoulli, 9*, 137–165.

Hallin, M., Paindaveine, D., & Verdebout, T. (2010). Optimal rank-based testing for principal components. *Annals of Statistics, 38*, 3245–3299.

Jupp, P., & Mardia, K. (1989). A unified view of the theory of directional statistics. *International Statistical Review, 57*, 261–294.

Kreiss, J. (1987). On adaptive estimation in stationary ARMA processes. *Annals of Statistics, 15*, 112–133.

Leong, P., & Carlile, S. (1998). Methods for spherical data analysis and visualization. *Journal of Neuroscience Methods, 80*, 191–200.

Ley, C., Swan, Y., Thiam, B., & Verdebout, T. (2013). Optimal R-estimation of a spherical location. *Statistica Sinica, 23*, 305–333.

Mardia, K. (1975). Statistics of directional data (with discussion). *Journal of the Royal Statistical Society: Series B, 37*, 349–393.

Mardia, K. V., & Edwards, R. (1982). Weighted distributions and rotating caps. *Biometrika, 69*, 323–330.

Mardia, K. V., & Jupp, P. (2000). *Directional statistics.* New York: Wiley.

Merrifield, A. (2006). An investigation of mathematical models for animal group movement, using classical and statistical approaches. *Ph.D. thesis.* The University of Sydney, Sydney, Australia.

Rao, C., & Mitra, S. (1971). *Generalized inverses of matrices and its apllications.* New York: Wiley.

Saw, J. G. (1978). A family of distributions on the m-sphere and some hypothesis tests. *Biometrika, 65*, 69–73.

Storetvedt, K. M., & Scheidegger, A. E. (1992). Orthogonal joint structures in the Bergen area, southwest Norway, and their regional significance. *Physics of the Earth and Planetary Interiors, 73*, 255–263.

Tsai, M., & Sen, P. (2007). Locally best rotation-invariant rank tests for modal location. *Journal of Multivariate Analysis, 98*, 1160–1179.

Watamori, Y. (1992). Tests for a given linear structure of the mean direction of the Langevin distribution. *Annals of the Institute of Statistical Mathematics, 44*, 147–156.

Watson, G. (1983). *Statistics on spheres.* New York: Wiley.

Some Extensions of Singular Mixture Copulas

Dominic Lauterbach and Dietmar Pfeifer

Abstract In Lauterbach (ZVersWiss, 101(5), 605–619, 2012) and Lauterbach and Pfeifer (Copulae in mathematical and quantitative finance, Springer, Dordrecht, 2013) the family of Singular Mixture Copulas was introduced. We present and discuss two extensions of these copulas. Both extensions are based on an approach introduced by Khoudraji (Contributions à l'étude des copules et à la modélisation des valeurs extrêmes bivariées. Ph.D. thesis, 1995). We study the dependence properties of the constructed copulas and show that the resulting copulas possess differing upper and lower tail dependence coefficients.

1 Introduction

Copulas are an effective and versatile tool for studying and modeling multivariate dependence. The term copula was first used in a mathematical sense by Sklar (1959), although the history of copulas can be traced back to Fréchet (1951) and Hoeffding (1940). In the 1970s, several authors rediscovered copulas under different names, among them Deheuvels (1978) who refered to them as dependence functions. Since then copulas have gained popularity in theory as well as in applications, see, e.g., Cherubini et al. (2004); Embrechts et al. (2003); Genest and MacKay (1986); Joe (1997); McNeil et al. (2005); Nelsen (2006); Wolff (1977).

In Durante and Sempi (2010) it was suggested that the "search for families of copulas having properties desirable for specific applications" should be one of the directions of future investigation in copula theory. It was also mentioned that these families of copulas should exhibit "different asymmetries, non-exchangeable copulas, copulas with different tail behavior, etc." As a contribution to this field of research, Lauterbach (2012) and Lauterbach and Pfeifer (2013) introduced a family of copulas—Singular Mixture Copulas. These copulas were constructed via a

D. Lauterbach · D. Pfeifer (✉)
Institut für Mathematik, Carl von Ossietzky Universität Oldenburg, Oldenburg, Germany
e-mail: dietmar.pfeifer@uni-oldenburg.de

D. Lauterbach
e-mail: dominic.lauterbach@uni-oldenburg.de

© Springer International Publishing Switzerland 2015

M. Hallin et al. (eds.), *Mathematical Statistics and Limit Theorems*,
DOI 10.1007/978-3-319-12442-1_15

convex sum[1] of certain singular copulas. It was also shown in Lauterbach (2012) that these copulas can be used to model the dependence between the flood levels of gauging stations along the German North Sea coast. In this paper, we want to present an extension of Singular Mixture Copulas and thus overcome some drawbacks of the aforementioned construction, such as the restricted support of Singular Mixture Copulas. To this end, we make use of an approach that was first studied by Khoudraji (1995) (see also Genest et al. (1998); McNeil et al. (2005)): Let C be an arbitrary copula, then C can be extended to a parametric family of copulas $C_{\alpha,\beta}$ by setting

$$C_{\alpha,\beta}(u, v) = u^{1-\alpha} v^{1-\beta} C(u^\alpha, v^\beta),$$

where $0 \leq \alpha, \beta \leq 1$. We study the resulting copulas and take a look at their mathematical properties, especially with respect to dependence.

This paper is organized as follows. In Sects. 2 and 3, we summarize the construction and some important properties of Singular Mixture Copulas. In Sects. 4 and 5, we present two extensions of Singular Mixture Copulas that are based on Khoudraji's device mentioned above.

2 Singular Copulas

In Lauterbach (2012); Lauterbach and Pfeifer (2013) we introduced a method of constructing singular copulas. This construction uses two distribution functions F and G on $[0, 1]$ which fulfill the equation

$$\alpha F(x) + (1 - \alpha)G(x) = x \tag{1}$$

for all $x \in [0, 1]$, where α is a constant in $(0, 1)$. The function G is given by

$$G(x) = \frac{x - \alpha F(x)}{1 - \alpha}. \tag{2}$$

Let X be a random variable with a continuous uniform distribution over $[0, 1]$, and let I be a random variable, independent of X, with a binomial $B(1, \alpha)$-distribution. Define the random variable Y via

$$Y := I \cdot F^{-1}(X) + (1 - I) \cdot G^{-1}(X). \tag{3}$$

Then the random variable Y also follows a continuous uniform distribution over $[0, 1]$. The distribution function of (X, Y) is the singular copula given by

$$C_{XY}(x, y) = \alpha \min(x, F(y)) + (1 - \alpha) \min(x, G(y)).$$

---

[1] See Nelsen (2006), Sect. 3.2.

The following lemma gives necessary and sufficient conditions for F to guarantee that G is also a distribution function.

Lemma 2.1 *Let F be an absolutely continuous distribution function on $[0, 1]$. Then G given by (2) is an absolutely continuous distribution function on $[0, 1]$ if and only if $F'(x) \leq \frac{1}{\alpha}$ for all $x \in [0, 1]$.*

Proof From $F(0) = 0$ and $F(1) = 1$, it follows immediately that $G(0) = 0$ and $G(1) = 1$. From Eq. (2), we have

$$G'(x) = \frac{1 - \alpha F'(x)}{1 - \alpha}, \tag{4}$$

so that $G'(x) \geq 0 \Leftrightarrow F'(x) \leq \frac{1}{\alpha}$, which completes the proof. $\qquad \square$

The assumption of absolute continuity of F is essential, as the following example shows.

Example 2.1 Let F be the distribution function of the Cantor distribution. This function is also known as the Cantor function.[2] Then F is an almost everywhere differentiable distribution function on $[0, 1]$ with $F'(x) = 0 \leq \frac{1}{\alpha}$ for all $x \in [0, 1]$ and any $\alpha \in (0, 1)$. However, F is not absolutely continuous. It holds that $F(x) = \frac{1}{2}$ for all $x \in [\frac{1}{3}, \frac{2}{3}]$. For $\alpha = \frac{3}{4}$, we can conclude that,

$$G\left(\frac{1}{3}\right) = \frac{\frac{1}{3} - \frac{3}{4} \cdot \frac{1}{2}}{\frac{1}{4}} = \frac{4}{3} - \frac{3}{2} = -\frac{1}{6} < 0.$$

Consequently, the function G is not a distribution function on $[0, 1]$.

We denote the class of functions that fulfill the properties in Lemma 2.1 by $\mathscr{F}_\alpha$, i.e.,

$$\mathscr{F}_\alpha := \{F : [0, 1] \rightarrow [0, 1] \mid F \text{ is abs. cont., } F(0) = 0, F(1) = 1, 0 \leq F'(x) \leq \frac{1}{\alpha}\}.$$

Remark 2.1 The copula C_{XY} is a special case of the construction presented in Durante (2009) for the choice of $f_1 = f_2 = id_{[0,1]}$, $g_1 = F$, $g_2 = G$, $A(u, v) = B(u, v) = \min(u, v)$ and $H(x, y) = \alpha x + (1 - \alpha)y$. In this setting Eq. (1) corresponds to the assumptions in Theorems 1 and 2 of Durante (2009).

The following statements show some properties of the copula C_{XY} which we will use later.

Proposition 2.1 *If α goes to zero then C_{XY} converges to the Fréchet-Hoeffding upper bound M^2.*

Proof For $\alpha = 0$ the function G is given by $G(x) = x$ and therefore C_{XY} is given by $C_{XY}(x, y) = \min(x, G(y)) = \min(x, y) = M^2(x, y)$. $\qquad \square$

---

[2] See Dovgoshey et al. (2006) for more information about the Cantor function.

Theorem 2.1 *For any $\alpha \in (0, 1)$ and any $F \in \mathscr{F}_\alpha$ the copula C_{XY} is positively quadrant dependent.*

Proof We have to show that $C_{XY}(x, y) \geq xy$ holds for all $(x, y) \in [0, 1]^2$. Due to the representation of C_{XY} we consider four cases.

Case 1:

$$C_{XY}(x, y) = \alpha x + (1 - \alpha)x = x \geq xy.$$

Case 2:

$$C_{XY}(x, y) = \alpha F(y) + (1 - \alpha)G(y) = \alpha F(y) + y - \alpha F(y) = y \geq xy.$$

Case 3:

$$C_{XY}(x, y) = \alpha x + (1 - \alpha)G(y) = y + \alpha(x - F(y)).$$

It is easily seen that $y + \alpha(x - F(y)) \geq xy$ is equivalent to

$$\frac{\alpha x - xy}{\alpha} \geq F(y) - \frac{y}{\alpha}. \tag{5}$$

For $y \leq \alpha$ the left-hand side of (5) is positive and the right-hand side is negative, since $F'(y) \leq \frac{1}{\alpha}$ for all $y \in [0, 1]$. For $y > \alpha$ the following holds

$$\frac{\alpha x - xy + y}{\alpha} = \frac{\alpha x + y(1 - x)}{\alpha} > \frac{\alpha x + \alpha(1 - x)}{\alpha} = 1 \geq F(y).$$

Case 4:

$$C_{XY}(x, y) = \alpha F(y) + (1 - \alpha)x = x + \alpha(F(y) - x).$$

It is easily seen that $x + \alpha(F(y) - x) \geq xy$ is equivalent to

$$F(y) \geq x \cdot \frac{y - (1 - \alpha)}{\alpha}. \tag{6}$$

For $y \leq 1 - \alpha$ the right-hand side of (6) is negative, therefore the desired inequality holds. For $y > 1 - \alpha$ we can conclude from $F'(y) \leq \frac{1}{\alpha}$ for all $y \in [0, 1]$ that

$$F(y) \geq \frac{y - (1 - \alpha)}{\alpha} \geq x \cdot \frac{y - (1 - \alpha)}{\alpha}. \qquad \square$$

3 Singular Mixture Copulas

Consider a family $\{F_\omega\} \subset \mathscr{F}_\alpha$ of distribution functions, then—using the construction above—for a fixed ω we can construct the singular copula $\check{C}_\omega$ given by

$$\check{C}_\omega(x, y) = \alpha \min(x, F_\omega(y)) + (1 - \alpha) \min(x, G_\omega(y)).$$

If Ω is a real-valued random variable and $F_\omega \in \mathscr{F}_\alpha$ for all observations ω of Ω, then the convex sum of $\{\check{C}_\omega\}$ is given by

$$\dot{C}(x, y) = \int \check{C}_\omega(x, y) \mathbb{P}^\Omega(d\omega)$$
$$= \alpha \int \min(x, F_\omega(y)) \mathbb{P}^\Omega(d\omega) + (1 - \alpha) \int \min(x, G_\omega(y)) \mathbb{P}^\Omega(d\omega).$$

These copulas were introduced in Lauterbach (2012), Lauterbach and Pfeifer (2013) as Singular Mixture Copulas. A special case considered the family of distribution functions F_ω given by

$$F_\omega(y) = \omega y^2 + (1 - \omega)y \tag{7}$$

with $\omega \in [-1, 1]$. Let $0 < \alpha \leq \frac{1}{2}$ then F_ω is an element of $\mathscr{F}_\alpha$ for all $\omega \in [-1, 1]$. Let Ω be a random variable with values in $[-1, 1]$ then the Singular Mixture Copula resulting from the family $\{F_\omega\}_{\omega \in [-1,1]}$ is given by

$$C_\alpha(x, y) = \mathbb{P}(X \leq x, Y \leq y)$$

$$= \begin{cases} x, & (x, y) \in A_1, \\ x + \alpha\left((x - y)(F_\Omega(\beta) - 1) + (y^2 - y)\int_\beta^1 \omega \mathbb{P}^\Omega(d\omega)\right), & (x, y) \in A_2, \\ \alpha\left((x - y)F_\Omega(\beta) + y + (y^2 - y)\int_\beta^1 \omega \mathbb{P}^\Omega(d\omega)\right) & \\ \quad + (1 - \alpha)(x + (y - x)F_\Omega(b)) + \alpha(y - y^2)\int_{-1}^b \omega \mathbb{P}^\Omega(d\omega), & (x, y) \in A_3, \\ \alpha(x - y)F_\Omega(\beta) + y + \alpha(y - y^2)\int_{-1}^\beta \omega \mathbb{P}^\Omega(d\omega), & (x, y) \in A_4, \\ y, & (x, y) \in A_5, \end{cases}$$
$$\tag{8}$$

where $\beta = \frac{x-y}{y^2-y}$, $b = \beta\frac{\alpha-1}{\alpha}$ and

$$A_1 = \left\{(x, y) \in [0, 1]^2 \mid x < y^2\right\},$$
$$A_2 = \left\{(x, y) \in [0, 1]^2 \mid y^2 \leq x < \frac{-\alpha}{1 - \alpha}(y - y^2) + y\right\},$$

$$A_3 = \left\{ (x, y) \in [0, 1]^2 \,\Big|\, \frac{-\alpha}{1-\alpha}(y - y^2) + y \le x < \frac{\alpha}{1-\alpha}(y - y^2) + y \right\},$$

$$A_4 = \left\{ (x, y) \in [0, 1]^2 \,\Big|\, \frac{\alpha}{1-\alpha}(y - y^2) + y \le x < 2y - y^2 \right\},$$

$$A_5 = \left\{ (x, y) \in [0, 1]^2 \,\Big|\, 2y - y^2 \le x \right\}.$$

The density of the copula is given by

$$
c_\alpha(x, y) = \begin{cases}
0, & (x, y) \in A_1, \\[4pt]
\alpha f_\Omega(\beta) \frac{y^2 - 2xy + x}{(y^2 - y)^2}, & (x, y) \in A_2, \\[4pt]
\frac{y^2 - 2xy + x}{(y^2 - y)^2} \left(\alpha f_\Omega(\beta) + \frac{(1-\alpha)^2}{\alpha} f_\Omega(b) \right), & (x, y) \in A_3, \\[4pt]
\alpha f_\Omega(\beta) \frac{y^2 - 2xy + x}{(y^2 - y)^2}, & (x, y) \in A_4, \\[4pt]
0, & (x, y) \in A_5.
\end{cases}
$$

Depending on the choice of the family of distribution functions, the resulting Singular Mixture Copula can be absolutely continuous, singular, or can possess an absolutely continuous part and a singular part. An example of an absolutely continuous Singular Mixture Copula was given above. If $F_\omega = F$ for all ω, then the resulting Singular Mixture Copula is singular and it is equal to the singular copula presented in Sect. 2. As another example, consider a family of distribution functions given by

$$
\tilde{F}_\omega(x) = \begin{cases}
\frac{1}{2} F_\omega(2x), & 0 \le x \le \frac{1}{2}, \\[4pt]
x, & \frac{1}{2} \le x \le 1,
\end{cases}
$$

where F_ω is given by (7). Then obviously $\tilde{F}_\omega \in \mathscr{F}_\alpha$ for all $\omega \in [-1, 1]$. Figure 1 shows a scatter plot of simulated points from this copula, which we denote with $\tilde{C}$, with a uniform mixing distribution and $\alpha = \frac{1}{2}$.

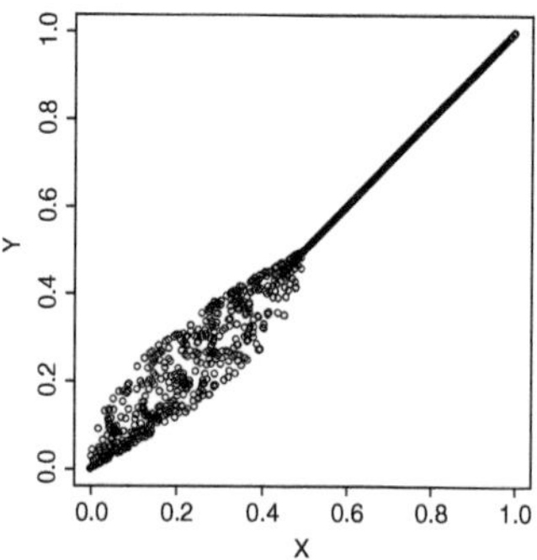

Fig. 1 Scatter plot of simulated points from the copula $\tilde{C}$

The copula $\tilde{C}$ clearly has a singular part and an absolutely continuous part. Moreover, it is the ordinal sum of the (absolutely continuous) Singular Mixture Copula presented above and the Fréchet-Hoeffding upper bound with respect to $\{[0, \frac{1}{2}], [\frac{1}{2}, 1]\}$.

The following propositions show some properties of Singular Mixture Copulas.

Proposition 3.1 *If α goes to zero then $\dot{C}$ converges to M^2.*

Proof The statement follows immediately from Proposition 2.1 and the construction of the copula $\dot{C}$. $\qquad\square$

Proposition 3.2 *The Singular Mixture Copula $\dot{C}$ is positively quadrant dependent.*

Proof In order to proof the statement, we have to show that

$$\dot{C}(x, y) \geq xy \text{ for all } x, y \in [0, 1].$$

By construction of $\dot{C}$ we have

$$\dot{C}(x, y) = \int \check{C}_\omega(x, y) \mathbb{P}^\Omega(d\omega) \geq \int xy \mathbb{P}^\Omega(d\omega) = xy \text{ for all } x, y \in [0, 1],$$

because all copulas $\check{C}_\omega$ are positively quadrant dependent (see Theorem 2.1). $\qquad\square$

Proposition 3.3 *The copula C_α has upper and lower tail dependence given by*

$$\lambda_U = 1 - \alpha \mathbb{E}(|\Omega|) = \lambda_L.$$

Proof The proof is straightforward. $\qquad\square$

4 First Extension

Figure 2 shows that the support of the copula C_α is very restricted. To overcome this problem of Singular Mixture Copulas, we now want to investigate an extension of the copula C_α that is based on the construction presented in Khoudraji (1995). Let a_1 and a_2 be two constants in $(0, 1]$ and let C_α be the Singular Mixture Copula defined in Sect. 3, then C_α^* given by

$$C_\alpha^*(u, v) = u^{1-a_1} v^{1-a_2} C_\alpha(u^{a_1}, v^{a_2})$$

is a copula. Of course, for $a_1 = a_2 = 1$ it holds that $C_\alpha^* = C_\alpha$, so we omit this case.

Remark 4.1 The above construction also works for $a_1 = 0$ and $a_2 = 0$, respectively. However in both cases, the resulting copula is the independence copula. Exemplary for $a_1 = 0$, we receive

$$C_\alpha^*(u, v) = uv^{1-a_2} C_\alpha(u^0, v^{a_2}) = uv^{1-a_2} v^{a_2} = uv.$$

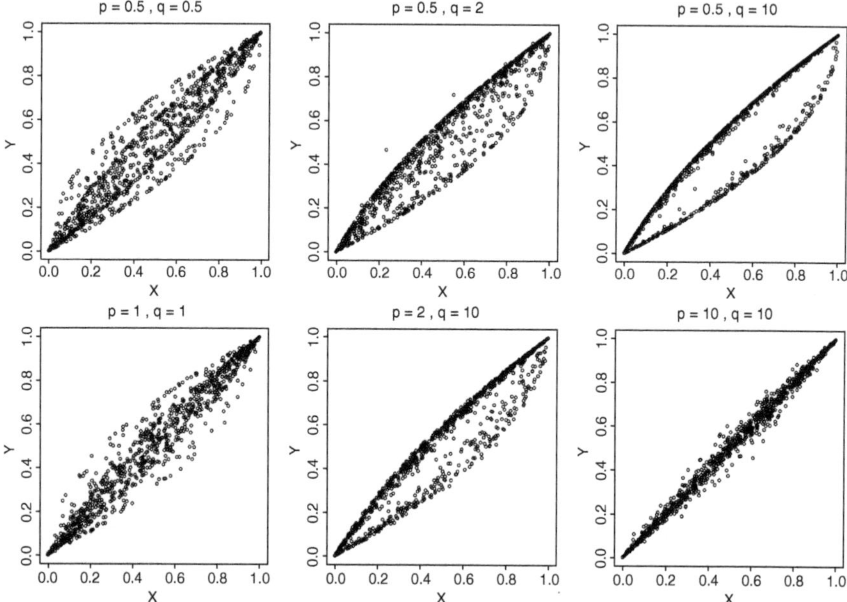

Fig. 2 Scatter plots of simulated points from a Singular Mixture Copula as in (8) for $\alpha = 0.3$ and with generalized beta mixing distribution (with shape parameters p and q)

The tail behavior of the C_α^* copulas differs from that of Singular Mixture Copulas, as the following theorems show.

Theorem 4.1 *For any* $(a_1, a_2) \in (0, 1]^2 \setminus \{1, 1\}$, *the tail dependence coefficient of the copula* C_α^* *(as defined above) equals 0.*

Proof By definiton,

$$\lambda_L(C_\alpha^*) = \lim_{u \searrow 0} \frac{C_\alpha^*(u, u)}{u} = \lim_{u \searrow 0} u^{1-a_1-a_2} C_\alpha(u^{a_1}, u^{a_2}).$$

Due to the piecewise representation of C_α, there are several cases to consider depending on the choice of a_1 and a_2. Instead of determining the choices of a_1 and a_2 that lead to a specific case, we will simply to calculate the above limit for all cases. This approach is more convenient, because—as we will see—most of the limits are the same—so there is no need for a distinction. We will denote the different cases by $A_1, \ldots, A_5$, as in the representation of C_α in Sect. 3.

A_1:

$$\frac{C_\alpha^*(u, u)}{u} = u^{1-a_1-a_2} \cdot u^{a_1} = u^{1-a_2} \longrightarrow 0 \text{ for } a_2 < 1.$$

For $a_2 = 1$ it would hold that $u^{a_1} \geq u^2$ for all $u \in [0, 1]$. Consequently, case A_1 cannot occur when $a_2 = 1$.

A_2:

$$\frac{C_\alpha^*(u, u)}{u} = u^{1-a_1-a_2}\left(u^{a_1} + \alpha\left((u^{a_1} - u^{a_2})\left(F_\Omega(\beta) - 1\right)\right.\right.$$
$$\left.\left. + (u^{2a_2} - u^{a_2})\int_\beta^1 \omega \mathbb{P}^\Omega(d\omega)\right)\right)$$
$$= u^{1-a_2} + \alpha\left((u^{1-a_2} - u^{1-a_1})(F_\Omega(\beta) - 1)\right.$$
$$\left. + (u^{1-a_1+a_2} - u^{1-a_1})\int_\beta^1 \omega \mathbb{P}^\Omega(d\omega)\right)$$
$$\longrightarrow 0 \text{ for } a_2 < 1.$$

When $a_1 = 1$, notice that $\beta = (u - u^{a_2})/(u^{2a_2} - u^{a_2}) = (u^{1-a_2} - 1)/(u^{a_2} - 1) \longrightarrow 1$. For $a_2 = 1$ case A_2 cannot occur: The right-hand derivative of u^{a_1} at $u = 0$ equals infinity, therefore $u^{a_1} > \frac{-\alpha}{1-\alpha}(u - u^2) + u$ for sufficient small (positive) u.

A_3:

$$\frac{C_\alpha^*(u, u)}{u} = u^{1-a_1-a_2}\left(\alpha\left((u^{a_1} - u^{a_2})F_\Omega(\beta) + u^{a_2}\right.\right.$$
$$\left.\left. + (u^{2a_2} - u^{a_2})\int_\beta^1 \omega \mathbb{P}^\Omega(d\omega)\right)\right.$$
$$+ (1-\alpha)\left(u^{a_1} + (u^{a_2} - u^{a_1})F_\Omega(b)\right.$$
$$\left.\left. + \alpha(u^{a_2} - u^{2a_2})\int_{-1}^b \omega \mathbb{P}^\Omega(d\omega)\right)\right)$$
$$= \alpha\left((u^{1-a_2} - u^{1-a_1})F_\Omega(\beta) + u^{1-a_1}\right.$$
$$\left. + (u^{1-a_1+a_2} - u^{1-a_1})\int_\beta^1 \omega \mathbb{P}^\Omega(d\omega)\right)$$
$$+ (1-\alpha)(u^{1-a_2} + (u^{1-a_1} - u^{1-a_2})F_\Omega(b)\right)$$
$$+ \alpha(u^{1-a_1} - u^{1-a_1+a_2})\int_{-1}^b \omega \mathbb{P}^\Omega(d\omega)$$
$$\longrightarrow 0 \text{ for } a_1, a_2 < 1.$$

For $a_1 = 1$ or $a_2 = 1$ case A_3 cannot occur: For $a_1 = 1$ the right-hand derivative of $\frac{-\alpha}{1-\alpha}(u^{a_2} - u^{2a_2}) + u^{a_2}$ at $u = 0$ equals infinity, therefore $\frac{-\alpha}{1-\alpha}(u^{a_2} - u^{2a_2}) + u^{a_2} > u$

for sufficient small (positive) u. For $a_2 = 1$ the right-hand derivative of u^{a_1} at $u = 0$ equals infinity, therefore $u^{a_1} > u + \frac{\alpha}{1-\alpha}(u - u^2)$ for sufficient small (positive) u.

A_4:

$$\frac{C_\alpha^*(u, u)}{u} = u^{1-a_1-a_2}\left(\alpha(u^{a_1} - u^{a_2})F_\Omega(\beta) + u^{a_2}\right.$$

$$\left. + \alpha(u^{a_2} - u^{2a_2})\int_{-1}^{\beta}\omega\mathbb{P}^\Omega(d\omega)\right)$$

$$= \alpha(u^{1-a_2} - u^{1-a_1})F_\Omega(\beta) + u^{1-a_1} + \alpha(u^{1-a_1} - u^{1-a_1+a_2})$$

$$\times \int_{-1}^{\beta}\omega\mathbb{P}^\Omega(d\omega)$$

$$\longrightarrow 0 \text{ for } a_1 < 1,$$

for $a_2 = 1$ notice that $\beta = (u^{a_1} - u)/(u^2 - u) = (u^{a_1-1} - 1)/(u - 1) \longrightarrow -\infty$. For $a_1 = 1$ case A_4 cannot occur: The right-hand derivative of $\frac{\alpha}{1-\alpha}(u^{a_2} - u^{2a_2}) + u^{a_2}$ at $u = 0$ equals infinity, therefore $\frac{\alpha}{1-\alpha}(u^{a_2} - u^{2a_2}) + u^{a_2} > u$ for sufficient small (positive) u.

A_5:

$$\frac{C_\alpha^*(u, u)}{u} = u^{1-a_1-a_2} \cdot u^{a_2} = u^{1-a_1} \longrightarrow 0 \text{ for } a_1 < 1.$$

For $a_1 = 1$ case A_5 cannot occur: The right-hand derivative of $2u^{a_2} - u^{2a_2}$ at $u = 0$ equals infinity, therefore $2u^{a_2} - u^{2a_2} > u$ for sufficient small (positive) u.

Since all limits exist and are equal to zero, the proof is complete. $\qquad\square$

Theorem 4.2 *The upper tail dependence coefficient of the copula C_α^* (as defined above) is given by*

$$\lambda_U(C_\alpha^*) = \begin{cases} a_2, & (a_1, a_2) \in B_1, \\ a_2 + \alpha(a_2 - a_1)(F_\Omega(\gamma) - 1) - \alpha a_2 \int_\gamma^1 \omega\mathbb{P}^\Omega(d\omega), & (a_1, a_2) \in B_2, \\ a_2 + (a_1 - a_2)(\alpha(1 - F_\Omega(\gamma)) + (1 - \alpha)F_\Omega(\delta)) & \\ \quad + \alpha a_2\left(\int_{-1}^\delta \omega\mathbb{P}^\Omega(d\omega) - \int_\gamma^1 \omega\mathbb{P}^\Omega(d\omega)\right), & (a_1, a_2) \in B_3, \\ a_1 + \alpha(a_2 - a_1)F_\Omega(\gamma) + \alpha a_2\int_{-1}^\gamma \omega\mathbb{P}^\Omega(d\omega), & (a_1, a_2) \in B_4, \end{cases}$$

where $\gamma := \frac{a_1-a_2}{a_2}$, $\delta := \gamma \cdot \frac{\alpha-1}{\alpha}$ *and*

$$B_1 = \{(a_1, a_2) \in (0, 1]^2 \mid a_1 > 2a_2\},$$

$$B_2 = \left\{(a_1, a_2) \in (0, 1]^2 \mid \frac{a_2}{1 - \alpha} < a_1 \leq 2a_2\right\},$$

$$B_3 = \left\{ (a_1, a_2) \in (0, 1]^2 \mid a_2 \frac{1 - 2\alpha}{1 - \alpha} < a_1 \le \frac{a_2}{1 - \alpha} \right\},$$

$$B_4 = \left\{ (a_1, a_2) \in (0, 1]^2 \mid a_2 \frac{1 - 2\alpha}{1 - \alpha} \ge a_1 \right\}.$$

Proof The upper tail dependence coefficient of C_α^* is given by

$$\lambda_U(C_\alpha^*) = 2 - \lim_{u \nearrow 1} \frac{1 - C_\alpha^*(u, u)}{1 - u} = 2 - \lim_{u \nearrow 1} \frac{1 - u^{2-a_1-a_2} C_\alpha(u^{a_1}, u^{a_2})}{1 - u}.$$

Due to the piecewise representation of C_α (see Sect. 3), we have to distinguish several cases. It is easily seen that $u^{a_1} < u^{2a_2}$ for $u \in [0, 1)$ if and only if $a_1 > 2a_2$. Therefore, if $(a_1, a_2) \in B_1$ then $C_\alpha(u^{a_1}, u^{a_2}) = u^{a_1}$, and consequently

$$\lambda_U(C_\alpha^*) = 2 - \lim_{u \nearrow 1} \frac{1 - u^{2-a_1-a_2} u^{a_1}}{1 - u} = a_2.$$

As a next step, we have to determine (a_1, a_2) such that $u^{a_1} < -\frac{\alpha}{1-\alpha}(u^{a_2} - u^{2a_2}) + u^{a_2}$ holds for $u \in (1 - \varepsilon, 1)$ for some $\varepsilon > 0$. Since both u^{a_1} and $-\frac{\alpha}{1-\alpha}(u^{a_2} - u^{2a_2}) + u^{a_2}$ are equal to 1 for $u = 1$ this can be done by comparing their derivatives at $u = 1$. It is $(u^{a_1})'(1) = a_1$ and $(-\frac{\alpha}{1-\alpha}(u^{a_2} - u^{2a_2}) + u^{a_2})'(1) = \frac{a_2}{1-\alpha}$, and consequently $u^{a_1} < -\frac{\alpha}{1-\alpha}(u^{a_2} - u^{2a_2}) + u^{a_2}$ holds for $u \in (1 - \varepsilon, 1)$ for some $\varepsilon > 0$ if and only if $\frac{a_2}{1-\alpha} < a_1$. Hence, if $(a_1, a_2) \in B_2$ then $C_\alpha(u^{a_1}, u^{a_2}) = u^{a_1} + \alpha \left((u^{a_1} - u^{a_2})(F_\Omega(\beta) - 1) + (u^{2a_2} - u^{a_2}) \int_\beta^1 \omega \mathbb{P}^\Omega(d\omega) \right)$ where β is given by

$$\beta = \frac{u^{a_1} - u^{a_2}}{u^{2a_2} - u^{a_2}} \quad \text{with} \quad \lim_{u \nearrow 1} \frac{u^{a_1} - u^{a_2}}{u^{2a_2} - u^{a_2}} = \frac{a_1 - a_2}{a_2} = \gamma.$$

Consequently,

$$\lambda_U(C_\alpha^*) = 2 - \lim_{u \nearrow 1} \frac{1 - u^{2-a_1-a_2} \left(u^{a_1} + \alpha \left((u^{a_1} - u^{a_2})(F_\Omega(\beta) - 1) + (u^{2a_2} - u^{a_2}) \int_\beta^1 \omega \mathbb{P}^\Omega(d\omega) \right) \right)}{1 - u}$$

$$= 2 - (2 - a_2) + \alpha(a_2 - a_1) \lim_{u \nearrow 1} (F_\Omega(\beta) - 1) - \alpha a_2 \lim_{u \nearrow 1} \int_\beta^1 \omega \mathbb{P}^\Omega(d\omega)$$

$$= a_2 + \alpha \left((a_2 - a_1)(F_\Omega(\gamma) - 1) - a_2 \int_\gamma^1 \omega \mathbb{P}^\Omega(d\omega) \right).$$

With analogous arguments, we can conclude that $u^{a_1} < \frac{\alpha}{1-\alpha}(u^{a_2} - u^{2a_2}) + u^{a_2}$ holds for $u \in (1 - \varepsilon, 1)$ for some $\varepsilon > 0$ if and only if $a_2 \frac{1-2\alpha}{1-\alpha} < a_1$. Therefore, if $(a_1, a_2) \in B_3$ then

$$\lambda_U(C_\alpha^*) = 2 - (2 - a_2) + \alpha(a_1 - a_2) + \alpha(a_2 - a_1) \lim_{u \nearrow 1} F_\Omega(\beta)$$

$$- \alpha a_2 \lim_{u \nearrow 1} \int_\beta^1 \omega \mathbb{P}^\Omega(d\omega) + (1 - \alpha)(a_1 - a_2) \lim_{u \nearrow 1} F_\Omega(b)$$

$$+ \alpha a_2 \lim_{u \nearrow 1} \int_{-1}^b \omega \mathbb{P}^\Omega(d\omega)$$

$$= a_2 + (a_1 - a_2)\left(\alpha(1 - F_\Omega(\gamma)) + (1 - \alpha)F_\Omega(\delta)\right)$$

$$+ \alpha a_2 \left(\int_{-1}^\delta \omega \mathbb{P}^\Omega(d\omega) - \int_\gamma^1 \omega \mathbb{P}^\Omega(d\omega) \right),$$

where $b = \beta \cdot \frac{\alpha - 1}{\alpha}$ with β as above and $\delta := \lim_{u \nearrow 1} b = \gamma \cdot \frac{\alpha - 1}{\alpha}$.

By comparing derivatives, we can conclude that $2u^{a_2} - u^{2a_2} \leq u^{a_1}$ holds for $u \in (1 - \varepsilon, 1)$ for some $\varepsilon > 0$ if and only if $a_1 \leq 0$ which would violate the aforementioned assumptions. Hence, if $(a_1, a_2) \in B_4$ then

$$\lambda_U(C_\alpha^*) = 2 - (2 - a_1) + \alpha(a_2 - a_1) \lim_{u \nearrow 1} F_\Omega(\beta) + \alpha a_2 \lim_{u \nearrow 1} \int_{-1}^\beta \omega \mathbb{P}^\Omega(d\omega)$$

$$= a_1 + \alpha(a_2 - a_1)F_\Omega(\gamma) + \alpha a_2 \int_{-1}^\gamma \omega \mathbb{P}^\Omega(d\omega). \qquad \square$$

Corollary 4.1 *If $a_1 = a_2 = a$, then the copula C_α^* has upper tail dependence given by*

$$\lambda_U(C_\alpha^*) = a\left(1 - \alpha \mathbb{E}(|\Omega|)\right) = a\lambda_U(C_\alpha).$$

Proof From Theorem 4.2 we can conclude

$$\lambda_U(C_\alpha^*) = a + \alpha a \left(\int_{-1}^0 \omega \mathbb{P}^\Omega(d\omega) - \int_0^1 \omega \mathbb{P}^\Omega(d\omega) \right) = a\left(1 - \alpha \mathbb{E}(|\Omega|)\right). \quad \square$$

Proposition 4.1 *The copula C_α^* is positively quadrant dependent.*

Proof By the fact that C_α is positively quadrant dependent, (see Proposition 3.2),

$$C_\alpha^*(u, v) = u^{1-a_1}v^{1-a_2}C_\alpha(u^{a_1}, v^{a_2}) \geq u^{1-a_1}v^{1-a_2}u^{a_1}v^{a_2}$$

$$= uv \text{ for all } u, v \in [0, 1]. \qquad \square$$

Figure 3 shows that the C_α^* copulas exhibit even more asymmetry than Singular Mixture Copulas. This is not surprising since the construction used was introduced by Khoudraji (1995) to construct asymmetric copulas from exchangeable copulas.

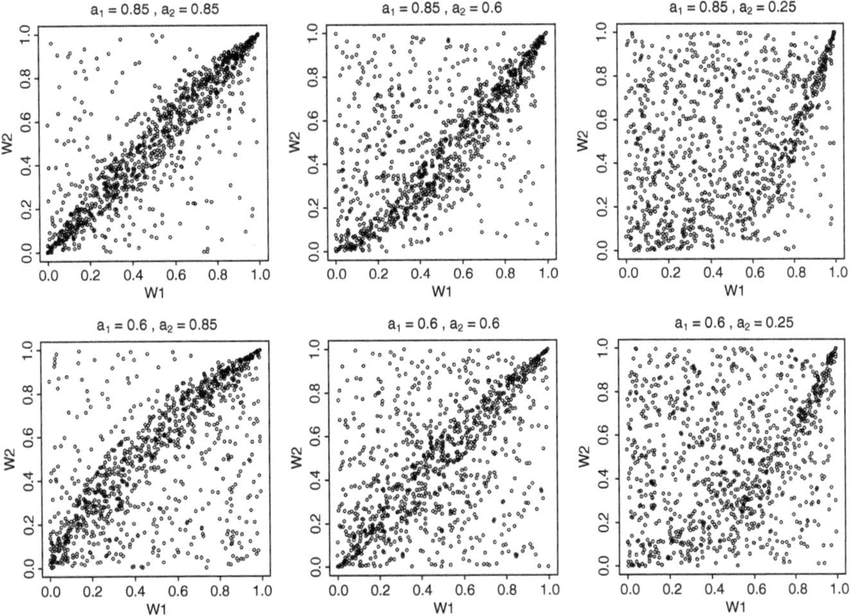

Fig. 3 Scatter plots of simulated points from the copula C_α^* for $\alpha = 0.3$ and different values for a_k. The underlying mixture distribution is a $\mathscr{U}(-1, 1)$-distribution

Moreover, this construction overcomes the drawback of a very restricted support (compare Fig. 2 with Fig. 3) which was a major disadvantage of Singular Mixture Copulas. Consequently, the copulas described in this section should find broader application.

The increased flexibility of the C_α^* copulas is also emphasized by the following proposition which shows that C_α^* copulas include both the Fréchet-Hoeffding upper bound and the independence copula as a limiting case.

Proposition 4.2 *The Fréchet-Hoeffding upper bound M^2 and the independence copula Π^2 are limiting cases of a series of C_α^* copulas.*

Proof Let C_{α,a_1,a_2}^* denote the copula given by $C_{\alpha,a_1,a_2}^*(u, v) = u^{1-a_1} v^{1-a_2} C_\alpha(u^{a_1}, v^{a_2})$, then clearly

$$\lim_{a_1 \to 0} \lim_{a_2 \to 0} C_{\alpha,a_1,a_2}^*(u, v) = uv C_\alpha(1, 1) = uv = \Pi^2(u, v).$$

On the other hand,

$$\lim_{a_1 \to 1} \lim_{a_2 \to 1} C_{\alpha,a_1,a_2}^*(u, v) = C_\alpha(u, v),$$

and Proposition 3.1 showed that $\lim_{\alpha \to 0} C_\alpha(u, v) = M^2(u, v)$. $\square$

5 Second Extension

Following the approach of Khoudraji (1995), it is also possible to construct a new copula using two Singular Mixture Copulas C_α and C_β via

$$C^\star(u, v) = C_\alpha(u^{1-a_1}, v^{1-a_2}) C_\beta(u^{a_1}, v^{a_2})$$

with $a_1, a_2 \in [0, 1]$.

Proposition 5.1 *The copula $C^\star$ is positively quadrant dependent.*

Proof By the fact that both C_α and C_β are positively quadrant dependent (see Proposition 3.2),

$$C^\star(u, v) = C_\alpha(u^{1-a_1}, v^{1-a_2}) C_\beta(u^{a_1}, v^{a_2}) \geq u^{1-a_1} v^{1-a_2} u^{a_1} v^{a_2}$$
$$= uv \text{ for all } u, v \in [0, 1]. \qquad \square$$

Like the C_α^* copulas, the $C^\star$ copulas include both the Fréchet-Hoeffding upper bound and the independence copula as a limiting case as the following proposition shows.

Proposition 5.2 *The Fréchet-Hoeffding upper bound M^2 and the independence copula Π^2 are limiting cases of a series of $C^\star$ copulas.*

Proof Let $C^\star_{\alpha,\beta,a_1,a_2}(u, v) = C_\alpha(u^{1-a_1}, v^{1-a_2}) C_\beta(u^{a_1}, v^{a_2})$, then clearly

$$\lim_{a_1 \to 0} \lim_{a_2 \to 1} C^\star_{\alpha,\beta,a_1,a_2}(u, v) = C_\alpha(u, 1) C_\beta(1, v) = uv = \Pi^2(u, v).$$

On the other hand,

$$\lim_{a_1 \to 0} \lim_{a_2 \to 0} C^\star_{\alpha,\beta,a_1,a_2}(u, v) = C_\alpha(u, v),$$

and Proposition 3.1 showed that $\lim_{\alpha \to 0} C_\alpha(u, v) = M^2(u, v).$ $\qquad \square$

As Fig. 4 shows, $C^\star$ copulas possess quite asymmetric shapes. This copula construction also overcomes—to some extent—the drawback of the restricted support. In contrast to the C_α^* construction, it is possible to create copulas which distribute probability mass only on a restricted area, but this area is much less restricted than the corresponding area in the Singular Mixture Copula approach.

At first glance, Fig. 4 might seem to show that $C^\star$ can possess a singular component. Nevertheless, this is not true. Since C_α and C_β are absolutely continuous copulas, it is apparent from its construction that $C^\star$ is absolutely continuous, too. What seems to be a singular component is in fact a very narrow band in which probability mass is distributed.

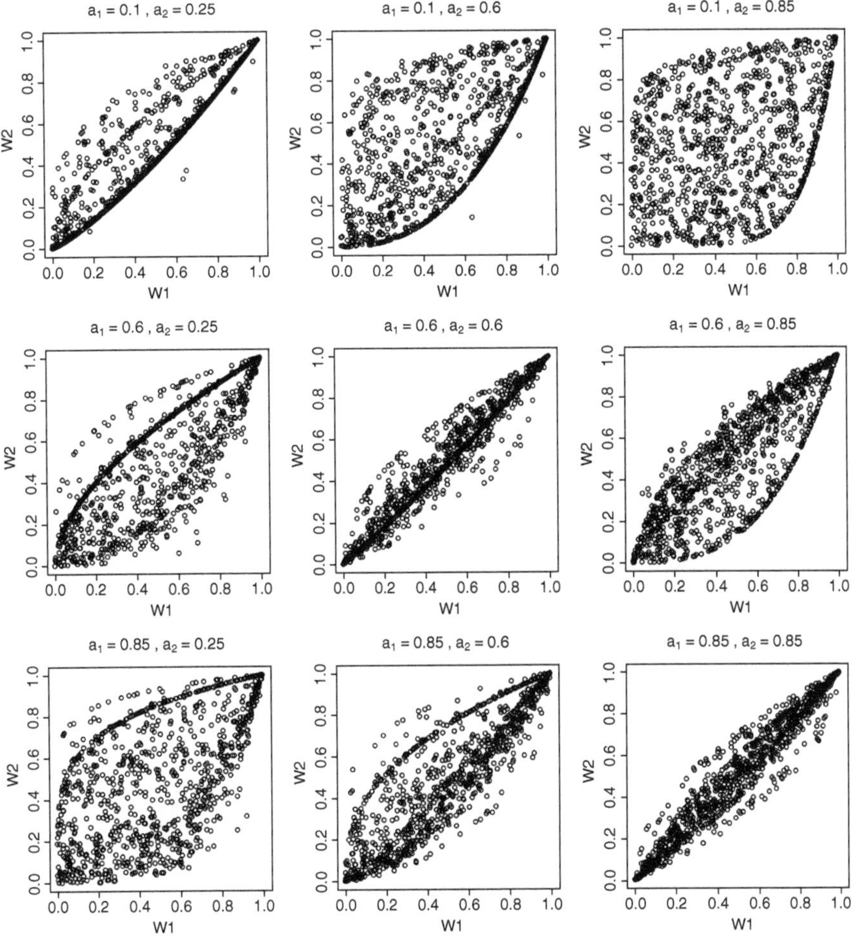

Fig. 4 Scatter plots of simulated points from the copula $C^\star$ for $\alpha = 0.3$, $\beta = 0.1$ and different values of a_k. The underlying mixture distributions are two $\mathscr{U}(-1, 1)$-distributions

6 Concluding Remarks

In this paper, we presented and discussed two extensions of Singular Mixture Copulas. These extensions are based on the approach introduced in Khoudraji (1995). We showed that the constructed copulas can overcome some drawbacks of Singular Mixture Copulas, and thus offer a more flexible tool for modeling stochastic dependence. We also showed that the copula C_α^* possesses a form of asymmetry in the way that it exhibits no lower tail dependence yet upper tail dependence.

Acknowledgments The authors would like to sincerely thank the reviewers for their detailed comments, which led to an improvement of the presentation of this paper.

References

Cherubini, U., Luciano, E., & Vecchiato, W. (2004). *Copula methods in finance*. Chichester: Wiley.

Deheuvels, P. (1978). Caractérisation complète des lois extrêmes multivariées et de la convergence des types extrêmes. *Publication Institute Statistics University of Paris* (Vol. 23, pp.1–37).

Dovgoshey, O., Martio, O., Ryazanov, V., & Vuorinen, M. (2006). The cantor function. *Expositiones Mathematicae, 24*, 1–37.

Durante, F. (2009). Construction of non-exchangeable bivariate distribution functions. *Statistical Papers, 50*, 383–391.

Durante, F., & Sempi, C. (2010). Copula theory: An introduction. In P. Jaworski, F. Durante, W. Härdle, & T. Rychlik (Eds.), *Copula theory and its applications* (pp. 3–31). Dordrecht: Springer.

Embrechts, P., Lindskog, F., & McNeil, A. J. (2003). Modelling dependence with copulas and applications to risk management. In S. T. Rachev (Ed.), *Handbook of heavy tailed distributions in finance* (pp. 329–384). Amsterdam: Elsevier.

Fréchet, M.: Sur les tableaux de corrélation dont les marges sont données. *Annales de l'Université de Lyon. Section A, 9*, 53–77 (1951).

Genest, C., Ghoudi, K., & Rivest, L.-P. (1998). Understanding relationships using copulas, by Edward Frees and Emiliano Valdez, January 1998. *North American Actuarial Journal, 2*(3), 143–149.

Genest, C., & MacKay, R. J. (1986). The joy of copulas: Bivariate distributions with uniform marginals. *The American Statistician, 40*(4), 280–283.

Hoeffding, W. (1940). Maßstabinvariante Korrelationstheorie. *Schriften des Mathematischen Instituts und des Instituts für Angewandte Mathematik der Universität Berlin, 5*, 179–233.

Joe, H. (1997). *Multivariate models and dependence concepts*. London: Chapman & Hall.

Khoudraji, A. (1995). Contributions à l'étude des copules et à la modélisation des valeurs extrêmes bivariées. Ph.D. thesis, Université de Laval, Québec (Canada).

Lauterbach, D. (2012). Singular mixture copulas: A geometric method of constructing copulas. *ZVersWiss, 101*(5), 605–619.

Lauterbach, D., & Pfeifer, D. (2013). Singular mixture copulas. In P. Jaworski, F. Durante, & W. Härdle (Eds.), *Copulae in mathematical and quantitative finance* (pp. 165–175). Dordrecht: Springer.

McNeil, A. J., Frey, R., & Embrechts, P. (2005). *Quantitative risk management: Concepts techniques and tools*. Princeton: Princeton University Press.

Nelsen, R. B. (2006). *An introduction to Copulas*. New York: Springer.

Sklar, A. (1959). Fonctions de répartition à n dimensions et leurs marges. *Publication Institute Statistics University of Paris* (Vol. 8, pp. 229–231).

Wolff, E.F. (1977). Measures of dependence derived from copulas. Ph.D. thesis, University of Massachusetts, Amherst.

Strong Laws of Large Numbers in an F^α-Scheme

Paul Doukhan, Oleg I. Klesov and Josef G. Steinebach

Abstract We study the almost sure limiting behavior of record times and the number of records, respectively, in a (so-called) F^α-scheme. It turns out that there are certain "dualities" between the latter results, that is, under rather general conditions strong laws for record times can be derived from the corresponding ones for the number of records, but in general not vice versa. The results extend, for example, the classical strong laws of Rényi (Annals Faculty Science University Clermont-Ferrand 8:7–12, 1962; Selected Papers of Alfred Rényi, vol. 3, pp. 50–65, Akadémiai Kiadó, Budapest 1976) for record times and counts.

Keywords Number of records · Record times · F^α-scheme · Almost sure convergence · Strong law of large numbers

1 Introduction

Consider a sequence $\{X_k, k \geq 1\}$ of independent, identically distributed (iid) random variables and assume that the distribution function of X_1 is continuous, so that events like $\{X_i = X_j\}$ only occur with probability 0 if $i \neq j$.

P. Doukhan
Université Cergy-Pontoise, UMR 8088 Analyse, Géométrie et modélisation,
2 avenue Adolphe Chauvin, 95302 Cergy-Pontoise Cedex, France
e-mail: doukhan@u-cergy.fr

O.I. Klesov
Department of Mathematical Analysis and Probability Theory,
National Technical University of Ukraine (KPI), pr. Peremogy 37,
Kyiv 03056, Ukraine
e-mail: klesov@matan.kpi.ua

J.G. Steinebach (✉)
Mathematisches Institut, Universität zu Köln, Weyertal 86–90,
50931 Cologne, Germany
e-mail: jost@math.uni-koeln.de

© Springer International Publishing Switzerland 2015
M. Hallin et al. (eds.), *Mathematical Statistics and Limit Theorems*,
DOI 10.1007/978-3-319-12442-1_16

Let $L(1) = 1$ and define recursively

$$L(n) = \inf\{k > L(n-1) : X_k > X_{L(n-1)}\}, \qquad n \geq 2, \tag{1}$$

where $\inf \emptyset := +\infty$. The random variables in the sequence $L = \{L(n), n \geq 1\}$ are called the *record times* constructed from the sequence $\{X_k, k \geq 1\}$.

We also define a counting process $\mu = \{\mu(n), n \geq 1\}$ via

$$\mu(n) = \#\{k : L(k) \leq n\}, \qquad n \geq 1, \tag{2}$$

i.e., the value $\mu(n)$ represents the *number of records* up to time n.

Rényi (1962) proved that

$$\lim_{n \to \infty} \frac{\mu(n)}{\log n} = 1 \quad \text{almost surely} \quad \text{(a.s.)}, \tag{3}$$

and

$$\lim_{n \to \infty} \frac{\log L(n)}{n} = 1 \quad \text{almost surely} \quad \text{(a.s.)} \tag{4}$$

(see also Gut 2005, pp. 307–308, for a modern presentation). In fact, it will turn out below that relations (3) and (4) are equivalent.

Our aim is to discuss the above "dualities" in a more general framework, that is, for a class of nonidentically distributed random variables called an F^α-scheme, which will be introduced below. A particular case of such a scheme, where the exponents $\alpha = \{\alpha_n\}$ form a geometric progression has been introduced by Yang (1975).

While the distributional results for F^α-schemes seem to have been studied in great detail (see, e.g. Nevzorov 2001, and the references therein), the almost sure convergence has found much less attention. Nevertheless some results are known.

Ballerini and Resnick (1987), for example, treat the case of $\alpha_n = [\lambda^{n-1}\alpha_1 + \frac{1}{2}]$, where $[\cdot]$ denotes the integer part and $\lambda > 1$ is a constant, and prove that

$$\lim_{n \to \infty} \frac{\mu(n)}{n} = 1 - \lambda^{-1} \qquad \text{a.s.} \tag{5}$$

This result shows the difference between the classical setting $\alpha_n \equiv 1$ reflected in (3) and an F^α-scheme with geometrically growing indices as in (5).

Weissman (1995) studies an F^α-scheme with exponents $\{\alpha_n\}$ for which

$$A_n := \sum_{j=1}^{n} \alpha_j \to \infty \quad \text{and} \quad \max\{\alpha_k : 1 \leq k \leq n\} = o(A_n) \quad \text{as} \quad n \to \infty. \tag{6}$$

Here it turns out that

$$\lim_{n \to \infty} \frac{\mu(n)}{\log A_n} = 1 \qquad \text{a.s.} \tag{7}$$

(see Weissman 1995). While (7) fits to the classical scheme, which is, in fact, a consequence of condition (6), this is not the case for the Ballerini and Resnick (1987) result (5). A more detailed discussion will follow below.

Some recent results on records in subsets of a random field, which are also related to F^α-schemes, have been obtained by Gut and Stadtmüller (2013). Further results concerning strong stability and asymptotic normality for $\mu(n)$ in F^α-schemes have also been discussed in Doukhan et al. (2013) under condition (6). We try to avoid the latter condition here. For example, relation (10) below does not assume (6). However, if (6) holds, then our results intersect with those given in Doukhan et al. (2013), but the methods presented in this paper are different from those used in Doukhan et al. (2013). Moreover, we also study the behavior of $L(n)$, the dual object to $\mu(n)$.

An even more general scheme, called a *random F^α-scheme*, has been introduced by Deheuvels and Nevzorov (1994). They extend the classical setting by assuming that $\{\alpha_n\}$ are independent, positive random variables. Deheuvels and Nevzorov (1994) are basically interested in normal approximations for the distribution function of $\mu(n)$ and do not discuss its almost sure convergence in much detail. For further discussion, we refer to Doukhan et al. (2013).

The paper is organized as follows. In Sect. 2, we recall some definitions concerning F^α-schemes. Then, in Sect. 3, we provide several almost sure asymptotics for the number $\mu(n)$ of records up to time n. It will be shown that, depending on the particular F^α-scheme, there is a variety of normalizations for $\mu(n)$ as well as for the corresponding limiting constants. In Sect. 5, we prove that (3) and (4) are equivalent. Moreover, it is shown that this remains true for arbitrary nonrandom functions μ and L, which are related via condition (30) below. We also retain (3) and (4) for the general F^α-scheme and prove that the first relation implies the second one, but that, in general, these relations are not equivalent. Finally, in Sect. 6, several strong laws for L are presented, which turn out to be corollaries of the corresponding results for μ in Sect. 3 in combination with the implication proved in Theorem 5.

2 F^α-Schemes

The class F^α of independent random variables we consider in the sequel has been introduced by Yang (1975). Let $\{X_k, k \geq 1\}$ be a sequence of independent random variables, $\alpha = \{\alpha_n, n \geq 1\}$ be a sequence of positive reals, and let F be a continuous distribution function. Assume that

$$\mathsf{P}\,(X_n \leq x) = F^{\alpha_n}(x), \qquad n \geq 1, \quad x \in \mathbf{R}.$$

Then we say that $\{X_n\}$ is an F^α-scheme.

We construct the sequences $\{L(n)\}$ and $\{\mu(n)\}$ as above and denote by I_n the indicator variable of a record at time n, i.e.

$$I_n = \begin{cases} 1, & \text{if } X_n > \max\{X_1, \ldots, X_{n-1}\}, \\ 0, & \text{otherwise.} \end{cases}$$

Note that the event $\{X_n = \max(X_1, \ldots, X_{n-1})\}$ occurs with probability 0.

Generalizing the Rényi (1962) classical result for iid random variables, Nevzorov (1985) showed for a general F^α-scheme that

(1) the random variables I_n, $n \geq 1$, are independent, and
(2) $\mathsf{P}(I_n = 1) = \alpha_n / A_n$, where $A_n = \alpha_1 + \cdots + \alpha_n$

(see also Borovkov and Pfeifer 1995). In fact, property (1) holds only if the sequence $\{X_n\}$ constitutes an F^α-scheme (see Nevzorov 1985).

Set

$$\xi_k = I_k - \frac{\alpha_k}{A_k}, \qquad S_n = \sum_{k=1}^{n} \xi_k.$$

It is clear that

$$\mathsf{E}\,\xi_k = 0 \quad \text{and} \tag{8}$$

$$\mathrm{var}\,[\xi_k] = \mathrm{var}\,[I_k] = \frac{\alpha_k}{A_k} \cdot \left(1 - \frac{\alpha_k}{A_k}\right). \tag{9}$$

3 Strong Laws for the Number of Records

3.1 Case of $\{A_n\}$ Bounded

Then obviously $\sum \mathrm{var}\,[I_k]$ converges and thus $\sum I_k$ is finite almost surely. This means that only a finite number of records occurs almost surely, so that the asymptotics have no meaning in this case.

3.2 Case of $\{A_n\}$ Unbounded

Since $A_n > 1$ for (say) $n \geq n_0$, one has $\int_{A_{n_0}}^{\infty} x^{-1}(\log x)^{-2} dx < \infty$, hence also

$$\sum_{k=n_0}^{\infty} \frac{\alpha_k}{A_k\,(\log A_k)^2} < \infty.$$

So, in view of (8) and (9), the random series

$$\sum_{k=1}^{\infty} \frac{\xi_k}{\log A_k}$$

converges almost surely (see, e.g., Billingsley 1995, Theorem 22.6).

Now, from Kronecker's lemma it follows that

$$\lim_{n\to\infty} \frac{1}{\log A_n} \sum_{k=1}^{n} \xi_k = 0 \quad \text{a.s.}$$

Since $\mu(n) = \sum_{k=1}^{n} I_k$, the number of records up to time n, we get from the latter relation that

$$\lim_{n\to\infty} \left[\frac{\mu(n)}{\log A_n} - \frac{1}{\log A_n} \sum_{k=1}^{n} \frac{\alpha_k}{A_k} \right] = 0 \quad \text{a.s.} \tag{10}$$

Various special cases will now be treated in more detail.

3.2.1 Case of $\alpha_n = o(A_n)$

Let

$$\lim_{n\to\infty} \frac{\alpha_n}{A_n} = 0. \tag{11}$$

Theorem 1 *Assume that* (11) *holds. Then*

$$\lim_{n\to\infty} \frac{\mu(n)}{\log A_n} = 1 \quad \text{a.s.} \tag{12}$$

Proof Indeed, for all $k \geq 2$,

$$\frac{\alpha_k}{A_k} \leq \int_{A_{k-1}}^{A_k} \frac{dx}{x} = \log A_k - \log A_{k-1}, \tag{13}$$

whence

$$\limsup_{n\to\infty} \frac{1}{\log A_n} \sum_{k=1}^{n} \frac{\alpha_k}{A_k} \leq 1. \tag{14}$$

On the other hand, if (11) holds, then $A_k/A_{k-1} \to 1$. We fix $\delta > 0$ and find a sufficiently large k_0 such that $(1+\delta)A_{k-1} \geq A_k, k \geq k_0$. Then, for $k \geq k_0$,

$$\frac{(1+\delta)\alpha_k}{A_k} \geq \frac{\alpha_k}{A_{k-1}} \geq \int_{A_{k-1}}^{A_k} \frac{dx}{x} = \log A_k - \log A_{k-1}. \tag{15}$$

Hence

$$\liminf_{n\to\infty} \frac{1}{\log A_n} \sum_{k=1}^{n} \frac{\alpha_k}{A_k} \geq \frac{1}{1+\delta}.$$

Since $\delta > 0$ can be chosen arbitrarily small, this together with (14) and (10) implies (12). $\qquad\square$

Corollary 1 *Let $c > -1$ and $\alpha_n = n^c \ell(n)$ for $n \geq 1$, where ℓ is a slowly varying function. Then*

$$\lim_{n\to\infty} \frac{\mu(n)}{\log n} = c + 1 \quad \text{a.s.}$$

Proof Note that in this case, by Karamata's theorem, $A_n \sim n^{c+1}\ell(n)/(c + 1)$, and thus $\log A_n \sim (c + 1) \log n$, since $\log \ell(n) = o(\log n)$. $\qquad\square$

The simplest case of Corollary 1 corresponds to $c = 0$ and $\ell(n) \equiv a > 0$, where all α_n are the same. If $a = 1$, then we are in the iid case and the result of Corollary 2 coincides with (3).

Corollary 2 *If $\alpha_n = a > 0$ for $n \geq 1$, then (3) holds.*

Remark 1 In fact, Corollary 2 coincides with the iid case even if $a \neq 1$. Indeed, the distribution function F^a is the same for all random variables X_n in this case. A more general setting corresponds to a finite number of possible values for $\{\alpha_n\}$ (cf. Doukhan et al. 2013, Example 2). Denoting by m and M the minimal and maximal values of $\{\alpha_n\}$, we have $nm \leq A_n \leq nM$, whence $A_n \to \infty$. Thus (11) holds and we obtain (12) from Theorem 1. Moreover, since $\log A_n \sim \log n$, we get (3).

An even more general result can be obtained as follows (see Doukhan et al. 2013, Example 3).

Corollary 3 *Assume that*

$$\inf\{\alpha_i \,|\, i \geq 1\} > 0 \quad and \quad \sup\{\alpha_i \,|\, i \geq 1\} < \infty.$$

Then (3) holds.

Another asymptotic corresponds to the case $c = -1$ extending Corollary 1.

Corollary 4 *Let ℓ be a slowly varying function such that the series $\sum \ell(n)/n$ diverges. Assume that*

$$\alpha_n = \frac{\ell(n)}{n}, \qquad n \geq 1.$$

Then (12) prevails.

Proof In view of Theorem 1 one only needs to show that (11) holds. For the latter result, we refer to Lemma 11.9.2 in Buldygin et al. (2012), which extends a result in Parameswaran (1961) from integrals to sums. Note that $\{A_n\}$ is also a slowly varying sequence (see Buldygin et al. 2012). $\square$

Corollary 4 provides various strong laws of large numbers for μ corresponding to specific choices of ℓ. Assume, for example, $\ell(x) = (\log x)^a$, $x \geq 3$, with some $a > -1$. Then $A_n \sim (\log n)^{a+1}/(a+1)$, whence

$$\lim_{n\to\infty} \frac{\mu(n)}{\log\log n} = a + 1 \quad \text{a.s.} \tag{16}$$

The asymptotic behavior changes if $a = -1$. Indeed, let $\ell(x) = (\log x)^{-1}$, $x \geq 3$. Then $A_n \sim \log\log n$, whence

$$\lim_{n\to\infty} \frac{\mu(n)}{\log\log\log n} = 1 \quad \text{a.s.} \tag{17}$$

In the case of $a < -1$ the series $\sum \ell(n)/n$ converges, which corresponds to the discussion in Sect. 3.1.

Now we study the case where condition (11) does not hold.

3.2.2 Case of $\alpha_n \sim \lambda A_n$, for Some $0 < \lambda < 1$

Let

$$\lim_{n\to\infty} \frac{\alpha_n}{A_n} = \lambda \quad \text{with} \quad 0 < \lambda < 1. \tag{18}$$

The normalizing sequence for $\{\mu(n)\}$ then changes as compared to Theorem 1.

Theorem 2 *Assume that* (18) *holds. Then*

$$\lim_{n\to\infty} \frac{\mu(n)}{\log A_n} = -\frac{\lambda}{\log(1-\lambda)} \quad \text{a.s.} \tag{19}$$

An equivalent form of (19) *is given by*

$$\lim_{n\to\infty} \frac{\mu(n)}{n} = \lambda \quad \text{a.s.} \tag{20}$$

Proof In case of (18),

$$\sum_{k=1}^{n} \frac{\alpha_k}{A_k} \sim n\lambda \quad \text{and} \quad \log A_n \sim n \log\left(\frac{1}{1-\lambda}\right). \tag{21}$$

Thus (19) and (20) follow from assertion (10).

Here, the first asymptotic in (21) is a straightforward consequence of (18), and the second one follows since (18) implies that $A_{n-1}/A_n \to 1 - \lambda$, whence $\log(A_n/A_{n-1}) \to -\log(1 - \lambda)$, which results in $\log A_n \sim -n \log(1 - \lambda)$. $\square$

Remark 2 The limit on the right-hand side of (19) equals 1 as $\lambda \downarrow 0$, which is "in agreement" with (12). Moreover, the limit in (19) equals 0 as $\lambda \uparrow 1$, which "agrees" with (23) below.

Theorem 2 generalizes the Ballerini and Resnick (1987) result (5) corresponding to the case of geometrically growing $\{\alpha_n\}$.

Corollary 5 *Let $\theta > 1$ and*

$$\alpha_n = \theta^n - \theta^{n-1}, \qquad n \geq 1.$$

Then

$$\lim_{n \to \infty} \frac{\mu(n)}{n} = 1 - \frac{1}{\theta} \quad \text{a.s.}$$

Proof Here, $\alpha_n/A_n \to 1 - \theta^{-1} =: \lambda$ as $n \to \infty$, whence $\log A_n \sim \log \alpha_n \sim n \log \theta = -n \log(1 - \lambda)$. $\square$

3.2.3 Case of $\alpha_n \sim A_n$

Let

$$\lim_{n \to \infty} \frac{\alpha_n}{A_n} = 1. \tag{22}$$

Theorem 3 *Assume that (22) holds. Then*

$$\lim_{n \to \infty} \frac{\mu(n)}{\log A_n} = 0 \quad \text{a.s.} \tag{23}$$

Proof Indeed, in case of (22),

$$\sum_{k=1}^{n} \frac{\alpha_k}{A_k} = o(\log A_n), \tag{24}$$

and (23) follows from (10). The argument for proving (24) is similar to that in the proof of the second part of (21). Note that (22) implies

$$\sum_{k=1}^{n} \frac{\alpha_k}{A_k} \sim n.$$

On the other hand, we obtain from (22) that $A_n/A_{n-1} \to \infty$. Let $\rho > 0$ be fixed. Then there exists an n_0 such that $A_n \geq \rho A_{n-1}$ for all $n \geq n_0$, whence $A_n \geq A_{n_0-1}\rho^{n-n_0+1}$, $n \geq n_0$. Passing to the limit results in

$$\liminf_{n\to\infty} \frac{1}{n} \log A_n \geq \log \rho.$$

Since ρ can be chosen arbitrarily large, this means that

$$\lim_{n\to\infty} \frac{1}{n} \log A_n = \infty,$$

which implies (24). $\qquad\Box$

Corollary 6 *Let $\alpha_1 = 1$ and*

$$\alpha_n = n! - (n-1)!, \quad n \geq 2.$$

Then

$$\lim_{n\to\infty} \frac{\mu(n)}{n \log n} = 0 \quad \text{a.s.} \tag{25}$$

Proof Note that $A_n = n!$, $\alpha_n/A_n = 1 - n^{-1} \to 1$, and $\log A_n \sim n \log n$. $\qquad\Box$

Remark 3 The second factor of the variance in (9) is of order $o(1)$ in case of (22). This can be used to improve the asymptotics of (23). Just for a brief demonstration, let us consider the F^α-scheme from Corollary 6. Then

$$\text{var}\, \xi_k = O(1)\left(1 - \frac{\alpha_k}{A_k}\right) = \frac{O(1)}{k}.$$

Fix $\gamma > 1/2$. By the three series theorem,

$$\sum \frac{\xi_k}{(\log k)^\gamma}$$

converges almost surely, whence

$$\lim_{n\to\infty} \frac{1}{(\log n)^\gamma}\left[\mu(n) - \sum_{k=1}^n \frac{\alpha_k}{A_k}\right] = 0 \quad \text{a.s.}$$

Now,

$$\sum_{k=1}^n \frac{\alpha_k}{A_k} = 1 + \sum_{k=2}^n \left(1 - \frac{1}{k}\right) = n - \log n + O(1).$$

Therefore, for all $\gamma > 1/2$,

$$\lim_{n \to \infty} \frac{1}{(\log k)^{\gamma}} \left[\mu(n) - n + \log(n) \right] = 0 \quad \text{a.s.} \tag{26}$$

In other words, there are very many records, since the "population grows too fast". In particular,

$$\lim_{n \to \infty} \frac{\mu(n)}{n} = 1 \quad \text{a.s.} \tag{27}$$

(compare with (25)).

An interesting question would be to investigate the case of $\gamma = 1/2$, which will be studied elsewhere.

4 Random F^{α}-Schemes

Deheuvels and Nevzorov (1993, 1994) introduced and investigated a (so-called) random F^{α}-scheme, in which the $\{\alpha_n\}$ are independent, positive random variables. We briefly discuss this scheme for the particular case of independent, identically distributed $\{\alpha_n\}$ possessing a positive first moment $\gamma = \mathsf{E}\,\alpha_1$, since this is a straightforward generalization of the classical iid case. For a more detailed discussion, we refer to Doukhan et al. (2013).

Denote by $\mathscr{A}$ the σ-algebra generated by the sequence $\{\alpha_n\}$. Then (see Deheuvels and Nevzorov 1994) the record indicators are conditionally independent given $\mathscr{A}$. Moreover,

$$\mathsf{P}\,(I_n | \mathscr{A}) = \frac{\alpha_n}{A_n} \quad \text{a.s.,} \tag{28}$$

where again $A_n := \alpha_1 + \cdots + \alpha_n$.

Limit theorems for conditionally independent random variables have been discussed in several papers (see, e.g., Beck 1974). For example, Kolmogorov's strong law of large numbers for nonidentically distributed random variables has been proved in Majerak et al. (2005) with a normalization $b_n = n$. For conditionally independent random variables, one can similarly treat the strong law of large numbers for the general case of increasing unbounded normalizations $\{b_n\}$. In particular, one can prove a conditionally independent version of Kolmogorov's strong law. As a consequence, we obtain for $\mu(n) = I_1 + \cdots + I_n$ that

$$\frac{\mu(n) - \mathsf{E}\,[\mu(n) | \mathscr{A}]}{\log n} \to 0 \quad \text{a.s.}$$

Combining the latter result with (28), we have

$$\frac{1}{\log n}\left(\mu(n) - \sum_{k=1}^{n}\frac{\alpha_k}{A_k}\right) \to 0 \quad \text{a.s.}$$

Now, by Kolmogorov's strong law of large numbers for iid random variables,

$$\frac{A_n}{n} \to \gamma \quad \text{a.s.},$$

whence we conclude that condition (6) holds almost surely. Denote the random event where (6) holds by Ω_1. Then, for any $\omega \in \Omega_1$, we can use (13)–(15) to obtain

$$\frac{1}{\log n}\sum_{k=1}^{n}\frac{\alpha_k}{A_k} \to 1 \quad \text{a.s.},$$

which finally results in

$$\frac{\mu(n)}{\log n} \to 1 \quad \text{a.s..}$$

Hence the limiting behavior of μ in the random F^α-scheme with integrable iid random variables $\{\alpha_n\}$ is the same as in the classical setting (confer (3)).

Remark 4 It should be noted that the result obtained above holds true in much more general settings. The case of iid $\{\alpha_n\}$ has just been chosen for the sake of an easy presentation.

5 An Auxiliary Result

By definition it is clear that

$$\mu(L(n)) = n, \quad n \geq 1. \tag{29}$$

Due to (29), the limit results (3) and (4) are equivalent (see Theorem 4 below). Moreover, this is true in more general cases, even for dependent and nonidentically distributed random variables $\{X_n\}$. We also like to mention that the equivalence holds for general numerical sequences μ and L, too, if they are only related via the relation

$$\lim_{n \to \infty}\frac{\mu(L(n))}{n} = 1 \tag{30}$$

instead of (29), so the randomness does not really matter. In fact, we only need this weaker condition for our asymptotics below.

In what follows in this section we assume that μ and L are two arbitrary positive sequences or functions.

Theorem 4 *Let μ be a nondecreasing sequence and L be a sequence of integers such that (30) holds. Then (31) and (32) are equivalent, where*

$$\lim_{n \to \infty} \frac{\mu(n)}{\log n} = 1 \tag{31}$$

and

$$\lim_{n \to \infty} \frac{\log L(n)}{n} = 1. \tag{32}$$

Proof First we prove that (31) implies (32). Note that $L(n) \to \infty$ as $n \to \infty$ in view of (30). We have

$$\frac{n}{\log L(n)} = \frac{n}{\mu(L(n))} \cdot \frac{\mu(L(n))}{\log L(n)}, \tag{33}$$

where the first factor on the right-hand side tends to 1 by (30), while the second one tends to 1 in view of (31). This proves (32).

Now we prove that (32) implies (31). Note that this implication does not follow via (33), since (33) proves (31) only for the subsequence $\{L(n)\}$. Instead we apply the following lemma.

Lemma 1 *Let f be a nondecreasing function and g be a positive function on $(0, \infty)$. If*

$$f(g(x)) \sim x \quad and \quad \log g(x) \sim x \quad as \quad x \to \infty, \tag{34}$$

then

$$f(x) \sim \log x. \tag{35}$$

We first complete the proof of Theorem 4 and then come back to the proof of Lemma 1. Set

$$f(x) = \mu([x]), \qquad g(x) = L([x]).$$

Obviously, (34) follows from (30) and (32). Thus (35) follows from Lemma 1, which yields (31). $\qquad\qquad\square$

Proof of Lemma 1. Let $0 < \varepsilon < 1$. It follows from (34) that there exists an x_0 such that

$$e^{(1-\varepsilon)x} \le g(x) \le e^{(1+\varepsilon)x}, \quad x \ge x_0,$$

and also

$$(1 - \varepsilon)x \le f(g(x)) \le (1 + \varepsilon)x, \quad x \ge x_0.$$

Since f is nondecreasing, $f(g(x)) \leq f\left(e^{(1+\varepsilon)x}\right)$, $x \geq x_0$, and

$$(1-\varepsilon)x \leq f(g(x)) \leq f\left(e^{(1+\varepsilon)x}\right), \qquad x \geq x_0,$$

whence

$$f(y) \geq \frac{1-\varepsilon}{1+\varepsilon} \log y, \quad y \geq e^{(1+\varepsilon)x_0}.$$

Similarly,

$$f\left(e^{(1-\varepsilon)x}\right) \leq f(g(x)) \leq (1+\varepsilon)x, \quad x \geq x_0,$$

and

$$f(y) \leq \frac{1+\varepsilon}{1-\varepsilon} \log y, \quad y \geq e^{(1+\varepsilon)x_0}.$$

Passing to the limit $y \to \infty$ in the latter two relations we get

$$\frac{1-\varepsilon}{1+\varepsilon} \leq \liminf_{y\to\infty} \frac{f(y)}{\log y} \leq \limsup_{y\to\infty} \frac{f(y)}{\log y} \leq \frac{1+\varepsilon}{1-\varepsilon}.$$

Since $\varepsilon > 0$ is arbitrary, this proves (32). $\qquad\square$

In the next section, we derive some strong laws of large numbers for L by using the results for μ obtained in Sect. 3. To do so we need the following generalization of the implication $(31) \Longrightarrow (32)$.

Theorem 5 *Let μ and L be two integer-valued sequences. Assume that μ is nondecreasing and that relation (30) holds. Let $\ell \in [0, \infty]$. If*

$$\lim_{n\to\infty} \frac{\mu(n)}{\log A_n} = \ell \tag{36}$$

for some sequence of positive numbers $\{A_n\}$, then

$$\lim_{n\to\infty} \frac{\log A_{L(n)}}{n} = \frac{1}{\ell}, \tag{37}$$

where $1/0 := \infty$ and $1/\infty := 0$.

Proof We only consider the case of $0 < \ell < \infty$. Note that $L(n) \to \infty$ as $n \to \infty$ in view of (30). Inserting $L(n)$ instead of n into (36), we get

$$\mu(L(n)) \sim \ell \log A_{L(n)} \quad \text{as} \quad n \to \infty,$$

whence (37) follows from (30). The proof for $\ell = 0$ or $\ell = \infty$ is similar. $\qquad\square$

6 Strong Laws for Record Times

Recall from Sect. 1 that $\mu = \{\mu(n)\}$ denotes the numbers of records up to time n and $L = \{L(n)\}$ are the nth record times, respectively. Using the results for μ proved in Sect. 3 together with Theorem 5, we now obtain the corresponding asymptotics for L.

6.1 Case of $\alpha_n = o(A_n)$

Theorem 6 *If* (11) *holds, then*

$$\lim_{n \to \infty} \frac{\log A_{L(n)}}{n} = 1 \quad \text{a.s.}$$

Proof Since $\mu(L(n)) = n$, Theorem 6 follows immediately from a combination of Theorems 1 and 5. $\qquad\square$

Corollary 7 *Let $c > -1$ and $\alpha_n = n^c \ell(n)$ for $n \geq 1$, where ℓ is a slowly varying function. Then*

$$\lim_{n \to \infty} \frac{\log L(n)}{n} = \frac{1}{c+1} \quad \text{a.s.} \tag{38}$$

Proof Condition (11) holds in this case, since $A_n \sim n^{c+1}\ell(n)/(c+1)$. Thus Theorem 6 implies that

$$\lim_{n \to \infty} \frac{\log(L(n))^{c+1}\ell(L(n))}{n} = 1 \quad \text{a.s.,}$$

whence (38) follows since $\log \ell(L(n))/\log L(n) \to 0$ on the random event where $L(n) \to \infty$. $\qquad\square$

The case of $c = -1$ is considered next.

Corollary 8 *Let $b \geq -1$, $a > 0$, and $\alpha_n = a(\log n)^b/n$. Then*

$$\lim_{n \to \infty} \frac{\log \log L(n)}{n} = \frac{1}{b+1} \quad \text{a.s.,}$$

if $b > -1$, and

$$\lim_{n \to \infty} \frac{\log \log \log L(n)}{n} = 1 \quad \text{a.s.,}$$

if $b = -1$.

Proof Condition (11) holds in this case, since $A_n \sim a(\log n)^{b+1}/(b+1)$, if $b > -1$, and $A_n \sim a \log \log n$, if $b = -1$. By an application of Theorem 6 the proof can be completed. $\qquad\Box$

The result of the next corollary with $a = 1$ coincides with (4) (confer also Remark 1).

Corollary 9 *If $\alpha_n = a > 0$ for $n \geq 1$, then (4) holds.*

Proof Combine Theorem 6 and Corollary 2. $\qquad\Box$

Corollary 10 *Assume that*

$$\min\{\alpha_i | i \geq 1\} > 0 \quad and \quad \max\{\alpha_i | i \geq 1\} < \infty.$$

Then (4) holds.

Proof Combine Theorem 6 and Corollary 3. $\qquad\Box$

6.2 Case of $\alpha_n \sim \lambda A_n$, for Some $0 < \lambda < 1$

Theorem 7 *Assume that (18) holds. Then*

$$\lim_{n \to \infty} \frac{L(n)}{n} = \frac{1}{\lambda} \quad \text{a.s.} \tag{39}$$

Proof Combine (20), with $L(n)$ replacing n, and (29). $\qquad\Box$

Corollary 11 *Let $\theta > 1$ and*

$$\alpha_n = \theta^n - \theta^{n-1}, \quad n \geq 1.$$

Then

$$\lim_{n \to \infty} \frac{L(n)}{n} = \frac{\theta}{\theta - 1} \quad \text{a.s.}$$

Proof Combine Theorem 7 and Corollary 5. $\qquad\Box$

6.3 Case of $\alpha_n \sim A_n$

Theorem 8 *Assume (22) holds. Then*

$$\lim_{n \to \infty} \frac{A_{L(n)}}{n} = \infty \quad \text{a.s.} \tag{40}$$

Proof Combine Theorem 3 and (29). □

Corollary 12 *Let $\alpha_1 = 1$ and $\alpha_n = n! - (n-1)!$, $n \geq 2$. Then*

$$\frac{L(n) \log L(n)}{n} \to \infty \quad \text{a.s.}$$

Proof Combine Theorem 8 and Corollary 6.

A more precise asymptotic in case of Corollary 12 follows from (27).

Corollary 13 *Let $\alpha_1 = 1$ and $\alpha_n = n! - (n-1)!$, $n \geq 2$. Then*

$$\lim_{n \to \infty} \frac{L(n)}{n} = 1 \quad \text{a.s.}$$

Proof Combine (27) and (29). □

Acknowledgments The research of P. Doukhan has been conducted as part of the project Labex MME-DII (ANR11-LBX-0023-01).

References

Ballerini, R., & Resnick, S. I. (1987). Embedding sequences of successive maxima in extremal processes with applications. *Journal of Applied Probability, 24*(4), 827–837.

Beck, A. (1974). Conditional independence. *Bulletin of the American Mathematical Society, 80*(6), 1169–1172.

Billingsley, P. (1995). *Probability and measure* (3rd ed.). New York: Wiley.

Borovkov, K., & Pfeifer, D. (1995). On record indices and record times. *Journal of Statistical Planning and Inference, 45*(1–2), 65–79.

Buldygin, V. V., Indlekofer, K.-H., Klesov, O. I., & Steinebach, J. G. (2012). *Regularly varying functions and generalized renewal processes*. TBiMC: Kiev (Ukrainian).

Deheuvels, P., & Nevzorov, V. B. (1993). Records in the F^α-scheme. I. Martingale properties. *Zap. Nauchn. Sem. St.-Peterburg. Otdel. Mat. Inst. Steklov. (POMI), 207*(10), 19–36, 158. (English transl. in *Journal of Mathematical Sciences, 81*(1): 2368–2378, 1996).

Deheuvels, P., & Nevzorov, V. B. (1994). Records in the F^α-scheme. II. Limit theorems. *Zap. Nauchn. Sem. St.-Peterburg. Otdel. Mat. Inst. Steklov. (POMI), 216*, 42–51, 162. (English transl. in *Journal of Mathematical Sciences, 88*(1): 29–35, 1998).

Doukhan, P., Klesov, O. I., Pakes, A., & Steinebach, J. G. (2013). Limit theorems for record counts and times in the F^α-scheme. *Extremes, 16*, 147–171.

Gut, A. (2005). *Probability: A graduate course*. New York: Springer.

Gut, A., & Stadtmüller, U. (2013). Records in subsets of a random field. *Statistics & Probability Letters, 83*(3), 689–699.

Majerak, D., Nowak, W., & Zieba, W. (2005). Conditional strong law of large numbers. *Journal of Pure and Applied Mathematics, 20*(2), 143–156.

Nevzorov, V. B. (1985). On record times and inter-record times for sequences of nonidentically distributed random variables. *Zap. Nauchn. Sem. (LOMI), 142*, 109–118.

Nevzorov, V. B. (2001). *Records: Mathematical theory*. Providence, RI: American Mathematical Society.

Parameswaran, S. (1961). Partition functions whose logarithms are slowly oscillating. *Transactions of the American Mathematical Society, 100*, 217–240.

Rényi, A. (1962). Théorie des éléments saillants d'une suite d'observations. *Annals Faculty Science University Clermont-Ferrand, 8*, 7–12 (French).

Rényi, A. (1976). On the extreme elements of observations. In *Selected Papers of Alfred Rényi* (vol. 3, pp. 50–65). Budapest: Akadémiai Kiadó.

Weissman, I. (1995). Records from a power model of independent observations. *Journal of Applied Probability, 32*(4), 982–990.

Yang, M. (1975). On the distribution of the inter-record times in an increasing population. *Journal of Applied Probability, 12*, 148–154.

On Two Results of P. Deheuvels

Ju-Yi Yen and Marc Yor

Abstract We highlight some works in the probabilistic literature which are closely related to two results by P. Deheuvels.

1 Introduction

The aim of this paper is to show the closeness between two results of P. Deheuvels (and G. Martynov) on one hand, and two results of M. Yor (and C. Donati-Martin) on the other hand, and more generally related works in the probabilistic literature.

- A first result, discussed in particular cases by the two authors is the existence of Brownian motions obtained as integral linear transforms of a given Brownian motion. One such transform originates very naturally from considerations of Brownian Bridges.
- A second result, again discussed in particular cases by the two authors, is that two quadratic functionals of Brownian motion may have the same law.

2 The Filtration of Brownian Bridges and Related Topics

2.1 The filtration

Given a Brownian motion $(B_t, t \geq 0)$, we consider, for fixed t, the $\sigma-$field generated by $B_u - \frac{u}{t} B_t$, for $u < t$. This family of $\sigma-$fields increases with t, and deserves to be called the filtration of Brownian bridges. It is easily shown that this filtration is

---

J.-Y. Yen (✉)
Department of Mathematical Sciences, University of Cincinnati,
Cincinnati, OH 45221-0025, USA
e-mail: ju-yi.yen@uc.edu

M. Yor
Laboratoire de Probabilités et Modèles Aléatoires, Université Pierre et Marie Curie,
Paris, France

© Springer International Publishing Switzerland 2015
M. Hallin et al. (eds.), *Mathematical Statistics and Limit Theorems*,
DOI 10.1007/978-3-319-12442-1_17

the natural filtration of a Brownian motion, obtained by taking $B_t - \int_0^t \frac{B_s ds}{s}$ (see also Jeulin and Yor (1990), Najnudel et al. (2012)). This process is precisely one of the transforms of Brownian motion considered by P.D. in Deheuvels (1982). Call $\mathcal{T}$ this transform. It preserves Wiener measure, and it is shown in Yor (1992), Chap. 1, that this transform is ergodic. Indeed, if we iterate $\mathcal{T}$, then the nth Brownian motion is identical to $\int_0^t dB_s L_n(\log \frac{t}{s})$, (see formula (2.n) in Jeulin and Yor (1990)), where L_n denotes the nth Laguerre polynomial, that is, the canonical sequence of orthonormal polynomials for the probability on $\mathbb{R}_+$, with exponential density. Moreover, considered up to time t, the nth Brownian motion is independent of the values at time t of the kth Brownian motions, for $k \leq n - 1$. The ergodicity of $\mathcal{T}$ follows from the fact that these polynomials form an orthonormal basis for the L^2 space of this probability.

We also note that in Najnudel et al. (2012) a continuous group of Wiener transforms extending the iterates of $\mathcal{T}$ has been exhibited.

2.2 A related computation

The computations by Chiu (1995), inspired by Lévy, show in particular that:

$$X_n(t) = \int_0^t P_n\left(\frac{u}{t}\right) dB_u, \quad \text{for} \quad P_n(t) = \frac{2n+1}{n} t^n - \frac{n+1}{n},$$

is also a Brownian motion. Integration by parts shows that:

$$X_n(t) = B_t - \int_0^t \frac{ds}{s}(2n+1) \int_0^s dB_u \left(\frac{u}{s}\right)^n.$$

In particular, for $n = 0$, we recover the Brownian motion we considered above, i.e.;

$$\beta(t) - \int_0^t \frac{ds}{s} \beta(s),$$

which we now denote by $\gamma(t)$. In fact, we may view that $X_n(t)$ is a Brownian motion from simple manipulations of both X_n and γ: take $h(s) = \frac{s^{2n+1}}{2n+1}$.

We leave it to the reader to establish that $\int_0^t s^n dX_n(s)$ is distributed as γ time-changed with $h(t)$, which is also distributed as $\int_0^t s^n d\gamma(s)$. Consequently X_n and γ have the same law, i.e. X_n is a Brownian motion. Thus, the Chiu-Lévy computations are not so far from the content of Sect. 2.1, which corresponds to $n = 0^+$. We also note that the covariance of X_n may be shown directly to be that of Brownian motion. Moreover, the same computation shows that the filtration of X_n is strictly contained in that of B because $\int_0^t s^n dB_s$ is independent of the past of X_n, up to time t.

3 Brownian Quadratic Functionals with the Same Law (Deheuvels and Martynov 2008), (Donati-Martin and Yor 1991)

The main aim of Deheuvels and Martynov (2008) was to enlarge the class of Gaussian processes for which an explicit Karhunen-Loeve decomposition may be computed (see also Pycke (2001)). This class is not so large. Then, this leads to identities in law for the integrals of the squares of such processes. It has been noticed by several authors, including P.D. and M.Y., that the variance of the Brownian path, on the unit time interval, is distributed as the integral of the square of the Brownian bridge, on the same unit interval. This result may be obtained as a consequence of a Fubini argument involving two independent Brownian motions, namely:

$$\int_0^1 \int_0^1 dB_u dC_s \psi(u, s) = \int_0^1 \int_0^1 dC_s dB_u \psi(u, s)$$

Taking characteristic functions of these two quantities, one finally obtains the identity in law:

$$\int_0^1 du \left(\int_0^1 dC_s \psi(u, s) \right)^2 \overset{\text{(law)}}{=} \int_0^1 du \left(\int_0^1 dC_s \psi(s, u) \right)^2$$

This may be called a Fubini-Wiener identity in law.

The identity in law between the variance of the Brownian path and the integral of the square of the Brownian bridge may be obtained as a particular case of a Fubini-Wiener identity in law. The origin of the interest by M.Y. in these identities in law comes from computations by physicists of laws related to radius of gyration (Duplantier 1989) (see also Chan (1994), Dean and Jansons (1992), Pycke (2001), Pycke (2005), Yen and Yor (2013)), as well as looking for some explanations of the celebrated Ciesielski-Taylor identities in law: the total time spent by Brownian motion in the unit sphere in $\mathbb{R}^{d+2}$ is distributed as the first hitting time of 1 by the radial part of Brownian motion in $\mathbb{R}^d$ (Ciesielski and Taylor 1962).

With the help of the corresponding Ray-Knight theorems for the local times of the Bessel processes involved, the Fubini-Wiener identity in law allows to recover the Ciesielski-Taylor identities. One may of course argue that the use of the Ray-Knight theorem is a formidable hammer to break this stone! However, from the purely one-dimensional Brownian viewpoint, this detour brings out remarkable identities in law which one might not have noticed otherwise. See Yor (1992), Chap. 4.

References

Chan, T., Dean, David S., Jansons, Kalvis M., Rogers, L.C.G. (1994). On polymer conformations in elongational flows. *Communications in Mathematical Physics, 160*(2), 239–257.

Chiu, Y. (1995). From an example of Lévy's. *Séminaire de Probabilités, XXIX* (pp. 162–165). New York: Springer.

Ciesielski, Z., & Taylor, S. J. (1962). First passage times and sojourn times for Brownian motion in space and the exact Hausdorff measure of the sample path. *Transactions of the American Mathematical Society, 103*, 434–450.

Dean, D. S., & Jansons, K. M. (1992). A note on the integral of a Brownian bridge. *Proceedings of the Royal Society of London Series A, 437*(1901), 729–730.

Deheuvels, P. (1982). Invariance of Wiener processes and of Brownian bridges by integral transforms and applications. *Stochastic Processes and Their Applications, 13*(3), 311–318.

Deheuvels, P., & Martynov, G. V. (2008). A Karhunen-Loeve decomposition of a Gaussian process generated by independent pairs of exponential random variables. *Journal of Functional Analysis, 255*(9), 2363–2394.

Donati-Martin, C., & Yor, M. (1991). Fubini's theorem for double Wiener integrals and the variance of the Brownian path. *Annales de l'Institut Henri Poincaré Probability and Statistics, 27*(2), 181–200.

Duplantier, B. (1989). Areas of planar Brownian curves. *Journal of Physics A, 22*(15), 3033–3048.

Jeulin, T., & Yor, M. (1990). Filtration des ponts browniens et équations différentielles stochastiques linéaires. *Séminaire de Probabilités, XXIV, 1988/89* (pp. 227–265). Berlin: Springer.

Najnudel, J., Stroock, D., & Yor, M. (2012). On a flow of transformations of a Wiener space. *Stochastic analysis and related topics* (pp. 119–131). Boston: Birkhäuser.

Pycke, J.-R. (2005). Sur une identité en loi entre deux fonctionnelles quadratiques du pont Brownien. *Comptes Rendus Mathematique, 340*(5), 373–376.

Pycke, J.-R. (2001). A generalization of the Karhunen-Loeve expansion of the Brownian bridge. *Comptes Rendus de l'Academie des Sciences Series I Mathematics, 333*(7), 685–688.

Yen, J.-Y., & Yor, M. (2013). On an identity in law between Brownian quadratic functionals. *Statistics & Probability Letters, 83*(9), 2015–2018.

Yor, M. (1992). *Some aspects of Brownian motion*. Part I. *Lectures in Mathematics ETH Zürich* Basel: Birkhäuser.

Some Topics in Probability Theory

Ju-Yi Yen and Marc Yor

Abstract We present a succinct discussion of a number of topics in Probability Theory which have been of interest in recent years.

1 The Set of Martingale Laws

Consider, on the Skorokhod space of càdlàg functions, all probabilities $\mathbb{P}$ which make the canonical process of coordinates a martingale. Call $\mathcal{M}$ this set. Clearly, it is a convex set, and it may be of interest to characterize its extremal points. An application of Hahn-Banach theorem (to the pair H^1-BMO, and the fact that a BMO martingale is locally bounded) allows to show that $\mathbb{P}$ in $\mathcal{M}$ is extremal if and only if any martingale under $\mathbb{P}$ may be written as the sum of a constant and of a stochastic integral with respect to the canonical (martingale) process. A particularly illustrative example is that of $\mathbb{P} = \mathbb{W}$, Wiener measure. Indeed, on one hand, from Lévy's martingale characterization of Brownian motion, it is easily shown that $\mathbb{W}$ is extremal in $\mathcal{M}$. On the other hand, it is a theorem (due to Itô) that all Brownian martingales may be written as the sum of a constant and of a stochastic integral with respect to Brownian Motion. That these two properties hold for $\mathbb{W}$ is not a mere coincidence, but is explained by the general statement above (Jacod and Yor 1977).

To our knowledge, the first author who tried to connect the two properties, namely: extremality of $\mathbb{P}$, and martingale representation property under $\mathbb{P}$ is Dellacherie (1974, 1975). Dellacherie (1975) corrects Dellacherie (1974) partially, but the local boundedness property which seems necessary for a correct proof is only found in Jacod and Yor (1977).

J.-Y. Yen (✉)
Department of Mathematical Sciences, University of Cincinnati,
Cincinnati, OH 45221-0025, USA
e-mail: ju-yi.yen@uc.edu

M. Yor
Laboratoire de Probabilités et Modèles Aléatoires, Université Pierre et Marie Curie,
Paris, France

© Springer International Publishing Switzerland 2015
M. Hallin et al. (eds.), *Mathematical Statistics and Limit Theorems*,
DOI 10.1007/978-3-319-12442-1_18

The use of the H^1-BMO duality in this topic is reminiscent to that of the L^1-L^∞ duality in the characterization of extremal probabilities, solutions of a (generalized) moment problem. In fact, it is a theorem, according to Douglas and Naimark (independently) the extremal points $\mathbb{P}$ of such a moment problem are those for which the vector space generated by the functions defining the problem and the constant function 1 is dense in $L^1(\mathbb{P})$. Yor (1978) explains how to relate the two frameworks and extremality results.

2 Strong and Weak Brownian Filtrations

We shall say that a filtration $\mathscr{F}_t$ is strongly Brownian if it is the natural filtration of a Brownian Motion. On the other hand, we shall say $\mathscr{F}_t$ is weakly Brownian if there exists a Brownian Motion B for this filtration such that all martingales for this filtration may be written as the sum of a constant and a stochastic integral with respect to B (but the integrand is predictable with respect to $\mathscr{F}_t$). Any strongly Brownian filtration is weakly Brownian (Itô's theorem recalled in Sect. 1). It is natural to ask whether any weakly Brownian filtration is strongly Brownian. The answer turns out to be negative:

- it is easily shown that on the canonical space of continuous functions, endowed with any probability $\mathbb{Q}$ equivalent to Wiener measure $\mathbb{W}$, the canonical filtration is weakly Brownian; however, it has been shown by Dubins et al. (1996) that there are infinitely many $\mathbb{Q}$'s such that $\mathscr{F}_t$ is not strongly Brownian under $\mathbb{Q}$;
- the filtration of Walsh's Brownian Motion with N rays, for $N \geq 3$, is weakly but not strongly Brownian, another result due to Tsirelson (1997). A posteriori, a clear explanation of this result emerged as it was shown that M. Barlow's conjecture holds: for g, the end of a predictable set in a strong Brownian filtration, the progressive σ-field up to g can only differ from the predictable one by, at most the addition of a set. This is clearly not the case for Walsh's Brownian motion with N rays, $N \geq 3$, and g the last zero of this process before time 1;
- there exist time changes of the canonical Brownian filtration such that the time changed filtration is weakly, but not strongly Brownian, a result due to Émery and Schachermayer (1999).

3 Weak Brownian Motions of Any Given Order

Although the adjective weak is used again here, this topic has nothing to do with topic in Sect. 2. It was suggested by a question of Stoyanov in his book of counter examples (Stoyanov 1987): does there exist, for a given integer k, a process which has the same k-dimensional marginals as Brownian Motion? The answer is yes, as was proven by Föllmer et al. (2000), by constructing probabilities $\mathbb{Q}$ equivalent to $\mathbb{W}$, the Wiener measure, such that the k-dimensional marginals of the canonical process under $\mathbb{Q}$ are those under $\mathbb{W}$.

4 Martingales with One-Dimensional Brownian Marginals

Note that this topic differs from Sect. 3, where the processes constructed there are not martingales, but, in general, semimartingales. For constructions of martingales, see Albin (2008), Baker et al. (2011), Hamza and Klebaner (2007), Madan and Yor (2002). There are at least two versions of these constructions, one where it is required that the martingale is continuous, e.g., Albin (2008); the other where discontinuity is allowed, e.g., Madan and Yor (2002).

5 Explicit Skorokhod Embedding

The problem is now well known: given a centered probability μ on $\mathbb{R}$, find a stopping time T of Brownian motion B, such that B_T is μ distributed and $B_{t \wedge T}$ is a uniformly integrable martingale. Although J. Obłój found 21 different solutions scattered in the literature (Obłój 2004), few of them are explicit, as in general, the authors proceed by finding solutions for simple μ's then pass to the limit.

Azéma-Yor found that if $T_\mu := \inf\{t : S_t \geq H_\mu(B_t)\}$, where $S_t = \underset{s \leq t}{\sup} B_s$, and the Hardy-Littlewood function $H_\mu(x)$ is defined as:

$$H_\mu(x) = \frac{1}{\mu([x, \infty))} \int_{[x,\infty)} t \, d\mu(t),$$

then T_μ solves Skorokhod problem for μ (Azéma and Yor 1979). To prove this result, Azéma and Yor (1979) use first-order stochastic calculus, whereas Rogers (1981) uses excursion theory. Madan and Yor (2002) remarked that for a family μ_t such that the corresponding Hardy-Littlewood family is pointwise increasing in t, the Brownian motion B taken at those stopping times is a martingale.

6 Peacocks and Associated Martingales

We say that a process X_t is a peacock (:PCOC) if, when composed with any convex function, the expectation of the obtained process is increasing in t. It is a consequence of Jensen's inequality that a martingale is a peacock. Conversely, it is a deep theorem due to Kellerer (1972) that a peacock is a process which has the same one-dimensional marginals as a martingale. Moreover, this martingale may be chosen Markovian. Thus, at least, two questions arise:

1. How to create peacocks in a systematic way? One answer is: the arithmetic average of a martingale is always a peacock. The original example of this seems to be due to Carr et al. (2008) who took for a martingale the geometric Brownian motion;

2. Given a peacock, how to associate to it a martingale with the same marginals? So far, there does not seem to exist a general answer. But, in their monograph, Hirsch, Profeta, Roynette, and Yor exhibit a number of general cases where some construction may be done (Hirsch 2011).

7 (Brownian) Penalisations

Consider $\mathbb{W}$, the Wiener measure and H_t, positive, a family of adapted probability densities (with respect to the canonical filtration). This allows to create a family $\mathbb{W}_t$ of probabilities on $\mathscr{F}_t$. The penalisation problem is to find whether, as $t \to \infty$, $\mathbb{W}_t$ when restricted to $\mathscr{F}_s$, for fixed s, converges weakly, and if so to describe the limit law. Two monographs have been devoted to this problem: Roynette and Yor (2009) and Najnudel et al. (2009), the first is a collection of examples, the second aims at finding general convergence criterions.

8 Martingales with the Wiener Chaos Decomposition

It is a well-known result, due to Wiener, that every L^2-martingale for the Brownian filtration may be written as the sum of a series of multiple integrals with respect to Brownian motion, with the series of squares of (deterministic) integrands, integrated with respect to Lebesgue measure on their corresponding sets of definitions, converging. A similar result is true for the martingale of the compensated Poisson process. For a long time, it was thought that these were the only two martingales with Wiener chaos decomposition. But, Émery (1989) showed that Azéma's martingale, i.e., the projection of Brownian motion on the filtration of Brownian signs up to time t, also satisfies this property. See also Azéma and Yor (1989) for another proof. Émery (1989) considered more generally some martingales solutions of so-called structure equations, some of which also enjoy the Wiener chaos decomposition; he also wrote a synthesis Émery (1991).

9 Asymptotics of Planar Brownian Windings

A number of limit theorems (in law) for additive functionals of one- or two-dimensional Brownian motion have been obtained throughout the years. This is in particular the case for the winding number of planar Brownian motion up to time t, which, when multiplied by $\frac{2}{\log(t)}$ converges in law toward a standard Cauchy variable Spitzer (1958). This result admits a number of multivariate extensions, in particular: with the same normalization $\frac{2}{\log(t)}$, the vector of n Brownian winding numbers around different points converges in law toward a random vector with (linked) Cauchy marginals Pitman and Yor (1986). The dependence between the different Cauchy marginals may be explained from the Kallianpur and Robbins (1953) asymptotic theorem: normalized by $\frac{1}{\log(t)}$, the time spent in an integrable Borel set by two-dimensional Brownian motion up to time t is asymptotically exponentially distributed.

10 How to Modify the Burkholder-Davis-Gundy Inequalities up to Any Time?

A version of the BDG inequalities is: for any positive p, the supremum of the absolute value of Brownian motion up to a stopping time T has L^p moment which is equivalent to that of $\sqrt{T}$. How could one modify this result when T is replaced by any random time L? A technique consists in making L a stopping time and to consider the semimartingale decomposition of Brownian motion stopped at L. Then, an extension of Fefferman's inequality allows to obtain the desired variants. For details, see Yor (1985).

References

Albin, J. M. P. (2008). A continuous non-Brownian motion martingale with Brownian motion marginal distributions. *Statistics & Probability Letters*, 78(6), 682–686.

Azéma, J., & Yor, M. (1989). Étude d'une martingale remarquable. *Séminaire de Probabilités, XXIII* (pp. 88–130). Berlin: Springer.

Azéma, J., & Yor, M. (1979). Une solution simple au problème de Skorokhod. *Séminaire de Probabilités, XIII (University of Strasbourg, Strasbourg, 1977/78)*, (pp. 90–115). Berlin: Springer.

Baker, D., Donati-Martin, C., & Yor, M. (2011). A sequence of Albin type continuous martingales with Brownian marginals and scaling. *Séminaire de Probabilités, XLIII* (pp. 441–449). Berlin: Springer.

Carr, P., Ewald, C.-O., & Xiao, Y. (2008). On the qualitative effect of volatility and duration on prices of Asian options. *Finance Research Letters*, 5(3), 162–171.

Dellacherie, C. (1974). Intégrales stochastiques par rapport aux processus de Wiener ou de Poisson. *Séminaire de Probabilités, VIII (University Strasbourg, année universitaire 1972–1973)*, (pp. 25–26). Berlin: Springer.

Dellacherie, C. (1975). Correction à: Intégrales stochastiques par rapport aux processus de Wiener ou de Poisson (*Séminaire de Probabilités, VIII (University Strasbourg, année universitaire 1972–1973)*, (pp. 25–26). Springer: Berlin, 1974). *Séminaire de Probabilités IX (Seconde Partie, University Strasbourg, Strasbourg, années universitaires 1973/1974 et 1974/1975)*, (pp. 494). Springer: Berlin.

Dubins, L., Feldman, J., Smorodinsky, M., & Tsirelson, B. (1996). Decreasing sequences of σ-fields and a measure change for Brownian motion. *Annals of Probability*, 24(2), 882–904.

Émery, M., & Schachermayer, W. (1999). Brownian filtrations are not stable under equivalent time-changes. *Séminaire de Probabilités, XXXIII* (pp. 267–276). Berlin: Springer.

Émery, M. (1989). On the Azéma martingales. *Séminaire de Probabilités, XXIII* (pp. 66–87). Berlin: Springer.

Émery, M. (1991). Quelques cas de représentation chaotique. *Séminaire de Probabilités, XXV* (pp. 10–23). Berlin: Springer.

Föllmer, H., Wu, C.-T. & Yor, M. (2000). On weak Brownian motions of arbitrary order. *Annales de l'Institut Henri Poincaré Probabilités et Statistiques, 36*(4), 447–487.

Hamza, K., & Klebaner, F. C. (2007). A family of non-Gaussian martingales with Gaussian marginals. *Journal of Applied Mathematics and Stochastic Analysis.* doi:10.1155/2007/92723.

Hirsch, F., Profeta, C., Roynette, B., & Yor, M. (2011). *Peacocks and associated martingales, with explicit constructions* (Vol. 3). *Bocconi & Springer Series Milan*: Springer.

Jacod, J., & Yor, M. (1977). Étude des solutions extrémales et représentation intégrale des solutions pour certains problèmes de martingales. *Zeitschrift für Wahrscheinlichkeitstheorie und Verw. Gebiete, 38*(2), 83–125.

Kallianpur, G., & Robbins, H. (1953). Ergodic property of the Brownian motion process. *Proceedings of the National Academy of Sciences of the United States of America, 39*, 525–533.

Kellerer, H. G. (1972). Markov-Komposition und eine Anwendung auf Martingale. *Mathematische Annalen, 198*, 99–122.

Madan, D., & Yor, M. (2002). Making Markov martingales meet marginals: With explicit constructions. *Bernoulli, 8*(4), 509–536.

Najnudel, J., Roynette, B., & Yor, M. (2009). *A global view of Brownian penalisations MSJ Memoirs.* (Vol. 19). Tokyo: Mathematical Society of Japan.

Obłój, J. (2004). The Skorokhod embedding problem and its offspring. *Probability Surveys, 1*, 321–390.

Pitman, J., & Yor, M. (1986). Asymptotic laws of planar Brownian motion. *Annals of Probability, 14*(3), 733–779.

Rogers, L. C. G. (1981). Williams' characterisation of the Brownian excursion law: Proof and applications. *Séminaire de Probabilités, XV* (pp. 227–250). Berlin: Springer.

Roynette, B., & Yor, M. (2009). *Penalising Brownian paths.* Berlin: Springer.

Spitzer, F. (1958). Some theorems concerning 2-dimensional Brownian motion. *Transactions of the American Mathematical Society, 87*, 187–197.

Stoyanov, J. M. (1987). *Counterexamples in probability.* Chichester: John Wiley & Sons Ltd.

Tsirelson, B. (1997). Triple points: From non-Brownian filtrations to harmonic measures. *Geometric and Functional Analysis, 7*(6), 1096–1142.

Yor, M. (1978). Sous-espaces denses dans L^1 ou H^1 et représentation des martingales. *Séminaire de Probabilités, XII.* (*University of Strasbourg, Strasbourg, 1976/1977*), (pp. 265–309). Berlin: Springer.

Yor, M. (1985). Inegalités de martingales continues arrétées à un temps quelconque: Le role de certains espaces BMO. *Grossissements de filtrations: Exemples et applications* (pp. 147–171). Berlin: Springer.

Publication List of Paul Deheuvels—2014

[1] Sur la convergence de sommes de minimums de variables aléatoires (1973). C. R. Acad. Sci. Paris Ser. A-B 276 309–312 [MR 48 #5156a]

[2] Sur la convergence de certaines suites de variables aléatoires (1973). C. R. Acad. Sci. Paris Ser. A-B 276 641–644 [MR 48 #5156b]

[3] Sur une application de la théorie des processus de renouvellement Ã l'estimation de la densité d'une variable aléatoire (1973). C. R. Acad. Sci. Paris Ser. A-B 276 943–946 [MR 48 #12710]

[4] Sur une famille d'estimateurs de la densité d'une variable aléatoire (1973). C. R. Acad. Sci. Paris Ser. A-B 276 1013–1015 [MR 48 #12711]

[5] Sur l'estimation séquentielle de la densité (1973). C. R. Acad. Sci. Paris Ser. A-B 276 1119–1121 [MR 48 #12712]

[6] Valeurs extrémales d'échantillons croissants d'une variable aléatoire réelle (1974). Annales de l'Institut Henri Poincaré Ser. B 10, f.1, 89–114 [MR 50 #11404]

[7] Majoration et minoration presque sûre des extrémums de suites de variables aléatoires indépendantes de même loi (1974). C. R. Acad. Sci. Paris Ser.A 278 513–516 [MR 50 #3317]

[8] Majoration et minoration presque sûre des extrema de processus Gaussiens (1974). C. R. Acad. Sci. Paris Ser. A 278 989–992 [MR 51 #1930]

[9] Conditions nécessaires et suffisantes de convergence ponctuelle presque sûre et uniforme presque sûre des estimateurs de la densité (1974). C. R. Acad. Sci. Paris Ser. A 278 1217–1220 [MR 49 #10032]

[10] Majoration et minoration presque sûre optimale des éléments de la statistique ordonnée d'un échantillon croissant de variables aléatoires indépendantes (1974). Rendi Conti della Academia Nazionale dei Lincei 8, 56, f.5, 707–719 [MR 52 #15625]

[11] Estimation non paramétrique de la densité par histogrammes généralisés (1977). Publications de l'Institut de Statistique de l'Université de Paris. 22, f.1, 1–24 [MR #81h:62071]

[12] Estimation non paramétrique de la densité par histogrammes généralisés (II) (1977). Revue de Statistique Appliquée 25, f.3, 5–42 [MR 58 #18876]

© Springer International Publishing Switzerland 2015

M. Hallin et al. (eds.), *Mathematical Statistics and Limit Theorems*,

DOI 10.1007/978-3-319-12442-1

[13] Caractérisation complète des lois extrêmes multivariées et de la convergence des types extrêmes (1978). Publications de l'Institut de Statistique de l'Université de Paris. 23, f.3, 1–36.

[14] Propriétés d'existence et propriétés topologiques des fonctions de dépendance avec applications à la convergence des types pour les lois multivariées. (1979). C. R. Acad. Sci. Paris, Ser. A 288 217–220 [MR0524771 (80e:60038)]

[15] Détermination complète du comportement asymptotique en loi des valeurs extrêmes multivariées d'un échantillon de vecteurs aléatoires indépendants (1979). C. R. Acad. Sci. Paris, Ser. A 288, f.3, 631–634 [MR #80c:60038]

[16] Détermination des lois limites jointes de l'ensemble des points extrêmes d'un échantillon multivarié (1979). C. R. Acad. Sci. Paris, Ser. A 288 631–644 [MR #80c:60034]

[17] Estimation non paramétrique de la densité compte tenu d'informations sur le support (1979). Revue de Statistique Appliquée 27, f.3, 47–68 (with P. Hominal) [MR #81e:62040]

[18] Estimation séquentielle de la densité (1979). In: Contribuciones en Probabilidad y Estadistica Matematica Enseñanza de la Matematica y Analysis. 156–168, Grindley, Granada, Spain [MR #81h:62072]

[19] La fonction de dépendance empirique et ses propriétés, un test non paramétrique d'indépendance (1979). Bulletin de l'Académie Royale de Belgique, Classe des Sciences (5) 65, f.6, 274–292 [MR #81h:62073]

[20] Non-parametric tests of independence (1980). In: Statistique Non Paramétrique Asymptotique (J. P. Raoult, Edit.) 95–107, Lecture Notes in Mathematics 821, Springer Verlag, Berlin [MR #82c:62061]

[21] Estimation automatique de la densité (1980). Revue de Statistique Appliquée 28, f.1, 23–55 (with P. Hominal) [MR #82j:62026]

[22] Some applications of dependence functions: nonparametric estimates of extreme value distributions and a Kiefer-type bound for the uniform test of independence (1980). In: Nonparametric Statistical Inference. Colloquia Math. Soc. Jànos Bolyai 32, 183–201, North Holland, Amsterdam [MR #85g:62064]

[23] The decomposition of infinite order and extreme multivariate distributions (1980). In: Asymptotic Theory of Statistical Tests and Estimation. (I. M. Chakravarti, Edit.), 259–286, Academic Press, New York [MR #82j:62032]

[24] A Kolmogorov-Smirnov test for independence (1981). Revue Roumaine de Mathématiques Pures et Appliquées 26, f.2, 213–226 [MR #83c:62064]

[25] La prévision des séries économiques, une technique subjective (1981). Archives de l'I.S.M.E.A. 34, f.4, 729–748.

[26] An asymptotic decomposition for multivariate distribution-free tests of independence (1981). Journal of Multivariate Analysis 11 102–113 [MR #82g:62067]

[27] Multivariate tests of independence (1981). In: Analytical Methods in Probability Theory. (D. Dugué, E. Lukacs and V. K. Rohatgi, Edit.), 102–113, Lecture Notes in Mathematics 861, Springer Verlag, Berlin [MR #83g:62074]

[28] A non-parametric test for independence (1981). Publications de l'Institut de Statistique de l'Université de Paris 26, f.2, 29–50.

[29] The strong approximation of extremal processes (1981). Zeitschrift für Wahrscheinlichkeitstheorie und Verwandte Gebiete 58 1–6 [MR0635267 (82k:60065)]

[30] Univariate extreme values - Theory and applications (1981). Proceedings of the 43d Session of the International Statistical Institute. 49, f.2, 837–858 [MR0820978]

[31] Strong limiting bounds for maximal uniform spacings (1982). Annals of Probability 10 1058–1065 [MR0672307 (84j:60039)]

[32] Spacings, record times and extremal processes (1982). In: Exchangeability in Probability and Statistics (G. Koch and F. Spizzichino, Edit.), North Holland, Amsterdam, 223–243 [MR0675978 (84e:62082)]

[33] A construction of extremal processes (1982). In: Probability and Statistical Inference (W. Grossmann, G. C. Pflug and W. Wertz, Edit.), 53–58, Reidel, Dordrecht [MR0668205 (84c:60112)]

[34] Sur des tests d'ajustement indépendants des paramètres (1982). In: Actas, II Coloquio de Estatistica, A Estatistica nos Processos Estocasticos, Departamente de Matematica, Universidade de Coimbra, 7–18.

[35] On record times associated with k-th extremes (1982). Proceedings of the 3rd Pannonian Symposium on Mathematical Statistics (J. Mogyoródi, I. Vincze and W. Wertz, Edit.), Akadémiai Kiadó, Budapest [MR0758999]

[36] Invariance of Wiener processes and Brownian bridges by integral transforms and applications (1982). Stochastic Processes and their Applications. 13, f.3, 311–318 [MR0671040 (83k:60044)]

[37] L'encadrement asymptotique des éléments de la série d'Engel d'un nombre réel (1982). C. R. Acad. Sci. Paris, Ser. A 295 21–24 [MR0669252 (83h:10095)]

[38] La Probabilité, le Hasard et la Certitude (1982). Presses Universitaires de France, Paris [MR0694839 (84i:60001)]

[39] Point processes and multivariate extreme values (1983). Journal of Multivariate Analysis. 13 257–272 [MR0705550 (85c:60069)]

[40] The complete characterization of the upper and lower class of the record and inter-record times of an i.i.d. sequence (1983). Zeitschrift für Wahrscheinlichkeitstheorie und Verwandte Gebiete. 62, 1–6 [MR0684204 (84f:60033)]

[41] The strong approximation of extremal processes (II) (1983). Zeitschrift für Wahrscheinlichkeitstheorie und Verwandte Gebiete. 62, 7–15 [MR0684205 (84c:60050)]

[42] Upper bounds for k-th maximal spacings (1983). Zeitschrift für Wahrscheinlichkeitstheorie und Verwandte Gebiete. 62, 465–474 [MR0690571 (84g:60055)]

[43] Strong bounds for multidimensional spacings (1983). Zeitschrift für Wahrscheinlichkeitstheorie und Verwandte Gebiete. 64, 411–424 [MR0717751 (85g:60038)]

[44] Indépendance multivariée partielle et inégalités de Fréchet (1983). In: Studies in Probabilities and Related Topics (Papers in Honour of Octav Ionicescu on

his 90th Birthday) (M. Demetrescu and M. Iosifescu, Edit.), 145–155, Nagard, Bucarest [MR0778216 (86j:60054)]

[45] Strong bounds for the maximal k-spacing when $k \leq c \log n$ (1984). Zeitschrift für Wahrscheinlichkeitstheorie und Verwandte Gebiete 66, 315–334 (with L. Devroye) [MR0751573 (85k:60043)]

[46] Strong limit theorems for maximal spacings from a general univariate distribution (1984). Annals of Probability 12, 1181–193 [MR0757775 (86a:60041)]

[47] How to analyze bioequivalence studies - The right use of confidence intervals (1984). Journal of Organizational Behavior and Statistics 1, f.1, 1–15.

[48] Asymptotic results for the pseudo-prime sequence generated by Hawkins's random sieve: twin primes and Riemann's hypothesis (1984). In: Proceedings of the 7th Conference in Probability Theory, Brasov (M. Iosifescu, Edit.), 109–115, Editura Academiei, Bucarest [MR0867422 (87m:11097)]

[49] Probabilistic aspects of multivariate extremes (1984). In: Statistical Extremes and Applications (J. Tiago de Oliveira, Edit.), 117–130, D. Reidel, Dordrecht [MR0784817 (86i:60064)]

[50] Strong approximations of records and record times (1984). In: Statistical Extremes and Applications (J. Tiago de Oliveira, Edit.), 491–496, D. Reidel, Dordrecht [MR0784838 (86h:60068)

[51] Strong approximation in extreme values, theory and applications (1984). In: Limit Theorems in Probability and Statistics (P. Révész, Edit.) Vol. 1, 369–404, Colloquia Math János Bolyai 36, North Holland, Amsterdam [MR0807566 (87g:60031)]

[52] The characterization of distributions by order statistics and record values - A unified approach (1984). Journal of Applied Probability. 21 326–334 (Corr. (1985). 22 997) [MR0741134 (85j:62009)]

[53] Point processes and multivariate extreme values (II) (1985). In: Multivariate Analysis VI (P. R. Krishnaiah, Edit.), 145–164, North Holland, Amsterdam.

[54] On the Erdős-Rényi theorem for random fields and sequences and its relationships with the theory of runs and spacings (1985). Zeitschrift für Wahrscheinlichkeitstheorie und Verwandte Gebiete. 70, 91–115 [MR0795790 (86k:60045)]

[55] Lois de type Pareto et applications à la théorie mathématique du risque (1985). Rend. Sem. Mat. Univers. Politecn. Torino. 43, f.1, 25–41 [MR0859847 (88a:62285)]

[56] Kernel estimates of the tail index of a distribution (1985). Annals of Statistics. 13 1050–1077 (with S. Csörgő and D. M. Mason) [MR0803758 (87b:62043)]

[57] Spacings and applications (1985). In: Probability and Statistical Decision Theory (F. Konecny, J. Mogyoródi et W. Wertz, Edit.), Vol. A, 1–30, Reidel, Dordrecht [MR0851016 (87k:60065)]

[58] The limiting behavior of the maximal spacing generated by an i.i.d. sequence of Gaussian random variables (1985). Journal of Applied Probability. 22, 816–827 [MR0808861 (87b:60045)]

[59] Exact convergence rate in the limit theorems of Erdős-Rényi and Shepp (1986). Annals of Probability. 14, 209–223 (with L. Devroye et J. Lynch) [MR0815966 (87d:60032)]

[60] Exact convergence rate of an Erdős-Rényi strong law for moving quantiles (1986). Journal of Applied Probability. 23, 355–369 (with J. Steinebach) [MR0839991 (87i:60033)]

[61] Strong laws for the k-th order statistic when $k \leq c \log n$ (1986). Probability Theory and Related Fields. 72, 179–186 [MR0835163 (87j:60048)]

[62] On the influence of the extreme values on the maximal spacing (1986). Annals of Probability. 14, 194–208 [MR0815965 (87e:60044)]

[63] A semigroup approach to Poisson approximation (1986). Annals of Probability. 14, 663–676 (with D. Pfeifer) [MR0832029 (87e:60043)]

[64] Simple random walk on the line in random environment (1986). Zeitschrift für Wahrscheinlichkeitstheorie und Verwandte Gebiete. 72, 215–230 (with P. Révész) [MR0836276 (87m:60193)]

[65] Operator semigroups and Poisson convergence in selected metrics (1986). Semigroup Forum. 34, 203–224 (with D. Pfeifer) [MR0868255 (88f:60092a-b)]

[66] Many heads in a short block (1987). In: Mathematical Statistics and Probability Theory (M. L. Puri, P. Révész et W. Wertz, Edit.), 53–67, D. Reidel, Dordrecht (with P. Erdős, K. Grill and P. Révész) [MR0922687 (89b:60073)]

[67] Weak laws for the increments of Wiener processes and Brownian bridges and applications (1987). In: Mathematical Statistics and Probability Theory (M. L. Puri, P. Révész and W. Wertz, Edit.), 69–87, D. Reidel, Dordrecht (with P. Révész) [MR0922688 (89a:60057)]

[68] Limit laws of the Erdős-Rényi-Shepp type (1987). Annals of Probability. 15, 1363–1386 (with L. Devroye) [MR0905337 (88f:60055)]

[69] Exact convergence rates in strong approximation laws for large increments of partial sums (1987). Probability Theory and Related Fields. 76, 369–393 (with J. Steinebach)[MR0912661 (88m:60081)]

[70] Semigroups and Poisson Approximation (1987). In: New Perspectives in Theoretical and Applied Statistics (M. L. Puri, J. Vilaplana and W. Wertz, Edit.) 439–448, Wiley, New York (with D. Pfeifer).

[71] Exact convergence rates in Erdős-Rényi-type theorems for renewal processes (1987). Annales de l'Institut Henri Poincaré. 23, 195–207 (with J. N. Bacro and J. Steinebach) [[MR0891710 (88k:60055)]

[72] An approximation of stopped sums with applications in queuing theory (1987). Advances in Applied Probability. 19, 674–690 (with M. Csörgő and L. Horvàth) [[MR0903542 (89c:60045)]

[73] The asymptotic behavior of sums of exponential extreme values (1988). Bulletin des Sciences Mathématiques. 112, 211–233 (with D. M. Mason) [MR0967146 (89k:60040)]

[74] Strong approximations of k-th records and k-th record times by Wiener processes (1988). Probability Theory and Related Fields. 77, 195–209 [MR0927237 (89b:60086)]

[75] Limit laws for the modulus of continuity of the partial sum process and for the Shepp statistic (1988). Stochastic Processes Appl. 29, 223–245 (with J. Steinebach) [MR0958501 (89i:60066)]

[76] The almost sure behavior of maximal and minimal k_n-spacings when $k_n = O(\log n)$ (1988). Journal of Multivariate Analysis. 24, 155–176 (with J. H. J. Einmahl, D. M. Mason and F. Ruymgaart) [MR0925137 (89a:60078)]

[77] Almost sure convergence of the Hill estimator (1988). Mathematical Proceedings of the Cambridge Philosophical Society. 104, 371–384 (with E. Haeusler and D. M. Mason) [MR0948921 (89i:62043)]

[78] Poisson approximations in selected metrics by coupling and semigroup methods with applications (1988). Journal of Statistical Planning and Inference. 20, 1–22 (with A. Karr, D. Pfeifer and R. Serfling) [MR0964424 (90a:60040)]

[79] Poisson approximations of multinomial distributions and point processes (1988). Journal of Multivariate Analysis. 25, 65–89 (with D. Pfeifer) [MR0935295 (89d:60063)]

[80] A new semigroup technique in Poisson approximation (1988). Semigroup Forum. 201, 1–13 (with D. Pfeifer and M. L. Puri) [MR0976202 (90b:60025)]

[81] On the relationship between Uspensky's theorem and Poisson approximations (1988). Annals of the Institute of Mathematical Statistics. 40, 671–681 (with D. Pfeifer) [MR0996692 (90m:60027)]

[82] Strong laws for the k-th order statistic when $k \leq c \log n$ (II) (1989). In: Extreme Value Theory (J. Hüsler and R. Reiss, Edit.), 21–35, Lecture Notes in Statistics 51, Springer-Verlag, Berlin [MR0992045 (90f:60062)]

[83] A Bahadur-Kiefer-type two sample statistic with applications to tests of goodness of fit (1989). In: Limit Theorems in Probability and Statistics (P. Révész, Edit.), Colloquia Mathematica Jànos Bolyai 57, 157–172 (with D. M. Mason) [MR1116785 (92i:62089)]

[84] Sharp rates for the increments of renewal processes (1989). Annals of Probability. 17, 700–722 (with J. Steinebach) [MR0985385 (90a:60058)]

[85] Asymptotic expansions of sums of non identically distributed binomial random variables (1989). Journal of Multivariate Analysis. 25, 65–89 (with M. L. Puri and S. Ralescu) [MR0991952 (90k:62039)]

[86] On the relationship between stability of the extreme order statistics and convergence of the maximum likelihood kernel density estimate (1989). Annals of Statistics. 17, 1070–1086 (with M. Broniatowski and L. Devroye) [MR1015138 (90k:62080)]

[87] On the non-parametric estimation of the bivariate extreme value distributions (1989). Statistics and Probability Letters. 9, 241–251 (with J. Tiago de Oliveira) [MR1028989 (92a:62077)]

[88] On the almost sure behavior of sums of extreme values from a distribution in the domain of attraction of a Gumbel law (1990). Bulletin des Sciences Mathématiques, Ser. 2, 114, 61–95 (with E. Haeusler and D. M. Mason) [MR1046701 (91b:60023)]

[89] Obituary; Daniel Dugué (1990). J. Roy. Statist. Ser. A 153, n°1, 99–100 [MR1050654 (91i:01071)]

[90] Bahadur-Kiefer-type processes (1990). Annals of Probability. 18, 669–697 (with D. M. Mason) [MR1055427 (91j:60055)]

[91] On the approximation of $P - P$ and $Q - Q$ plot processes by Brownian bridges (1990). Statistics and Probability Letters. 9, 241–251 (with J. Beirlant) [MR1045191 (91i:60081)]

[92] On the sample path behavior of the first passage time process of a Brownian motion with drift (1990). Annales de l'Institut Henri Poincaré. 26, 145–179 (with J. Steinebach) [MR1075443 (91i:60083)]

[93] Nonstandard functional laws of the iterated logarithm for tail empirical and quantile processes (1990). Annals of Probability. 18, 1693–1722 (with D. M. Mason) [MR1071819 (91j:60056)]

[94] On some alternative estimates of the adjustment coefficient in risk theory (1990). Scandinavian Actuarial Journal. 135–159 (with J. Steinebach) [MR1157782 (93f:62139)]

[95] Bahadur-Kiefer theorems for uniform spacings processes (1991). Teoria Verojatnosti i ee Primienienia. 36, 724–743 (with J. Beirlant, J. H. J. Einmahl and D. M. Mason) [MR1147173 (93g:60061)]

[96] Laws of the iterated logarithm for density estimators (1991). In: Nonparametric Functional Estimation and Related Topics (G. Roussas, Edit.), 19–29, Kluwer, Dordrecht [MR1154317 (93b:62058)]

[97] Estimating the quantile-density function (1991). In: Nonparametric Functional Estimation and Related Topics (G. Roussas, Edit.), 213–224, Kluwer, Dordrecht (with M. Csörgő and L. Horvàth) [MR1154330 (93a:62052)]

[98] A tail empirical process approach to some nonstandard laws of the iterated logarithm (1991). Journal of Theoretical Probability. 4, 53–85 with D. M. Mason) [MR1088393 (92e:60061)]

[99] On the limiting behavior of the Pickands estimator for bivariate extreme-value distributions (1991). Statistics and Probability Letters. 12, 429–439 [MR1142097 (93g:62069)]

[100] Statistical analysis and modelling of slug lengths (1991). In: Multi-Phase Production (A. P. Burns, Edit), 80–112, Proceedings of the 5-th International Conference on Multi-Phase Production, Elsevier, London (with H. Dhulesia and M. Bernicot).

[101] Pointwise Bahadur-Kiefer-Type theorems (II) (1992). In: Nonparametric Statistics and Related Topics (A. K. Md. E. Saleh, Edit.), 331–346, North Holland, Amsterdam [MR1226733 (94g:60044)]

[102] On the limiting behavior of the Bahadur-Kiefer statistic for partial sums and renewal processes when the fourth moment does not exist (1992). Statistics and Probability Letters. 13, 179–188 (with J.Steinebach) [MR1158856 (93d:60041)]

[103] Functional Erdős-Rényi laws (1991). Studia Scientiarum Math. Hungarica. 26, f.2–3 261–295 [MR1180494 (93i:60056)]

[104] Deux exemples d'estimation de la fiabilité de matériels à partir d'informations fournies par les maintenances corrective et préventive. (1992). Revue de Statistique Appliquée. 40, f.2 77–90 (with J. Lannoy and C. Chevalier).

[105] Large deviations by Poisson approximations (1992). Journal of Statistical Planning and Inference. 32, 75–88 [[MR1177360 (93k:60068)]

[106] A functional law of the iterated logarithm for tail quantile processes (1992). Journal of Statistical Planning and Inference. 32, 63–74 [MR1177359 (93i:60057)]

[107] Pointwise Bahadur-Kiefer-type theorems. I. (1992). Probability Theory and Applications. Essays to the Memory of Joszef Mogyoródi. (J. Galambos et I. Katai edit.) 235–256, Math. Appl. 80, Kluwer, Dordrecht [MR #94a:60028]

[108] A functional LIL approach to pointwise Bahadur-Kiefer theorems. (1993). Probability in Banach Spaces. 8, Proceedings of the Eight International Conference. R. M. Dudley, M. G. Hahn and J. Kuelbs Edit. 255–266. Progress in Probability 30 Birkhäuser, Boston (with D. M .Mason) [MR1227623 (94e:60029)]

[109] Functional laws of the iterated logarithm for the increments of empirical and quantile processes (1992). Annals of Probability. 20, 1248–1287 (with D. M. Mason) [MR1175262 (93h:60046)]

[110] Functional laws of the iterated logarithm for large increments of empirical and quantile processes (1992). Stochastic Processes and Applications. 43, 133–163 [MR1190911 (93m:60057)]

[111] Approximations and two-sample tests based on P-P and Q-Q plots of the Kaplan-Meier estimators of lifetime distributions. (1992) Journal of Multivariate Analysis. 43, 200–217. (with J. H. J. Einmahl) [MR1193612 (93k:60053)]

[112] Some results on the influence of extremes on the bootstrap (1993). Annales de l'Institut Henri Poincaré. 29, 83–103. (with D. M. Mason and G. R. Shorack) [MR1204519 (94a:62068)]

[113] On fractal dimension and modelling of bubble and slug flow processes. (1993). Multi-Phase Production. BHR Group Conference Series 4, 95–16, A. Wilson Edit, Mechanical Engineering Publications Ltd, London (with M. Bernicot and H. Dhulesia).

[114] Records in the F^α-scheme. I. Martingale properties. Zapiski Nauchnykh Seminarov POMI 207, 19–36. (in Russian) (with V. B. Nevzorov) [MR1227973 (94g:60096)]

[115] On the coverage of Strassen-type sets by sequences of Wiener processes (1993). Journal of Theoretical Probability. 6, 427–449. (with P. Révész) [MR1230339 (94k:60063)]

[116] Strassen-type functional laws for strong topologies (1993). Probability Theory and Related Fields. 97, 151–156 (with M. A. Lifshits) [MR1240720 (94i:60048)]

[117] Limit laws for k-record times (1994). Journal of Statistical Planning and Inference. 38, 279–307 (with V. B. Nevzorov) [MR1261805 (95e:60024)]

[118] Random fractals generated by oscillations of processes with stationary and independent increments. (1994). Probability in Banach Spaces 9, J. Hoffman-Jørgensen, J. Kuelbs, and M. B. Marcus Edit., Birkhäuser, Boston, 73–89 (with D. M. Mason) [MR1308511 (96b:60191)]

[119] Functional laws of the iterated logarithm for local empirical processes indexed by sets (1994). Annals of Probability 22, 1619–1661 (with D. M. Mason) [MR #96e:60048]

[120] Necessary and sufficient conditions for the Strassen law of the iterated logarithm in non-uniform topologies. (1994). Annals of Probability. 22, 1838–1856 (with M. A. Lifshits) [MR #96d:60051]

[121] A unified model for slug flow generation (1995). Revue de l'Institut FranÃ§ais du Pétrole. 50, 219–236 (with M. Bernicot).

[122] Nonstandard local empirical processes indexed by sets (1995). Journal of Statistical Planning and Inference. 45, 91–112 (with D. M. Mason [MR #97c:60011]

[123] On the fractal nature of empirical increments (1995). Annals of Probability. 23, 355–87 (with D. M. Mason) [MR #96e:60064]

[124] A new law of the iterated logarithm for arrays (1996). Studia Scientiarum Mathematicarum Hungarica. 31, 145–185 (with H. Teicher) [MR #96k:60073]

[125] Strong laws for extreme values of sequences of partial sums (1996). Studia Scientiarum Mathematicarum Hungarica. 31, 121–144 (with J. Steinebach) [MR1367703 (96k:60072)]

[126] Cramér-Von Mises-type tests with applications to tests of independence for multivariate extreme-value distributions (1996). Communications in Statistics - Theory and Methods. 25, n°4, 871–908 (with G. V. Martynov) [MR #97d:62087]

[127] On the strong limiting behavior of local functionals of empirical processes based upon censored data (1996). Annals of Probability. 24, 504–525 (with J. H. J. Einmahl) [MR #98f:62088]

[128] Functional laws for small increments of empirical processes (1996). Statistica Neerlandica. 50, 261–280 [MR1406305 (97e:62083)]

[129] On the Hausdorff dimension of the set generated by exceptional oscillations of a Wiener process (1997). Studia Scientiarum Mathematicarum Hungarica. 33, 75–110 (with M. A. Lifshits) [MR1454103 (99b:60137)]

[130] On random fractals (1997). Supplemento ai Rendiconti del Circolo Matematico di Palermo, Ser. II 50, 111–132 [MR1603109]

[131] Strong laws for local quantile processes (1997). Annals of Probability. 25, 2007–2054 [MR #99c:60064]

[132] Protuberance effect in the generalized functional Strassen-Révész law (1998). Journal of Mathematical Sciences. 88, 22–28 (with M. A. Lifshits) [translated from: Zapiski Nauchnykh Seminarov St. Petersburg. Otdel. Mat. inst. Steklov. POMI. 216, 33–41 (1994)] [[MR #96m:60071]

[133] Records in the F^α-scheme. II. Limit laws (1998). Journal of Mathematical Sciences. 88, 29–35 (with V. B. Nevzorov) [translated from : Zapiski Nauchnykh Seminarov St. Petersburg. Otdel. Mat. inst. Steklov. POMI. 216, 42–51 (1994)] [MR #96e:60091]

[134] Random fractals and Chung–type functional laws of the iterated logarithm (1998). Studia Scientiarum Mathematicarum Hungarica. 34, 89–106 (with D. M. Mason) [MR1645150 (99g:60061)]

[135] On local oscillations of empirical and quantile processes (1998). Asymptotic Methods in Probability and Statistics. 127–134. Edit. B. Szyszkowicz, North Holland, Amsterdam [MR #99i:60076]

[136] On the approximation of quantile processes by Kiefer processes (1998). Journal of Theoretical Probability. 11, 997–1018 [MR #2000f:60045]

[137] Fluctuations du processus de risque et coefficient d'ajustement (1998). Atti del XXI Convegno Annuale A.M.A.S.E.S. Appendice. 41–60.

[138] Bootstrap for maxima and records (1999). Zapiski nauchny seminarov POMI (Notes of Sci. Semin. of Peterburg Branch of Math. Inst.). 260, 119–129 (with. V. A. Nevzorov) (in Russian) [Translated in: J. Math. Sci. (New York) 109, (2002), n°6, 2115–2121] [MR #2001h:62077]

[139] Chung-type functional laws of the iterated logarithm for tail empirical processes. (2000). Annales de l'Institut Henri Poincaré. 36, 583–616 [MR #2001m:60065]

[140] Functional Limit laws for the Increments of Kaplan-Meier Product-Limit Processes and Applications (2000). Annals of Probability. 28, 1301–1335 (with J. H. J. Einmahl) [MR #2001k:62039]

[141] Nonstandard Strong Laws for Local Quantile Processes (2000). Journal of Statistical Planning and Inference. 91, 239–266 [MR #2001m:62067]

[142] Une approche nouvelle pour l'exploitation des empreintes chromatographiques. (2000). Ann. Fals. Exp. Chim. 92, 455–470 (with Richard, J. P. and Pierre-Loti-Viaud, D.).

[143] Uniform limit laws for kernel density estimators on possibly unbounded intervals. (2000). In: Recent Advances in Reliability Theory: Methodology, Practice and Inference. N. Limnios and M. Nikulin, Edit. 477–492, Stat. Ind. Technol. Birkhäuser, Boston [MR #2001j:62042]

[144] Strong Approximation of Quantile Processes by Iterated Kiefer Processes (2000). Annals of Probability. 28, 909–945 [MR #2001j:60038]

[145] Asymptotic independence of the local empirical process indexed by functions (2001). High dimensional Probability II (Seattle, WA, 1999). 183–206. Progress in Probability, Vol. 47 (E. Giné, D. M. Mason, and J. A. Wellner Edit.), Birkhäuser, Boston (with D. M. Mason and U. Einmahl) [MR #2002f:60057]

[146] Limit Laws for kernel density estimators for kernels with unbounded supports (2001). In: Asymptotics in Statistics and Probability. Papers in Honor of George Gregory Roussas. 117–131. M. L. Puri Edit. VSP International Science Publishers, The Netherlands.

[147] Probabilities of hitting shifted small balls by a centered Poisson processes (2001). Zapiski Nauchn. Seminarov St. Petersburg. Otdel. Mat. Inst. Steklov POMI. (In Russian) 278, 63–85 (with M. A. Lifshits) [Verojatn. i Stat. 4, 310–311] (In Russian) [MR #2002i:60100]

[148] Estimation non-paramétrique de la régression dichotomique - Application biomédicale (2002). C. R. Acad. Sci. Paris, Ser. I. 334, 59–63 (with G. Derzko) [MR #2002m:62054]

[149] Tests of Independence with Exponential Marginals (2002). In: Goodness-of-Fit Tests and Model Validity. 463–476 (C. Huber-Carol, N. Balakrishnan, M. S. Nikulin and M. Mesbah edit.), Birkhäuser Verlag, Basel [MR1901856].

[150] Exact laws for sums of logarithms of uniform Spacings (2003). Austrian Journal of Statistics. 32, n°1 & 2, 29–47 (with G. Derzko).

[151] Karhunen-Loève expansions for weighted Wiener processes and Brownian bridges via Bessel functions. (2003). High dimensional Probability III. Progress in Probability, 57–93 (M. Marcus and J. A. Wellner edit.), Birkhäuser, Boston (with G. V. Martynov) [MR2033881 #2005i:62074]

[152] General confidence bounds for nonparametric functional estimators (2004). Statistical Inference for Stochastic Processes. Vol. 7, 225–277 (with D. M. Mason) [MR2111291#2005k:62131]

[153] An extension of Lévy's formula to weighted Wiener processes (2004). In: Parametric and Semiparametric Models wit Applications to Reliability, Survival Analysis, and Quality of Life. 519–531. Stat. Ind. Technol., Birkhäuser, Boston [MR2091632 #2006f:60079]

[154] Tests of fit based on products of spacings (2006). (with G. Derzko). In: Probability, Statistics and Modelling in Public Health, 119–135, Springer, New York [MR2230727#2007h:62042]

[155] Weighted multivariate Cramér-von Mises type statistics (2005). Afrika Statistika 1 n°1, 1–14 [MR2298871 #2008k:62103]

[156] Limit laws for kernel smoothed histograms (with G. Derzko and L. Devroye). Proceedings of the St.Petersburg Conference on limit theorems in Probability.

[157] On quadratic functionals of the Brownian sheet and related functionals (2006) (with G. Peccati and M. Yor). Stochastic Processes Applications. Vol. 116, 493–538 [MR2199561 #2007k:60064]

[158] Karhunen-Loève expansions of mean-centered Wiener processes (2006). In: IMS Lecture Notes Monograph Series High Dimensional Probability. Vol. 51, 62–76, Institute of Mathematical Statistics [MR2387761 #2009e:62184]

[159] Weighted multivariate tests of independence (2007). Comm. Statist. Theory Meth. Vol. 36, 1–15 [MR2413589];

[160] A Karhunen-Loève expansion for a mean-centered Brownian bridge (2007). Statist. Probab. Letters. Vol. 77, 1190–1200 [MR2392790 #2009b:62089].

[161] A Karhunen-Loève decomposition of a Gaussian process generated by independent pairs of exponential random variables (2008). Journal of Functional Analysis. Vol. 255, n°9, 2363–2394. (with G. Martynov) [MR2473261].

[162] Topics on Empirical Processes (2007). EMS Ser. Lect. Math, 93–190., Eur. Math. Soc., Zürich [MR2347004].

[163] Asymptotic certainty bands for kernel density estimators based upon a bootstrap resampling scheme (2008). Statistical models and methods for biomedical and technical systems. 171–186, Stat. Ind. Technol., Birkhäuser, Boston [MR2462913].

[164] A Glivenko-Cantelli-type theorem for the spacing-ratio empirical distribution functions (2008). Ann. I.S.U.P. Vol. 52, 25–38 [MR2435038#2010a:62037].

[165] A multivariate Bahadur-Kiefer representation for the empirical copula process (2009). Journal of Mathematical Sciences. Vol. 163, n°4, 382–398 [MR2749128 (2012c:60083)].

[166] Spacings-ratio empirical processes (with G. Derzko) (2010). Periodica Mathematica Hungarica. Vol. 61, 121–164 [MR2728435 (2011g:60040)]

[167] Non-uniform spacings processes (2011). Statistical Inference Stoch. Processes. Vol.14, n°2, 141–175 [MR2794959 (2012f:60110)].

[168] One bootstrap suffices to generate sharp uniform bounds in functional estimation (2011). Kybernetika, Vol. 47, n°6, 855–865 [MR2907846]

[169] Hommage à Marc Hallin. (2012). J. SFdS 153, no. 1, 2–4 [MR2930286]

[170] Uniform in bandwidth functional limit laws (with S. Ouadah) (2013). J. Theoretical Probab. Vol. 26, 697–721 [MR3090547]

[171] On testing stochastic dominance by exceedance, precedence and other distribution-free tests, with applications (2014). In: Statistical Models and Methods for Reliability and Survival Analysis (Edit. V. Couallier, L. Gerville-Réache, C. Huber-Carol, N. Limnios and M. Mesbah), John Wiley & Sons, Inc., Hoboken, USA, 145–159.